# Praise for *Grand Coulee*

Paul Pitzer has produced a volume that rivals Joseph E. Stevens's prize-winning study of Hoover Dam in its discussion of the construction of Grand Coulee. His work merits a place beside Charles McKinley's *Uncle Sam in the Pacific Northwest* as one of the most significant studies examining the Pacific Northwest yet to be published.

> Richard Lowitt, author
> *The New Deal and the West*

Paul Pitzer's splendid study of Grand Coulee Dam and the Columbia Basin reclamation project will be required reading for anyone interested in the future of the Inland Empire or making long-term plans for land redevelopment. It is sure to be consulted by a generation of economists, engineers, politicians, historians, and social critics.

> Murray Morgan, author
> *The Dam* and *Skid Road*

Paul Pitzer's marvelous book on Grand Coulee is at once the most complete history of the dam's construction and the best account of its importance to Pacific Northwest history. Pitzer has addressed an impressive range of subjects, from the details of building the massive structure, to the local and national political decisions that led to its construction, to the environmental and social changes it wrought on the landscape. Pitzer's impressive research and analysis is an important contribution to Northwest historiography.

> William L. Lang, Director
> Center for Columbia River History

Paul Pitzer's history of Grand Coulee Dam and the Columbia Basin Project is thoroughly researched, highly readable, authoritative in the extreme, and a major contribution to our understanding of the development of the modern American West.

Robert E. Ficken, author
*Washington: A Centennial History*


This illuminating book for the first time brings together the controversies and triumphs of the dam's construction and operation, the political issues of hydroelectricity and irrigation, and the effects on surrounding communities and the fish in the "Great River of the West." . . . A fascinating portrayal based on thorough scholarly research, this convincing account reveals the significance of Grand Coulee for the West and America as a whole.

David H. Stratton, editor
*Washington Comes of Age*

PAUL C. PITZER

# GRAND COULEE

## HARNESSING A DREAM

Washington State University Press
Pullman, Washington

Washington State University Press, Pullman, Washington, 99164-5910
First printing 1994
Printed and bound in the United States of America on pH neutral, acid-free paper.

Cover design based on a concept by Erik Sturdevant.

Portions of the present work appeared in "The Atmosphere Tasted Like Turnips: The Pacific Northwest Dust Storm of 1933," *Pacific Northwest Quarterly,* Vol. 79 (1988): 50-55, and "A Farm-In-A-Day: The Publicity Stunt and the Celebrations That Initiated the Columbia Basin Project," *Pacific Northwest Quarterly,* Vol. 82 (1991): 2-7.

*Library of Congress Cataloging-in-Publication Data*
Pitzer, Paul C.
    Grand Coulee : harnessing a dream / Paul C. Pitzer.
        p.    cm.
    Includes bibliographical references and index.
    ISBN 0-87422-113-7 (alk. paper).—ISBN 0-87422-110-2 (pbk. : alk. paper)
    1. Grand Coulee Dam (Wash.)—History.    2. Columbia Basin Project (U.S.)—History.    3. Water resources development—Columbia River Watershed—History.    I. Title.
TC557.W22G737    1994
333.91'0097972—dc20                                                  94-27155
                                                                          CIP

To Richard Maxwell Brown

It is easier to do a thing when someone you admire
and trust believes you can do it.

Some day at Grand Coulee Dam there will be harnessed 2,000,000 wild horses—energy that today is wasting itself away day by day as it flows down the Columbia River, the wildest big stream in the civilized world.

*Rufus Woods*

# Contents

Statue of Franklin Roosevelt looks out onto the back of Grand Coulee Dam. *Courtesy Grace Pitzer.*

# Introduction

> Such a power if developed would operate railroads, factories, mines, irrigation pumps, furnish heat and light in such measure that all in all it would be the most unique, the most interesting, and the most remarkable development of both irrigation and power in this age of industrial and scientific miracles.
> *Rufus Woods*

In the 1950s the American Society of Civil Engineers identified the seven civil engineering wonders of the United States. Selection committee members dismissed size and looked at uniqueness and pioneering design as their main criteria. They selected Chicago's Sewage Disposal System, the Colorado River Aqueduct, the Empire State Building, the Panama Canal, the San Francisco-Oakland Bay Bridge, and Hoover Dam. They also included Grand Coulee Dam and the Columbia Basin Project. Seeing significance in what the popular press had dubbed "The Greatest Structure in the World," "The World's Greatest Engineering Wonder," "The Eighth Wonder of the World," and "The Biggest Thing on Earth" hardly surprised anyone. Through the 1930s and 1940s the dam generated sensational nationwide publicity. It collared so much attention that freelance journalist Richard L. Neuberger wrote in 1942, "Everyone in America has heard of Grand Coulee."[1]

But few Americans, then or now, know much about the Columbia Basin Project—the irrigation network that Grand Coulee Dam makes possible, even though it is the largest single reclamation project ever undertaken in the United States.[2] In all, the project area of over 2,500,000 acres is roughly twice the size of the state of Delaware.[3] Grand Coulee Dam, once the largest concrete structure on the planet, is its key feature but only one of its many parts. In all, the total includes 333 miles of main canals, 1,993 miles of laterals (smaller distribution canals) 3,498 miles of drains and wasteways, and four large dams besides Grand Coulee. In addition there is an enormous pump-generating plant beside Franklin D. Roosevelt Lake—the reservoir formed by Grand Coulee Dam.[4]

These irrigation features represent a construction effort larger than Grand Coulee Dam itself.

As of September 1986, the total Columbia Basin Project, since 1933, including Grand Coulee Dam, cost $1,687,000,000.[5] Of that amount, well over $500,000,000 was for the pump plants, reservoirs, canals, laterals, and other irrigation works. Over $500,000,000 covered the third Grand Coulee powerhouse. In return, the complex can produce more than 6.18 million kilowatts of electrical energy and now irrigates more than 556,000 acres—roughly one-half of the ultimate 1,029,000 acres possible for the entire project. Government officials today estimate that it will take at least another two billion dollars to finish the job.[6]

The power and the irrigation provided by the Columbia Basin Project make it an important element in the West's economy. Grand Coulee Dam is famous because of the electricity it has generated since 1942 and it is a popular attraction visited by thousands annually. But seldom do the tourists realize that the fields of potatoes, corn, and other crops that they see on their way to Grand Coulee rely on the dam and the sale of its power. The critical link between power and reclamation remains obscure to most Americans.

The businessmen and professionals of Wenatchee, Ephrata, Spokane, and Pasco understood the link. They imagined that damming the river could provide cheap electricity and abundant water transforming the region into an agricultural/industrial empire. On May 14, 1919, Rufus Woods wrote a headline for his *Wenatchee Daily World* stating that a dam on the Columbia River "Would furnish [the] Power to Run all Industries in [a] Washington Empire." Selling the electricity, he theorized, would eventually pay the costs covered at the outset by the government. That dream of almost-free irrigation, supported by power ratepayers, is one that has plagued the project since the first water arrived on the land. More than other farmers in the West, Columbia Basin Project boosters saw no reason why they should not have the same conditions as the wetter regions to their east. They demanded irrigation to compensate for that lack and they wanted someone else to pay the bills.

Replication and accommodation drove Western Hemisphere expansion and settlement. Individuals who came to new land brought ideas about how to use it based on the place or places they left behind. They aimed to establish New England, New France,

New Spain, or New Amsterdam. They wanted a fresh start but, plagued by the twin diseases of culture shock and homesickness, also worked to recreate familiar surroundings. None of the participants lost their desire to replicate what was familiar to them. Those in the West have always wanted to make it as much like the East as possible, while at the same time keeping it vigorous and untainted. The result is a West that is both a continuation and a place unique. Look at the people who came, the ideas they carried with them, and the changes the new environment and association with different peoples forced them to make, and you can understand American history.

The arid West did not easily accede to the goals of its settlers. To accommodate the differences, they made subtle and dramatic changes in their lifestyles. In the process they became the democrats that Frederick Jackson Turner saw when he wrote his famous thesis in 1893. As settlement expanded, leavened by racial, cultural, and economic diversity, the American character changed. It showed continuity with the past, but through a multifaceted dialectic between different peoples with different ideals meeting each other in a variety of new places and conditions over time, it also produced something unique and changing. What happened at Grand Coulee and on the Columbia Basin Project is one tiny piece of a much larger mosaic which, taken together, exhibits this process.

Farmers, businessmen, professionals, and promoters looked to irrigation to make the dry parts of the West bloom. This lure would bring industry, development, and self-sufficiency. Increasingly they expected the federal government to pay the costs. Seldom did they see the paradox as they demanded financial support from the East and at the same time resented their colonial status. They wanted independence, growth, development, and a successful agricultural base to make their regions autonomous, independent, and prosperous. They wanted to remake the East and they accommodated themselves to local conditions and federal largess only enough to accomplish that end. It led them to decry both the stingy support they felt they received from the government and the strings attached to the money they fought so hard to garner.

All of this drove the construction of Grand Coulee Dam and the Columbia Basin Project. Settlers who came to the arid Columbia Basin in the 1870s and 1880s dreamed of irrigation.[7] The

Columbia Basin Project is the result of many overlapping and diverse visions, all aiming toward that end, which emerged from the late-nineteenth century through the present. The goal was always reclamation to compensate for "nature's failure." Once irrigated, the promoters felt certain that the land would support thousands of farmers who in turn would provide the human base for an industrial empire. The dam's power would turn machines, illuminate cities, and bring prosperity to an area avoided as a no-man's-land by those with lesser vision. The dam itself would be the biggest thing on earth, man's greatest engineering undertaking, and a demonstration of modern civilization. It would symbolize the West's bigness. It would make a part of the West like the East—the same, only better, and different.

When Franklin Roosevelt's New Dealers began the Columbia Basin Project in 1933, they added the concept of planning. They hoped to create a "Planned Promised Land."[8] On small farms of around eighty acres each, displaced Dust Bowl refugees would find homes. Through a controlled economy the government would guarantee the success of those settlers. In the late 1930s, the New Deal planners and others in the region debated how best to achieve their goals.[9] Then, before any of the land had received water, World War II and the rapid changes that it brought altered the vision. The project, as it emerged in the 1950s, differed from the blueprint drawn two decades earlier. The changes continued into the 1980s and 1990s.

Every society leaves monuments that tell us what that society held as important. Egyptian tombs show preoccupation with death, and the Great China Wall indicates concern with boundary and security. Electrical power preoccupied the twentieth century. It could unburden our lives and improve our standard of living both physically and spiritually.[10] We leave behind as our monuments to this obsession the thousands of dams that barricade our rivers.

This history of Grand Coulee Dam and the Columbia Basin Project is not unlike a biography. As with any person, the dam, too, has its many facets, all complicated and inextricably interconnected. Thousands of people played larger and smaller roles in the political maneuvering that led to construction and more helped to build the dam and irrigation works. Some readers may find the landscape cluttered with personalities while others might regret omission of this or that character. Grand Coulee is a big story and there is insufficient space to include everyone and every detail. Some peripheral issues, such as the struggle between the Bureau of Reclamation and the Army Corps of Engineers or a complete survey of the uses for the dam's power receive mention but are not treated in depth, as they are in themselves expansive topics. The focus here is on the dam and its attendant irrigation works.

The two most prominent aspects of Grand Coulee Dam and the Columbia Basin Project are their political history and their physical construction and operation. In separating the two there is of necessity the need to cover one topic, then drop back to look at the other. Grand Coulee Dam did not come into being easily. The struggle to have the government undertake and then finance the dam and the project stretched across the first five decades of the twentieth century and its roots went back even farther. Often those who promoted the vision squabbled among themselves. The dam, the power, and the irrigation exist today as much in spite of as because of the people who worked to accomplish them.

It is fashionable in the late twentieth century to decry the dams and the environmental damage they brought. But, through the first half of the century, dam builders and irrigation backers saw their work as promoting conservation. To them, taming the rivers, stopping erosion and floods, and reclaiming land outweighed any harm that might follow, and they focused on the promise of power and prosperity. Despite the tragic loss seen in retrospect, people in the Northwest are not likely to tear down Grand Coulee or many other hydroelectric projects.[11] The challenge now is to find a way to accommodate the works of the past with the new visions of the future.

Dry Falls, in the middle of the Grand Coulee, a remnant of the Missoula Floods. *Courtesy Bureau of Reclamation.*

# Chapter One

# The Land and the People

From its birth in the womb of the great Columbia Ice Field in British Columbia to its disappearance in the Pacific Ocean off the Oregon-Washington coast, the Columbia graces some of the most beautiful scenery in North America.

*Ralph W. Johnson*[1]

It is not a beautiful land, nor is the country surrounding it. . . . Grand Coulee and the plains below it are ugly and lifeless, yet today they command the interest of the nation.

*Arthur W. Baum*[2]

Although large and formidable, the Columbia is not the greatest river in North America.[3] Its strength lies in its volume and its rapid descent as it moves toward the sea. From source to mouth, it falls 2,600 feet. In the state of Washington alone it drops more than 1,000 feet over about 400 miles.[4] Consequently, along its course lie many potential hydroelectric power sites.

Paradoxically much of the region through which the Columbia flows is arid. The Cascade Mountains to its west leave a rain shadow that limits precipitation across most of eastern Washington. There the river cuts through a deep canyon leaving dry the plateau above it. The river makes a large curve around the plateau nearly encircling the region often called the "Big Bend." With an area of 12,780 square miles, that expanse is a curious mixture of rich grassland and occasional barren scablands. Where they exist, the soils are rich, deep, and gray or dark brown in color.[5] The addition of water makes them excellent for farming.

Various geologists attempted to explain the forces that left the deep coulees and scarred channels on the plateau surface. Most agree that in the earliest geologic epochs of the Tertiary period, over thirty million years ago, oceans covered the area that is now eastern Washington.[6] The rise of land to the west created a huge inland sea that slowly changed to a string of freshwater lakes and then to one river, the antecedent of the modern Columbia. That river ran in a southerly direction from what is today the northeast corner of Washington to the mouth of the Snake River. The land surface, with its deep, fertile soil and rich vegetation, resembled the present-day Palouse region, south and east of the Big Bend.[7] Ginkgo, sequoia, oak, elm, cypress, and pine trees, found today in fossil form, give evidence of the once warm, humid climate.[8]

The Cascade Range rose during the mid-Miocene epoch, from ten to twenty million years ago, causing a rain shadow and leaving the land to their east increasingly arid. At the same time great fissures opened in the plateau, and over hundreds of thousands of years enormous lava flows gushed from them. Resembling the activity of Kilauea volcano in modern-day Hawaii, these extrusions of Yakima basalt eventually covered well over 250,000 square miles and in places accumulated to a thickness of more than 4,000 feet.[9]

Heavy basaltic lava on the underlying granite tilted the plateau so that its northeast corner is 2,500 feet higher than the southwest section near present-day Pasco. Deformations in the cooling surface produced the Saddle Mountains, the Frenchman Hills, and the Horse Heaven Hills which run generally east-to-west across the southwestern section of the plateau.[10] The lava also pushed the Columbia River to the west, up against the Cascade Mountains, forming the Big Bend.[11] Consequently, today the river flows south into the United States, curves abruptly to the north, and then swings around in a great arc, much like the letter "C," at which point it flows briefly toward the east. Suddenly it turns west through the Cascades at the Columbia Gorge and on to the Pacific Ocean.

Starting about one to two million years ago, during the Pleistocene Epoch, or Ice Age, great glaciers covered the northern parts of North America. Geologists agree that for well over one million years the ice flows came and went, disrupting the normal drainage patterns. The great Cordilleran Ice Sheet crept south over Canada into the United States, and its western lobes moved into Washington state. The Okanogan lobe crossed the Columbia River and forced it into a new channel. Other lobes crossed the Columbia or its tributaries creating temporary lakes that existed for hundreds or thousands of years. Over the millennia, ground by the glaciers, some of the lava decayed into a fine dust that settled across the region building the thick loessal soils, remnants of which form the famed Palouse.[12]

Geologists agree on most of this, but there has been controversy. The traditional opinion held that the Columbia River, dammed by the ice, slowly created a new channel—the Grand Coulee or "great canyon"—with its magnificent stair-step at Dry Falls. Occasional overflows accounted for the other unique features of the area such as the soil-less and barren Channeled Scablands.[13] There water dug away the earth and cut deep coulees, or ditches, into the once flat, undulating plateau.[14]

In the 1920s and 1930s professor Joseph Harlan Bretz of the University of Chicago investigated the Grand Coulee and its surroundings.[15] Bretz doubted that a slow process formed the Grand Coulee. He suggested that the ice flows created large dams and enormous lakes. During warm periods the dams collapsed, releasing huge volumes of water that rushed toward the ocean as devastating floods. These cataclysms, Bretz maintained, dug the Grand Coulee and swept away the loess in a series of catastrophes scattered down the centuries.[16] Now named in his honor, the Bretz Floods repeatedly poured out of ancient, glacial-lake Missoula—the remnant of which is today's Lake Pend Oreille. With a volume of perhaps 500 cubic miles of water, these geologic disasters produced the greatest floods yet documented in earth's history.[17]

It is difficult to imagine such volumes of water, in flows perhaps twenty miles wide and 600 feet deep, rushing across the land, gouging up huge blocks of basalt, digging the Grand Coulee, Moses Coulee, Lind Coulee, the Channeled Scablands, Crab Creek Channel, and the other unique and intricate features that mar what had once been a nearly flat landscape.[18] But each time the velocity of rushing water doubles, its ability to dislodge and propel material increases sixty-four times. The Bretz Floods moved at speeds up to fifty miles an hour and the work they did was at once magnificent and terrifying.

Bretz speculated that no humans witnessed the catastrophic floods.[19] The last inundation occurred approximately 15,000 years ago and the first signs of man date back only about 12,000 years.[20] Others argue that Amerinds—Indians—predated the last of the floods and could have been among its casualities.[21] If that happened, they heard a puzzling mounting rumble for perhaps thirty minutes to an hour and then saw an advancing, churning mass of water carrying ten times the flow of all the world's rivers, full of sediment, debris, and huge ice blocks, an onslaught that first paralyzed them with fear and then enveloped them.

Bretz concluded that the Grand Coulee was the last channel formed by the catastrophic floods. Diversion of the Columbia by the Okanogan lobe of the ice sheet also helped dig the channel, but the floods did most of the work.[22] The floods left a mammoth coulee—the "Grand" Coulee, so named by later French explorers— that today opens over 500 feet above the bed of the Columbia at a point where the river turns to the north just before entering its Big Bend curve. Dry Falls divides the fifty-mile-long channel into the upper and the lower coulees. That ancient waterfall is over 400 feet high and three to four miles wide. The coulee is between one and six miles wide and its walls tower up to 900 feet.

At first, geologists ridiculed notions of great catastrophes and openly scoffed at Bretz's findings. Undaunted, he held fast despite derision from his peers. Slowly, over three decades from the 1940s through the 1970s, accumulating evidence, including photographs from Nimbus satellites, verified Bretz's thesis and scientists grudgingly accepted his findings.[23] When Bretz died in 1981, at the age of ninety-eight, the honors given him by his profession at last recognized his singular achievement.[24] He alone had unraveled the riddle of the unique and spectacular Grand Coulee.

Either before or not long after the Bretz Floods, North America's first residents arrived. When they came, the Indians avoided most of the Columbia Plateau because of its dryness. They believed that the awesome Grand Coulee harbored spirits, so they skirted it much of the time. In the spring some tribes went to Moses Lake to spear carp. Sweat houses stood along the banks of the Columbia, near where Grand Coulee Dam is today.[25] Europeans explored late in the eighteenth century. Thirteen years after Captain Robert Gray crossed the bar of the Columbia River in 1792, Meriwether Lewis and William Clark floated down the Snake River into the Columbia near present-day Pasco. They observed the southern tip of the Big Bend country, then moved on to the Pacific.

Mapmaker David Thompson, in the employ of the North West Company, journeyed down the Columbia in 1811, the first European to make the trip. He passed Kettle Falls and later the site of Grand Coulee Dam going on to Fort Astoria. A year later Alexander Ross passed through Grand Coulee country, describing it as "one of the most romantic picturesque and marvelously formed chasms west of the Rocky Mountains."[26] And he added "No one traveling in these parts ought to resist paying a visit to the wonder of the West."[27]

In 1836 Samuel Parker, possibly the first American to enter and settle the Columbia plain, noted the fertile upper prairies. But only a small handful of pioneers resided there ten years later. The Whitman Massacre and the Indian wars through the late 1840s and 1850s slowed settlement. Although Parker and others signaled the arrival of farmers, the 1850s brought prospectors and more conflict with the Indians. The rush for gold on the Upper Columbia drew Americans and Chinese laborers into the region, provoking the Shoshones, Yakimas, Spokanes, Colvilles, and others to protect their land.[28] Wars raged from 1855-1856 and again in 1858.[29] The government ultimately subdued all except the Nez Perce and established a period of peace that also marked the start of real settlement on the Columbia plateau.[30]

Cattle, sheep, wheat, and gold, with a slow shift toward wheat, were the main products around the Grand Coulee as the Timber Culture Act of 1873 and the Desert Land Act of 1877 encouraged movement onto dry Western land.[31] Even so, few came to the Big Bend and even then only the area south of the scablands saw serious development. Depending on the year, farmers or stockmen often suffered water shortages. In 1871 the land that became Douglas County had only one permanent resident. A few other ranchers joined him in 1883.

Three events accelerated settlement in the 1880s. First, a series of unusually wet years convinced farmers that rainfall followed the plow and that the Big Bend offered better prospects than previously indicated.[32] Second, in 1883 the Northern Pacific Railroad, building on its government land grant, completed its track across the Big Bend and James J. Hill's Great Northern followed in the early 1890s. These transcontinental railroads ended the region's isolation and allowed pioneers easy access and a way to market their crops. Finally, a series of hard winters from 1880 through 1890 ruined many cattlemen when their stock froze to death. This began the era of the wheat farmer.

By the late 1880s wheat dominated the Columbia plateau. In 1890 it grew on more than 250,000 acres, and by 1910 that increased to nearly three million acres.[33] Farmers favored the rich lands of the eastern Columbia plateau where rainfall averaged or exceeded ten inches annually. There farming succeeded. Most of the drier land to the west remained in the hands of the government or the railroads. The area around Soap Lake and Moses Lake had almost no inhabitants.[34] Where a few farmers grew wheat, the plantings quickly absorbed whatever water the land held and reduced the organic content of the soil, diminishing its fertility and making it susceptible to erosion. To conserve moisture the luckless settlers planted half of their land annually while the other half lay fallow, absorbing water from rain and snow. In the first decade of the twentieth century some around Moses Lake turned to fruit orchards. When the moisture finally gave out, most left.

From 1910 through 1930 the number of farms in the Big Bend declined. The only exception occurred in the places where limited irrigation, mostly by wells, brought water. Residents of the Big Bend, both on its farms and ranches, and in its few small towns, increasingly concluded that their region would be permanently settled and prosperous only when people somehow found a steady, reliable water source. From the 1880s through the 1930s, eastern Washington residents sought a way to do just that. They looked at the Columbia, the largest river in the state, and they wondered how to lift its water up onto their land. And they noted the amazing geologic site near the Grand Coulee. In the valley below the canyon's opening, steady erosion had reached the ancient bedrock of solid granite and formed an extraordinary potential dam site.[35] There the river canyon is 4,300 feet wide with cliffs on either side that rise over 600 feet. Upstream, the canyon walls provide a nearly watertight receptacle for a deep reservoir. Few locations are as perfectly suited to host a hydroelectric project and irrigation source as the Columbia River at the Grand Coulee. To build it required only three things: promoters with vision, the right technology, and a great deal of money.

William "Billy" Clapp. *Courtesy Bureau of Reclamation.*

# Chapter Two

# The Visions

With irrigation properly conducted, it is safe to say that nearly every foot of land now classed as desert will be found as productive as the regions more favored by rain.

*First Lieutenant Thomas William Symons, 1882*

Irrigation came to eastern Washington with the first settlers, and small projects speckled the region by the last half of the nineteenth century.[1] First Lieutenant Thomas William Symons surveyed the Columbia River and the land surrounding it for the United States Army Corps of Engineers in 1882. "It is a desert pure and simple, an almost waterless, lifeless desert," he wrote.[2] But Symons noted the richness of the soil and added that irrigation would allow it to produce bountiful harvests.

Both the Northern Pacific Railroad and the Great Northern Railroad conducted surveys in eastern Washington in the 1880s and 1890s, investigating ways to irrigate and attract settlers. Their slogans promoted promising ventures that filled landowners with hope—hope that vanished because of their expense or because the champions were more often charlatans than honest entrepreneurs.[3] Scores lost money on evanescent schemes that looked good on paper but disintegrated after backers realized their true costs.

Increasingly the government saw empty dry land and the need for irrigation as problems it might attack. In 1879 the United States Geological Survey replaced the Geographical and Geological Survey. Frederick H. Newell headed the hydrography section of the revamped agency and secured a $350,000 government appropriation for surveys to identify feasible irrigation projects.[4] In Washington Territory, settlers on the Columbia plateau argued that the government should pay to irrigate their land.

An act, passed by the Washington state legislature in 1890, allowed irrigation districts to organize and sell bonds. The state joined the federal government in paying for a survey of the Big Bend area to determine how best to irrigate it. Professor Israel Cook Russell, an engineer-geologist from the University of Michigan working for the Geological Survey, oversaw much of the effort that included drilling test wells in Douglas County. They failed to produce water in usable amounts. Russell reported unfavorably on the prospects for artesian wells due to the high cost and the lack of water, but he did suggest diverting water from the Spokane River to irrigate land in Lincoln and Douglas counties, or a ditch through the Saddle Mountains that might carry water somehow lifted out of the Columbia River.[5]

An 1892 copy of the *Coulee City News* proposed irrigating the Big Bend with water diverted into the Grand Coulee from behind a large dam on the Columbia.[6] The *Spokane Spokesman* of September 28, 1892 ran the same story with more details. Laughlin MacLean proposed a ninety-five-mile-long canal crossing the Big Bend carrying the entire flow of the Columbia River. Later, MacLean added the idea of a 1,000 foot-high dam to divert the river directly into the Grand Coulee. That dam and canal, enormous even by today's standards, was perhaps the first proposal for what became Grand Coulee Dam.[7] A more modest suggestion by an anonymous dreamer proposed a dam, at Albeni Falls on the Pend Oreille River, diverting its water onto the Columbia plateau.[8] These two ideas from the 1890s—a dam at the Grand Coulee or a canal starting at some upstream location—were the ancestors of the two schemes that, through the 1920s, polarized Washington state and national reclamation politics. But through the 1890s and early 1900s realists always discounted such plans when they calculated the high costs and technical difficulties.

Dramatic events changed the fortunes and hopes of irrigation promoters. First, the assassination of President William McKinley moved Theodore Roosevelt, a reclamation champion, into the White House. The new president formed a committee that included Wesley Livsey Jones, then a Republican representative from Washington and a Yakima resident. That committee drew up the bill that subsequently became the Reclamation Act or so-called Newlands Act of

1902 (for Francis Griffith Newlands, then Democratic representative and soon-to-be-senator from Nevada). It set the stage for the Reclamation Service (later the Bureau of Reclamation), which the government created that July as a branch of the Geological Survey.[9] The bill allowed the federal government to undertake self-liquidating irrigation projects that produced crops for local consumption and helped establish farms to relieve crowded urban areas. The new agency incorporated the progressive notion of efficient multipurpose river development yielding the greatest good for the largest number of people—ideals not fully implemented until thirty years later under Franklin Roosevelt's New Deal.[10] Eastern and Midwestern legislators dissented, unsuccessfully arguing that such unfair aid to Western farmers would increase agricultural surpluses and violate the principle of states' rights.[11] This conflict between Western farmers with arid land seeking government-supplied irrigation and established Eastern agriculture, already oversupplying national needs, was from the start a contentious debate that haunted the development of Grand Coulee Dam and the Columbia Basin Project.

Frederick Haynes Newell, Chief of the Hydrographic Branch of the Geological Survey, became the chief engineer of the new Reclamation Service, with Arthur Powell Davis, nephew of explorer John Wesley Powell, his assistant. Under their direction, civil engineer Theron A. Noble of Seattle and others began investigating irrigation schemes in eastern Washington. They analyzed proposals for irrigating the whole region with water from the Pend Oreille River.[12] The *Spokane Spokesman-Review*, gushed, "Evidently the homestead rush is nearing its end. It will be succeeded by the age of irrigation."[13] That newspaper had dubbed the effort the "Big Bend Project," and noted that Newell compared its size with the Panama Canal. A front-page editorial a year later saw the preliminary work as a sign that the government would soon start construction, an observation premature by three decades.[14]

In its first annual report, the infant Reclamation Service suggested artesian wells as the most probable way to irrigate the dry Columbia plateau.[15] Engineer Noble explored ways to divert water from the Spokane, Palouse, and Columbia rivers and a plan to divert water from the Pend Oreille River into the Spokane River and then by canal to the Grand Coulee, that would act as a storage reservoir.[16] Frederick Newell himself visited the Columbia Basin in 1903 and later testified before a congressional committee that the Big Bend, where the Service had spent over $8,000, would someday be the greatest irrigation project in the world, but that its cost would be "staggering."[17]

Businessmen statewide and locals pushed both Washington and federal officials to consider and reconsider each irrigation idea. "That the Palouse irrigation project will ultimately be pushed to completion by the national government, there is not the slightest doubt in my mind," said Washington Republican representative William Ewart Humphrey in 1905.[18] Around 500 hopeful settlers staked claims that year on land that needed only water to make it bloom. But a year later, in the spring of 1906, as Reclamation Service officials rejected irrigation plans and prepared to leave, the locals gathered and threw rocks at them. In a prophetic statement, the Reclamation Service argued that the Big Bend might serve as a project during some future "hard times" as a large public works effort.[19]

In 1906, David R. McGinnis, chief of immigration for the Great Northern Railroad and landowner in Wenatchee, approached Rufus Woods, then a fledgling newspaper publisher and secretary of the Wenatchee Commercial Club, with a plan to irrigate 500,000 acres around Quincy. McGinnis envisioned using water from the Clark Fork River (now called the Pend Oreille). Later he switched and suggested drawing water from Lake Wenatchee and taking it across the Columbia in a huge inverted tube called a siphon. The $12 million idea quickly withered and died.[20] A reevaluation of the Palouse Project met a similar fate when the Reclamation Service announced that the estimated cost of around $6 million would not produce a successful result.[21]

The Reclamation Service promoted strong local water user organizations and the state statutes needed to legalize them.[22] The Service then used these bodies to govern water distribution on its projects and, more importantly, to make contracts with the government to cover repayment of costs. In 1907 the people around Quincy, Washington, founded the Quincy Valley Water Users Association. Three years later that association established the Quincy Valley Irrigation District and, directed by engineer Joseph Jacobs, formulated a $44 million dollar plan to water 435,000 acres at just over $100 an acre.[23] Jacobs also suggested going up the Columbia River to the "head of the Grand Coulee" for some sort of installation.[24] Startled at the cost, Quincy residents asked the Reclamation Service for help, but Director Newell turned them down citing the expense. Undaunted, the promoters and farmers approached congressmen, chambers of commerce, and even wrote to President Roosevelt directly, but without result.

The Army Corps of Engineers surveyed the Columbia River looking for dam sites to aid navigation; they also considered irrigation as a possible side benefit. The engineers reported river commerce as insufficient to warrant such expensive improvements and noted that any dams would have to be primarily for power.[25] Here was the start of the long conflict between the Bureau of Reclamation and the Corps of Engineers at Grand Coulee. At first each looked only after its own interests, with the Bureau concentrating on irrigation and the Army focusing on navigation and power production. Later, when it became evident that any successful project would need to combine both power and reclamation, the two agencies clashed. But in the early 1900s, that was still in the future.

"It may not be in this generation or the next, but the time will come when an immense irrigation project will be carried through for the reclamation of the Big Bend country," stated Frederick Newell when he visited Spokane in May 1909. Newell talked about the reclamation of 500,000 acres using water from the Spokane River.[26]

Although the Reclamation Service pointed to successes throughout the West, irrigation was expensive and many farmers receiving government-subsidized water went broke. In 1914, Congress extended the ten-year repayment period to twenty years, allowing farmers to meet their obligations more easily. Congress also then limited the Service to projects it approved. That year Frederick Newell, demoted because of pressure from reclamation opponents in the East, left the Service, replaced by Arthur Powell Davis. The bright promise of reclamation dimmed and many, especially in the East, firmed their opposition to allowing the government to subsidize irrigation of land that competed with farms already in production.

Through 1912 and 1913, plans for the Quincy and the Palouse projects repeatedly resurfaced and the Quincy plan received considerable attention in Olympia. The Quincy Valley Water Users Association approached the state legislature in 1913 requesting $100,000 for a thorough survey of their project. This passed the state Senate and became Referendum Measure Number Two that appeared on the statewide ballot in 1914. That ambitious venture asked the voters to support a $40 million bond issue to build the Quincy Project and establish a board of governors to operate it. In the November election, only Chelan County residents approved the measure.[27]

The Washington voters acted wisely. That year farmers defaulted on millions of dollars worth of irrigation securities in Colorado, Wyoming, Utah, Idaho, Oregon, and Washington. Through 1917 Washington state refinanced over forty of its irrigation districts facing bankruptcy.[28] Plagued by unanticipated construction expenditures and by huge maintenance and replacement costs, irrigation repeatedly ruined those who put money into various schemes. Of course every dreamer, visionary, and promoter assured doubters that his project would be different, but few realized that hope. From 1903 until 1915 the Reclamation Service spent over $86,000 attempting to find ways to irrigate the Big Bend country. The State of Washington added $10,000 to the Palouse venture alone. The nearly $100,000 total produced no results.

World War I reversed the sagging fortunes of reclamation nationally and rekindled interest in irrigating the Big Bend. Columbia Basin wheat prices rose dramatically. As forty million new acres entered production, the country conserved food and families augmented their diets with "wheatless" or "meatless" days. Patriotic slogans like "If you can't fight, farm: Food will win the war!" and "Plow to the Fence for National Defense," urged farmers to increase output.[29] "Food, not money may win the war," ran an editorial headline in the *Wenatchee Daily World*.[30] Farmers, business owners, and promoters in the Columbia Basin looked at their two million unproductive acres awaiting only water. How to get that water became the general topic of conversation in Pasco, Moses Lake, Wenatchee, Ephrata, and throughout the Big Bend. The two basic proposals—a dam at the Grand Coulee or water diverted from somewhere in Idaho or Montana—resurfaced. Each generated its camp of supporters and they engaged in a debate that lasted almost fifteen years.

One of the great legends surrounding Grand Coulee Dam is the story of Billy Clapp. Quiet, mild-mannered William M. Clapp operated a modest legal practice in Ephrata. As the story goes, he and his friends occasionally gathered and discussed matters of local importance. One spring day in 1917, Clapp, A. A. Goldsmith of Soap Lake, Paul D. Donaldson and Warren Gale Matthews of Ephrata, and possibly Samuel Billingsley Hill (a Democrat who later became a United States representative from Washington) lamented the failure of the 1914 bond measure that would have irrigated land and could have aided the war effort. Then Donaldson talked about his recent trip with Dr. Henry Landes, state geologist and professor at the University of Washington.[31] Landes had worked since the early 1900s with the United States Geological Survey on a topographic survey of eastern Washington and Donaldson often went along when he took measurements. In passing conversations Landes explained the theory that glaciers once crossed the Okanogan highlands and dammed the Columbia somewhere below the Grand Coulee, diverting the flow of the river into that now-dry canyon. Clapp entered the conversation and suggested that if nature once dammed the Columbia, nothing could stop men from doing the same thing—this time with concrete. That dam could again divert water into the ancient coulee and onto land in Grant, Adams, and Franklin counties, including points as far south as the Snake River. The men agreed that Clapp offered an interesting idea, and they ruminated over it for a few months.

Norval Enger, deputy Grant County engineer, attended some of those discussions. One of the men asked Enger if he could measure the distance from the Columbia River up to the mouth of the Grand Coulee to determine how high a dam diverting the river would need to be. Enger apparently wanted official authorization, so late in the summer of 1917 Clapp, Donaldson, Goldsmith, and Matthews approached the Grant County commissioners. They told Enger to conduct something of an investigation, provided that everyone kept quiet about the whole thing. Afraid of being laughed at for wasting county money on such a wild adventure, the men all agreed to the terms. That winter Enger examined the Columbia at the Grand Coulee and later reported that the idea for a dam had merit, but that the engineering would be costly—expensive well beyond anything that Grant County could afford.

The Ephrata men played with the idea of the big dam but it would have eventually died had not Rufus Woods come to town. Woods's *Wenatchee Daily World* circulated throughout the Big Bend. It constantly promoted the growth and development of the region. "Sees Mighty Empire on Upper Columbia River," a headline that ran in the spring of 1918, was typical.[32] Woods scoured the countryside for stories on that theme and used them to fill the pages of his newspaper. In that quest he often drove around in his car, which carried a typewriter bolted to a stand in the back seat. Wherever he found an interesting story, there he sat and wrote it.

On a hot July day in 1918, Rufus Woods drove into Ephrata, approached Gale Matthews, and asked if he had any news. Matthews brushed off Woods saying he was busy, and then playfully added that Woods should look up Billy Clapp if he wanted a story about something really big. It would be an idea about a dam on the Columbia River, Matthews hinted. Woods found Clapp and the lawyer sheepishly explained his idea.

"FORMULATE BRAND NEW IDEA FOR IRRIGATION GRANT, ADAMS, FRANKLIN COUNTIES, COVERING MILLION ACRES OR MORE," ran the headline that Woods wrote for his story that appeared on page seven of the July 18, 1918 issue. News of General Foch's aggressive actions in the European war filled the front pages while the small story about Billy Clapp's proposal appeared at the top of a back page.[33] "The latest, newest; the most ambitious idea in the way of reclamation and development of water power ever formulated is now in process of development," stated Woods in his flamboyant style.

The article in the *Wenatchee Daily World* was not the first time someone suggested building a dam on the Columbia River near the Grand Coulee. And it was not the first time that a newspaper printed such an idea but, in contrast to its predecessors, it was the story that people remembered, took seriously, and which caused them to act. In that sense it was the genesis of Grand Coulee Dam.

Rufus Woods himself probably discounted the story and only used the suggestion for the big dam as something of a jest to liven up his newspaper. Superior Judge R. J. Steiner of Waterville wrote a letter to the Wenatchee editor poking fun at the suggestion and he ended with the comment "Baron Munchausen, thou art a piker."[34] Commenting on the story in 1930, Woods wrote, "It was a joke to a certain extent but it began to 'take' all over this territory."[35] There is no evidence that Woods took the suggestion seriously and if he did, he did not actively crusade on its behalf at that time.[36] No other newspapers in the area even picked up the story.[37]

The next mention of the proposed dam in the *Wenatchee Daily World* came that December, just after World War I ended.[38] In an article proposing development of the Quincy Project as a place to house soldiers returning from Europe, the *World* included the dam among the possible options. It pointed out, however, that Grant County Engineer Enger estimated the height of such a structure at over 550 feet. The article indicated a preference for a gravity system carrying water from a distant point.[39] Woods might later have regretted that opinion when he fought creation of just such a system, but that was still ahead.

Failure of the Quincy Project referendum in 1914 taught irrigation backers that their success would depend on federal funding. Long lists of those delinquent in paying charges on land already irrigated appeared in the *Wenatchee Daily World* throughout 1918. Promoters had to persuade the government that investment in the Big Bend was worthwhile and the government had to build the project with a minimal financial burden on farmers. To that end the plan proposed by Elbert F. Blaine carried greater immediate promise.

Like the legend of Billy Clapp, the story of Elbert F. Blaine is difficult to document precisely. In Blaine's story, one thing is certain, however. Although hardly the first to think of it, he is the person generally credited with suggesting what everyone later called the "gravity plan." In 1917 and 1918, Blaine, a resident of Grandview and a landowner in the Yakima Valley, was chairman of the Washington State Public Service Commission. An eminent Republican attorney, horticulturist, and real estate developer, he also served, at the behest of Governor Ernest Lister, as chairman of the state Railroad Commission.[40] In his work Blaine frequently crossed the dry eastern areas of the state.[41] In later recountings, Blaine explained that during one trip he lamented the aridity of the Big Bend country. It struck him that water, diverted from the Pend Oreille River at or near Albeni Falls (then commonly called Albany Falls) and carried by gravity flow out of northern Idaho, could successfully irrigate most of the land east of the Columbia River and west of the Palouse.

On a trip to Washington, D. C., Blaine said he looked up land surveys and geological survey maps and checked his plan's feasibility.[42] He found the route predominantly downhill and pronounced the scheme workable.[43] Back in Spokane, Blaine announced his idea at a meeting of the Chamber of Commerce, probably on November 9,[44] although Blaine later stated that he first considered the plan as early as 1917.[45] Because the competition between the Clapp plan, known as the "pumping plan," and the Blaine plan, or "gravity plan," became so heated through the 1920s, it is possible that over time both sides exaggerated their stories. Backers of Clapp argued that Blaine and his friends developed the gravity plan only after people took the dam seriously. Blaine's people charged just the reverse. Certain only is that late in 1918 two old ideas for irrigating the Columbia Basin re-emerged, each with its avid backers.

The Spokane Chamber of Commerce actively supported Blaine. For his part, Blaine immediately promoted his suggestion. He spoke about it around the state wherever and whenever he could and gained the ear of both Governor Lister and Seattle Mayor Ole Hanson. Hanson and Lister traveled to Spokane and elsewhere speaking actively in favor of Columbia Basin irrigation. During the fall and early winter of 1918, the gravity plan appeared as the most likely method for Big Bend reclamation. Billy Clapp's idea remained in the background, receiving little attention.

Neither idea originated in 1918. But the war, food shortages, the need to resettle veterans, and the desire of eastern Washington residents to see their area grow all converged and heightened interest in reclamation. Motivated by the twin desires of making money and seeing their region grow and prosper, residents grasped the promise of irrigation with renewed vigor. But none of them could have guessed that it would take almost fifteen years for the government to decide between the plans, and over thirty years before any water poured onto Columbia Basin land. First came the debate. It was long, hard-fought, contentious, bitter, and it divided eastern Washington. It was so fierce that it weakened the effectiveness of those who advocated the project and almost prevented it from happening at all.

Block no. 26 of the Columbia Basin Project before irrigation. *Courtesy Bureau of Reclamation.*

# Chapter Three
# The Columbia Basin Survey
# Commission

The ruthless pursuit of private gain was the driving force behind most western enterprises.

*Rodman Paul*[1]

Through the late fall and winter of 1918, and on into early 1919,
Elbert F. Blaine's "gravity plan" gained attention and support in
eastern Washington and beyond. In the late spring the *Wenatchee
Daily World* rekindled interest in the large dam. Billy Clapp's proposed structure, high enough to divert the Columbia directly into
the Grand Coulee, would have been incredibly huge and would
have backed water well into Canada. Local engineers suggested
that instead the river generate power to pump water uphill from a
reservoir behind a more modest dam. This evolved into what became the "pumping plan."[2] Blaine and Clapp became minor players
in the contest between the two sides. After the first flurry of interest Blaine disappeared entirely and Clapp, modest and somewhat
shy, appeared from time to time but essentially stayed out of the
fray.

Spokane's Chamber of Commerce backed the gravity plan,
while people scattered throughout smaller Big Bend towns favored
the pumping plan. A number of factors intensified the protracted
debate between the two sides. First, local pride pitted Spokane
against Wenatchee. Businessmen in each area hoped, if not to dominate, at least to influence the direction and economy of the region.
Spokane saw the Big Bend as its own private commercial domain
and the city's Chamber of Commerce talked as if it owned the region. Residents of smaller cities resented the second-class status

to which Spokane's economic clout relegated them.[3] This rivalry and the suspicions it generated added color and, at times, acrimony, clouding the real issues.

Second, as it developed, the idea for a dam on the Columbia River required both federal funding and hydroelectric power production on a large scale. This would have resulted in direct government competition with the Washington Water Power Company,[4] Spokane's largest employer and the most influential supporter of that city's Chamber of Commerce. The company's need to protect its economic interests understandably fueled its desire to suppress the big dam despite the plan's merits or advantages.

Third, the gravity plan included diversion of irrigation water into the Spokane River during the off-season. This would have benefited the Washington Water Power Company by increasing output at its generating facilities located there.[5] If Washington Water Power people killed plans for the big dam and implemented the gravity project, the company would doubly benefit.

Fourth, the years after World War I brought the "Red Scare" era during which nationalists and capitalists saw the threat of socialism and foreign radicals everywhere. The thought of a large government project in direct competition with free enterprise was something the public, especially its entrepreneurial sector, viewed with alarm. The notion of the government not only managing the project but even using it to enter the power market as a competitor with private companies became the bugaboo that gravity plan backers successfully used against the pumpers for over a decade. They thought the government should pay the bills but not usurp the resulting project, which represented huge potential profits for whoever managed it. There was no danger, they reasoned, of those things happening under their plan.[6]

Finally, the 1920s saw expansion of irrigation slow to a standstill nationally. Except for occasional small acreages developed by the Reclamation Service, little new land came under the ditch during that decade.[7] The Reclamation Service, attacked by conservatives and Easterners as expensive, wasteful, and unnecessary, struggled to maintain itself as a viable government agency and it looked for dramatic projects to revitalize its image. The two proposed Columbia Basin options appeared huge, expensive, and burdened by difficult engineering problems that hardly excited government engineers. Through the 1920s the gravity people vied with the pumpers for recognition by the Reclamation Service (later the Bureau of Reclamation) while that agency acted interested but actually eschewed involvement. Over the years from 1918 until 1933, overt and often covert activity including surveys, debates, organizing, lobbying, fund-raising, deceit, and acrimony were all part of the protracted struggle to irrigate the Big Bend. Each side had self-serving motives and each used devious tactics to advance its cause.

During the late fall and winter of 1918, Governor Ernest Lister became the gravity plan's most visible backer.[8] He wrote to Secretary of the Interior Franklin K. Lane and offered to let Washington state hydraulic engineer Marvin Chase and state geologist Dr. Henry Landes help the Reclamation Service conduct detailed investigations of the Columbia Basin. Lane replied that the project, estimated at $250 million, would cost too much.[9] At a reclamation conference, held in Seattle late in November, the Big Bend Project suffered a second setback. Farmers advocating smaller developments elsewhere in the state actively opposed it, voicing fears that it would reduce their chances of gaining irrigation.

Undaunted, in December Landes, who for years lent his name, position, and time to back the scheme, and Marvin Chase laid out preliminary plans for the gravity project. They pictured an intake at Albani Falls, a canal in the valley of the Little Spokane River, an aqueduct two miles east of Spokane carrying water over the Spokane River, a tunnel from there to Hangman Valley, a second tunnel taking water to Rock Creek Valley north of Bonnie Lake, and then more canals directing the water onto 2.5 million irrigable acres. Professor Landes suggested calling the plan the Columbia Basin Project rather than the Big Bend or Pend Oreille Project, and his name stuck.[10]

On December 15, with barometers in hand, Chase, Landes, and a few other men, mostly engineers with the Washington Water Power Company, drove along the main canal routes. For three days they bounced over the distance from Albani Falls to Davenport.[11] On December 22 at a dinner in the Davenport Hotel in Spokane they pronounced the plan as practical both physically and economically. Everything was downhill all the way.

The Spokane Chamber of Commerce formed a Columbia Basin Committee headed by its secretary, James A. Ford, who became the constant force behind and orchestrator of activity emanating from Spokane. As such, although he avoided publicity, he was a major figure, albeit something of a gray eminence, in the skirmishes between the gravity people and the pumpers. Armed with Landes's findings, Ford turned his attention to the state government in Olympia.

Governor Lister asked the legislature to appropriate $100,000 for a complete survey of the project,[12] and Spokane's representative introduced a bill to cover the request. It passed the state House of Representatives by a vote of 54 to 39 with opposition coming mostly from the western parts of the state.[13] On February 20 the bill passed the state Senate more comfortably at 34 to 5.[14] It provided the money and created the Columbia Basin Survey Commission to oversee its use.[15] Due to Governor Lister's increasing ill-health, lieutenant governor—soon to be acting governor—Louis F. Hart faced the difficult task of naming members to that commission.[16]

On March 1, Hart signed the legislation, then named state hydraulic engineer Marvin Chase as chairman of the commission. As members he appointed Osmar Waller, Spokane investor and realtor Arthur D. Jones, former state senator and rancher from Adams County Daniel A. Scott, and state commissioner of agriculture Elbert F. Benson. When Scott declined, Hart replaced him with wheat and stock rancher and Spokane Federal Reserve Bank director Peter McGregor.[17] Reluctantly the Reclamation Service named its engineer and consultant, David C. Henny of Portland, and engineer James Munn of Denver as its representatives. In late March the Columbia Basin Survey Commission hired its staff. J. C. Ralston, then a consulting engineer with the Washington Water Power Company, became the chief consulting engineer, Arthur J. Turner chief engineer, Ivan E. Goodner office engineer, Lars Langloe field engineer, and Fred A. Adams educational director. All except Adams had been or were then engineers with the Washington Water Power Company. The selection of the commission and the personnel the commission hired, all linked to Spokane and its private power concerns, destined the pumping plan to receive little serious consideration.

Rufus Woods and his *Wenatchee Daily World* looked closely at the commission, its composition, and its direction. Woods realized that the men represented Spokane interests who might justify their project rather than conduct a more objective survey, including the idea for a dam at the Grand Coulee. Woods felt both the dam and gravity plan merited investigation. On April 21, 1919, the *Wenatchee Daily World* headline read, "Next Big Thing After Panama Canal," followed by an article about a dam costing around $30 million.[18] "A dam on [the] Columbia would yield untold electrical energy," continued the article, and it would provide irrigation at a maximum cost of about $75 an acre. In conclusion the article explained that the *Wenatchee Daily World* had asked James O'Sullivan to summarize the facts in a series that would appear soon.

James Edward O'Sullivan immediately became one of the most important players in the gravity plan/pumping plan drama. Born on June 26, 1876 in Port Huron, Michigan, where his Irish-Catholic family ran a construction business, he graduated from the University of Michigan in 1902 with a law degree and passed the state bar one year later. In 1905 he moved to Seattle and worked as a concrete foreman; then he taught for three years in Bellingham. In 1909 he visited Ephrata and one year later opened a law office and bought land there. In 1914 he went back to Michigan to help with his father's business.[19] Five years later, in 1919, he returned to Ephrata.

Magniloquent and often contentious, O'Sullivan eventually adopted Grand Coulee Dam as a cause that became his obsession. Through the 1920s and 1930s he jeopardized his health and his family in the fight with the gravity people, whom he saw as rich businessmen conspiring to dominate eastern Washington. O'Sullivan stood for the small farmer and private landowners such as himself, but more than that, he visualized the great dam as a way to transform the state into an agricultural and industrial success story unequaled nationally. He and Rufus Woods shared this vision. While many people had a part in getting Grand Coulee Dam built, none made it as much the focal point in their lives as did James O'Sullivan.

Researching articles for Rufus Woods, O'Sullivan read *Principles of Irrigation Engineering* by Frederick Haynes Newell and

Daniel William Murphy. The two authors discussed the idea of using water power to irrigate land at higher elevations. The excess power sold to light and heat towns would subsidize projects otherwise too expensive.[20] O'Sullivan made a quantum leap. If done correctly with a big dam generating enough power, the irrigation would be almost free.[21] And O'Sullivan knew that the Reclamation Act of 1906 authorized the Secretary of the Interior to sell surplus power.[22] This was the concept that James O'Sullivan preached continually for over a decade and which animated the entire pumping plan effort. The gravity plan involved Columbia Basin farmers making repayments which, in bad years, might ruin them financially, just as they had irrigators elsewhere. With the dam and its abundant cheap power to pay all the bills, the irrigation would cost nothing. That overriding idea, hypothetical and without precedent on such a large scale, became the great promise that motivated O'Sullivan and others in their fight against the Spokane interests.

This concept appealed to people in Wenatchee, Quincy, Ephrata, Moses Lake, and Pasco. Presenting the specter of power surpluses and intense government activity in the private sector, it appalled businessmen, especially those with the Washington Water Power Company in Spokane. A huge block of electricity generated by the government would compete directly with their monopoly and eliminate their control over the economic development of eastern Washington, to say nothing of slashing their profits. The idea struck at the heart of their financial well being, their leadership position regionally, and their conservative political and economic sensitivities. Not surprisingly they pursued every way they could imagine, fair and foul, to quash the proposed dam and pumping plan.

On May 3, 1919, O'Sullivan's first article appeared in the *World* on page seven. He pushed the dam as the best alternative for the Columbia Basin Project because its construction would save millions of dollars over the gravity plan. O'Sullivan's May 14 article stated that the Columbia River Dam (nobody specifically called it the Grand Coulee Dam until sometime later) would save $60 million over the gravity plan.[23] He urged the Columbia Basin Survey Commission to spend at least 10 percent of its $100,000 investigating the pumping plan. "The revenue from sale of electric energy alone would surely pay all the upkeep, interest on the investment; and provide a sinking fund for the liquidation of the cost of the project itself."[24] The O'Sullivan articles heightened interest in the pumping plan, and Rufus Woods later said, perhaps exaggerating, that they doubled his newspaper's circulation.[25] After writing the articles, O'Sullivan returned to Michigan.

The commission gave only passing attention to the Columbia Dam suggestion. On May 5 its members left Spokane on a four-day inspection trip to the Pend Oreille River and a quick look at the Columbia River near the Grand Coulee. Rufus Woods accompanied the group. A few days later he wrote that they investigated "the Greatest Project yet conceived by man—for that is what the Columbia Basin Project is."[26] Irked by their short stay at Grand Coulee, O'Sullivan sent his articles to Arthur Powell Davis, head of the Reclamation Service. Davis consequently wired commission members and urged them to look more closely at the dam. Since project backers knew that the federal government would ultimately have to finance the project, they could not ignore the suggestion.[27] Grudgingly they investigated the pumping plan, emphasizing why it could not succeed.

Through the spring of 1920 the commission rushed to complete its report.[28] It decided against any dam on the Columbia River saying, "There is little question but that it is infeasible to place a dam in the Columbia River at [the] head of [the] Grand Coulee."[29] Issued on July 5, 1920, the report placed the gravity plan cost at $171 an acre.[30] An appendix quoted sections of the August 5, 1846 treaty with Great Britain guaranteeing the Hudson's Bay Company free navigation of the Columbia River, concluding that it prohibited construction of any dam on the Columbia River unless it included expensive locks.[31] Finally, the report also argued that the Columbia River at the Grand Coulee contained deep deposits of glacial material with bedrock anywhere from 150 to 250 feet below the surface, far too deep to support a dam.[32]

The 177-page report proposed irrigating 1,733,000 acres with water from the Pend Oreille River. Canals spanning 134 miles would provide irrigation for 44,000 families living on forty-acre farms. A $200,000 aqueduct east of Spokane would allow excess water to run into the Spokane River, boosting power capacity at the

Washington Water Power Company facilities. The report did not suggest charging the company for the benefits.

Startled by the report, David C. Henny, the Reclamation Service's consulting engineer, wrote to Reclamation Service Chief Engineer Franklin E. Weymouth, "There are many single items in the estimate which in themselves would constitute the equivalent of a large Reclamation Service project."[33] Weymouth replied that the commission's work was preliminary, sketchy, and that the project was infeasible.[34] In his recommendation to his boss, Arthur Davis, Weymouth wrote,

> Neither Mr. Henny nor any other Reclamation official has been given opportunity to make other than a cursory examination and study of the project and it has seemed to me that nothing more has been desired by the Commission. On the other hand, in newspaper articles and other ways, it has seemed to me, an earnest effort has been made by the State Commissioners to make it appear to the public that the Reclamation Service is cooperating fully, while as a matter of fact the Service has had little opportunity to become familiar with the project. . . . For these reasons and for the reason that the acreage cost is so high and the total cost so great, I believe that the Service should at the first favorable opportunity withdraw from further participation in regard to this project.[35]

This decision, firmly held by the Reclamation Service for years, meant that neither the pumping plan nor the gravity plan had any chance of success. But project backers knew little of those feelings for neither Davis, Weymouth, or Henny openly shared them. If the commission members or Spokane people had any inkling of the truth, they too kept it a secret and gave the public the opposite impression. But disengaging from the project proved more difficult than Weymouth anticipated, and despite many efforts the Reclamation Service never successfully distanced itself from Columbia Basin Project proponents.

Meanwhile, to oppose the Spokane forces, a group of pumping plan supporters spent the summer and fall of 1920 organizing the Columbia River Dam, Irrigation, and Power Association. They held their first formal meeting in Ephrata on Friday, November 26 and they elected Nathaniel Willis "Nat" Washington of Ephrata as their president.[36] Had he lived longer, Nat Washington might well have been one of the familiar heroes of the Grand Coulee Dam story, and he did have considerable impact over the next few years. His organization, and later Rufus Woods's newspaper, kept alive the effort to publicize the pumping plan when few gave it serious consideration. As most of the Ephrata group tended to be Democrats and since Woods was a Republican, they did not always work together comfortably.[37] Indeed, support by Woods was intermittent and at times nonexistent.

In late September the Northwest Reclamation and Land Development Congress opened in Seattle.[38] James O'Sullivan attended and managed an introduction to A. P. Davis who came from Washington, D. C. In their conversation, O'Sullivan invited Davis to look over the proposed dam site and Davis accepted, visiting the Grand Coulee on September 23. On October 2 a large headline on the first page of the *Wenatchee Daily World* declared, "Columbia Dam is Feasible," followed by a smaller line that read, "Thinks A. P. Davis, Head of U.S. Reclamation Service."[39] O'Sullivan later claimed that Davis declared the Grand Coulee the greatest damsite in the world. The amount of enthusiasm that Davis actually showed remains unclear, but the story brought howls from gravity plan backers in Spokane. Members of the Quincy Valley Irrigation District wrote Davis and asked if he said the things attributed to him. "This statement is an error," Davis replied. "I have never had the data or made the study upon which to pronounce on the feasibility of this structure, and have not done so."[40]

Fred Adams, educational director for the Columbia Basin Survey Commission, reflecting the heated protest in Spokane, wrote a bitter letter to Davis criticizing his "unauthorized" visit to the damsite and asking Davis "to set the record straight."[41] "I am greatly surprised at the belligerent tone of the letter," Davis replied. "I received copies of newspapers with exaggerated accounts of my visit to the Grand Coulee, and I wrote a letter . . . disclaiming the articles."[42] In another letter Davis wrote:

> Newspapers in the vicinity of the Big Bend Country publishing some very fulsome laudatory accounts of my visit to the dam site "did so" without my knowledge or consent and goes so far as to be rather humiliating.[43]

O'Sullivan apologized to Davis for anything he wrote that might have caused discomfort for the Reclamation Service.[44]

On December 13, 1920 the Reclamation Board of Review issued its analysis of the Columbia Basin Survey Commission report. The board stated that the commission estimated gravity plan costs 9 percent too low and the pumping plan $5^{1}/_{2}$ percent too high. Review board member David Henny wrote privately, "I believe [it] inadvisable to show that [the] commission made such [a] glaring mistake and prefer [to] leave it implied."[45] Privately, Reclamation officials felt that the survey commission had stacked, if not outright falsified, its report to favor the gravity plan. Unwilling to affront the apparently influential Spokane forces, however, they hedged and refused to say so directly. This timidity was partly responsible for the Reclamation Service's inability to divorce itself from the project. Reclamation officials instead recommended that the state spend $50,000 for drilling to check the feasibility of the dam and pumping plan in which they did have some interest.[46] A measure to appropriate the needed money wound its way through the state legislature despite everything from lukewarm support to open hostility from Spokane. Benjamin E. Thomas, the representative from Soap Lake, later recalled that unnamed people offered him $1,000 to vote against the core drilling appropriation.[47] Nevertheless, the bill finally passed the legislature.

On March 13 the measure—that not only provided the money but also abolished the Columbia Basin Survey Commission—appeared before the governor. The bill transferred authority for investigation and promotion of the plan to the state's Director of Conservation and Development. Governor Hart appointed Daniel A. Scott of Ritzville to that position and Scott in turn named Fred A. Adams as supervisor of the Columbia Basin Project Division. Adams, a past employee of the Washington Water Power Company and educational director for the Columbia Basin Survey Commission, came strongly recommended by the Columbia Basin Committee of the Spokane Chamber of Commerce.[48] A week later Scott announced that his offices would remain in Spokane and that Ivan Goodner and Arthur J. Turner would be his chief engineers. In other words, the Spokane people again dominated and the new organization remained in roughly the same hands that had promoted the gravity plan.[49] Adams, who had openly opposed the core drilling plan and had spoken to the legislature against it, awarded the contract to the International Diamond Drill Company of Spokane on June 20, 1921.

Hedging its bet in the event that a dam might someday be built at the Grand Coulee, on June 30, 1921 the Washington Water Power Company filed with the Federal Power Commission for a permit to place a dam at Kettle Falls. Located 110 miles upstream from the Grand Coulee site, a dam at that location would have limited the height of any dam downstream. When the move became public on July 7, pumping plan backers saw it as a deliberate ploy to block a high dam and extensive power development at Grand Coulee.[50] The Washington Water Power Company, lamented Rufus Woods, is a "soulless corporation."[51] Despite repeated objections frequently expressed over the next year by the pumpers, who really had almost no political clout at the state level, let alone with national agencies, the Federal Power Commission granted the preliminary permit on July 26, 1922. Significant because of the threat it posed through its ability to limit the size of any dam at Grand Coulee, the Kettle Falls project was a nuisance that plagued the pumpers for years.

The *Spokane Spokesman-Review* began an editorial attack on the core drilling project, focusing on Arthur Davis and David Henny:[52]

> If D. C. Henny, chairman of the reviewing board appointed by Director Davis of the United States reclamation service, wanted to wear out the state and defeat the Columbia basin project he could not have devised a more effective means than the recommendations in his report.[53]

Spokane people reasoned that if they could discredit the drilling and possibly cause its cancellation, nobody would ever know the nature of the proposed dam site. They were willing to risk angering Reclamation Service high officials, their possible future benefactors, in that effort. Despite the negative publicity, the International Diamond Drill Contracting Company set up its equipment that May and began work. In July, Fred Adams reported that they had not found satisfactory bedrock, saying specifically that the drilling showed a "flimsy foundation" of soft rotten rock.[54] And the *Spokane Daily Chronicle* reported, "Drilling to a depth of 156 feet at Grand Coulee . . . has failed to reveal any rock formation upon which any kind of masonry work might be constructed."[55] On July 26 the *Spokesman-Review* reported, "No rock was encountered."

Speaking for the pumpers, Nat Washington urged David Henny to visit the drilling site and investigate. Henny went and then wrote to Arthur Davis, "The first hole drilled on tentative center line . . . hit gumbo or clay to a depth of 172 feet when granite was struck. Second hole . . . found granite at 157 feet."[56] Since this data differed markedly from the adverse news reports, Henny ordered Ivan Goodner and Fred Adams to release accurate information or he would expose them. In response the *Spokesman-Review* increased its personal attacks on Henny.[57] But on August 25 the *Spokesman-Review* reported, on a back page, "Solid rock has been struck at a depth of 62 feet below low water level in the fifth hole to be drilled at the Columbia basin dam site." Increasingly favorable reports continued from the drillers as the work progressed through the fall and into the winter.

The successful core drilling heated up the debate as each side looked for ways to strengthen its position or discredit the opposition. The state's Department of Conservation and Development investigated two possible ploys. First, the state hired consulting engineer Willis Tryon Batcheller to investigate the power potential at Foster Creek, more than thirty miles downstream from the Grand Coulee, a site that could generate and transmit enough electricity to operate pumps upstream.[58] Such a plan would eliminate the threat of a large block of power posed by a high dam. On August 8, 1921, Batcheller, who had worked for Seattle City Light, took the job.[59] Later, as drilling revealed a satisfactory damsite at the Grand Coulee, his job description unofficially grew to include his evaluation of a possible dam and pump plant there also.[60] Batcheller quietly went to work and produced little news.

Second, hoping to find a prominent public figure to support the gravity plan, end the debate, and draw national attention to the project, officials in the Department of Conservation and Development courted General George Washington Goethals, well-known heroic builder of the Panama Canal.[61] On December 19 the news about Goethals broke in the *Wenatchee Daily World* and his coming appeared certain. Goethals demanded a $25,000 consultation fee, but the state had only around $15,000 left from the $50,000 core drilling appropriation. After negotiations with Spokane Chamber of Commerce Secretary Ford, Goethals dropped his price to $20,000[62] and Spokane businessmen, aided by a number of chambers of commerce around the state, donated the remaining $5,000.[63] On paper, then, it appeared that the government, rather than a promotional agency, hired the general.[64] Spokane interests, however, not only provided the needed $5,000, but they also paid most of the incidental bills involved with housing, feeding, and entertaining the general.

"The judgment of as distinguished an engineer as Goethals should settle [the matter] once for all, and would have great weight in winning federal support if the plan is approved," declared the *Spokesman-Review*.[65] Goethals arrived in Spokane on February 1, 1922. For nine days he surveyed the Columbia Basin, largely from private railroad cars supplied by the Milwaukee and Great Northern Railroads. Engineers formerly with the Columbia Basin Survey Commission always traveled along. At every stop, local public officials, members of the Spokane Chamber of Commerce, or both, wined and dined Goethals in grand style.[66] "The Basin Project is fascinating," said Goethals, "but years of work will be needed and it might cost $170,000,000." At Albani Falls he wore galoshes and waded around in two feet of snow. He spent most of his time investigating the gravity plan, a few days going over facts and figures in the Spokane Office of the Department of Conservation and Development, and a few hours looking at the pumping plan site. He finished in ten days.

Then Goethals left the Pacific Northwest, but his son remained behind to complete the anxiously awaited report. Although the newspapers heightened anticipation, those in the inner circle knew exactly what Goethals found. Like many consultants, the general first discerned what his employer wanted and then produced that product. In this case it was easy since the officials, who constantly hovered around him, spelled out exactly what they wished. He knew that the Spokane people demanded a favorable view of the gravity plan and he gave it to them, or rather, he allowed them to give it to themselves.[67]

On March 30, 1922, Goethals submitted his report and the state made it public on April 7.[68] The document rejected the pumping plan, Goethals conveniently finding that no market existed for the anticipated power. It praised the gravity plan, which he put "at

$145.56 an acre, $26.14 less than the original estimate."[69] The gravity plan would irrigate 1,753,000 acres as compared to 1,403,000 acres for the pumping plan. Goethals added that the gravity plan involved no international treaty complications.[70] The data in his report came almost directly from the Columbia Basin Survey Commission. Rufus Woods later charged that state engineer Ivan E. Goodner wrote it.[71] In an interview years later, Bob McCann of the Spokane Chamber of Commerce admitted that Goethals did nothing more than "say amen" to the report of the Columbia Basin Survey Commission.[72] Then he took his money and left.

With the Goethals report in hand, James Ford and the Spokane Chamber of Commerce began selling the project to Congress. People around the state lined up behind the gravity plan, and while a few continued the struggle for the pumping plan, even the *Wenatchee Daily World* temporarily spoke favorably of the gravity venture. From this time in 1922 until 1928, the *World*, often credited with avidly and continually backing the pumpers, carried few serious references to Grand Coulee Dam.[73] For the moment, the field belonged to the Spokane forces.

Privately, the Reclamation Service counted itself among those unimpressed by Goethals's report, feeling it full of errors.[74] In his work, however, Goethals briefly mentioned the study being quietly done by Willis Batcheller. Arthur Davis, who knew about that study through David Henny, who had indirectly supervised Batcheller, used Goethals's comment as a reason to request a copy of Batcheller's findings.

Batcheller had worked alone since hired by the Department of Conservation and Development in August 1921. In December, Dan Scott, of that office, wondered what Batcheller was doing. He feared that Batcheller, whom the state paid by the day, had expanded and prolonged his job to increase his salary. As Scott quickly discovered to his dismay, David Henny, in his capacity with the Reclamation Service, had directed Batcheller away from Foster Creek and to the Grand Coulee site. There Batcheller explored the feasibility of pumping water into the Grand Coulee from behind dams of different heights. This clearly went beyond Batcheller's instructions from the state and pointed him in a direction the gravity people wanted ignored. The Department of Conservation and Development moved to stop Batcheller. First they reminded him that the state hired him and not Henny or the Reclamation Service, and that his drawings, plans, notes, and computations belonged to the state.[75] Then, on February 1, 1922, Fred Adams fired Willis Batcheller.

Forced to leave his papers behind, Batcheller left for Seattle with only a draft of his report that he all but smuggled out of the office. Ten days later he sent the finished product to the state.[76] It showed the pumping plan as dramatically cheaper than the gravity plan even if power sales did not offset costs![77] The Department of Conservation and Development refused to pay Batcheller for any work after February 1, or reimburse him for the cost of binding his report. Then it consciously suppressed Batcheller's study.

Eventually the Reclamation Service, and later James B. Cavanaugh, working on a survey for the Federal Power Commission,[78] successfully obtained copies (it took Arthur Davis two tries), and these influenced subsequent investigations done by other agencies. Only years later, however, did the significance of Batcheller's effort become apparent. He had further convinced the Reclamation Service that it should not seriously consider the gravity plan. In that contribution, the work of Willis Batcheller was more important than this minor incident might indicate. Throughout his later life, Batcheller never forgave Rufus Woods, among others, for not granting him credit as one of the leading figures who helped bring about the great dam.

It is noteworthy that during this period, and throughout the 1920s for that matter, few of the key players in the Columbia Basin drama lived on or farmed those lands; they were all lawyers, businessmen, professionals, promoters, and politicians. Like Woods, they believed that irrigation would provide a sound agricultural base supporting industrial and urban growth. Woods called his vision the "agricultural-industrial empire," and he worked tirelessly promoting it.[79] As for the farmers, they were few and seldom actively participated in the debate. During those ten years irrigators abandoned 2,000 miles of canals in Washington state, and 30,000 acres that once received water went dry. The number of farms on what became the Columbia Basin Project dropped from 1,524 in 1919 to 1,141 in 1929.[80] More and more basin land returned to the state, banks, or land companies as drought added insult to financial injury and farmers abandoned their holdings. Only those on the east side, which received more rain, many of them German-Russians familiar with dry steppe conditions, stayed in appreciable numbers and they did not actively push for irrigation.

The people in Spokane and the towns throughout the Big Bend, however, did work for irrigation. The Goethals study closed one chapter in that effort. Local and state advocates of the project now at least realized that the federal government would not finance the project without its own surveys. Goethals's work gave the gravity plan a temporary edge. With it in hand, project backers organized and approached the government, urging it to investigate. To that end they pushed government officials with a vengeance.

Major John S. Butler. *Courtesy U. S. Army Corps of Engineers.*

# Chapter Four
# The Government Investigations

To a great part of the nation, which knew nothing of irrigation and cared less, the whole project was a giant "pork barrel" scheme of the Pacific Northwest.

*William D. Miner*[1]

The Grand Coulee leaders called the opponents of the dam, "private power pickpockets," and they were called "desert rat land speculators."

*Clarence C. Dill*[2]

Suppression of the Batcheller study and a high profile for the Goethals report temporarily subdued advocates of the pumping plan. Backers of the gravity plan now assembled a broad coalition to lobby their project before the federal government. They would find controlling the agencies they encountered, and Congress, more difficult than manipulating state and local officials. And although they may not have realized it at the time, by involving those entities they risked losing control of their project, something they very much feared. Of immediate concern, however, was cajoling the government into endorsing and then financing the gravity plan.

On April 21, 1922, after a series of preliminary gatherings, the Spokane Chamber of Commerce sponsored a meeting in Pasco. The event drew over 500 people from around the state and beyond. They formed the Columbia Basin Irrigation League (CBIL) and elected John A. Gellaty of Wenatchee as its chairman with James A. Ford of the Spokane Chamber of Commerce as the secretary, again in the background but, as usual, directing events. On the spot they raised $72,180, with $53,596 coming from Spokane, and they vowed to collect $200,000 to promote the "Goethals Plan" as the way to irrigate the Columbia Basin.[3] Choosing "It Shall Be

Done," as their motto, they dispatched Ford to Washington, D. C., to open an office from which he could lobby members of Congress. The CBIL located its headquarters in Spokane, in rooms provided by the Chamber of Commerce.[4] At Ford's behest, Senator Miles Poindexter introduced a bill on June 14 calling for the creation of a national Columbia Basin Commission within the Department of the Interior. Modified, it passed the Senate but the House of Representatives adjourned without considering it.

In the November 1922 election, Democrat Clarence Cleveland Dill unseated Poindexter.[5] Although the *Spokesman-Review* opposed Dill because it claimed that Dill did not back the gravity plan,[6] Dill refuted the claims and vowed to press for the Columbia Basin Project. He eventually worked with both sides during the controversy, switching to whatever position he thought would bring irrigation to the Columbia Basin or, more importantly, whichever side was ahead. He became a central figure if for no other reason than because he always presented himself as being on the winning team, and he took credit for every accomplishment regardless of how much he had actually done to promote or, in some cases, discourage it. Consequently both sides sought his help but also suspected him. His role in backing the project is debatable. As someone who always gave himself considerable credit for bringing about Grand Coulee Dam, the colorful and enigmatic Dill exaggerated his own achievements and importance.

Early in 1923 Congress, on Dill's urging, reconsidered the Columbia Basin irrigation project, passed the bill, and created a Federal Columbia Basin Commission with $100,000 appropriated for an investigation.[7] Secretary of the Interior Hubert Work appointed Assistant Secretary of the Interior, Francis Goodwin, and the head of the Reclamation Service, Arthur Davis, to that commission and they hired Homer Gault, an engineer who had been with the Reclamation Service since 1906, to conduct the study.[8] Gault, who actually supervised the work (Goodwin and Davis were figureheads) established an office in Spokane.[9]

On April 21 the Columbia Basin Irrigation League went to Pasco for its second meeting. A special train from Spokane carried 130 delegates, 61 came from Seattle, 20 from Tacoma, and in all over 1,000 attended the event. They elected Harvey Lindley of

Seattle as president, a job he held for the next seven years.[10] Roy Gill, another Spokane personality, became chairman of the executive committee and in effect manager of the CBIL. Gill, like Ford, was a moving power who directed CBIL activities, and his lobbying abilities and relationships with government figures greatly increased the organization's influence. The delegates celebrated the Gault investigation, then in progress, and they paraded carrying signs that read, "Basin or Bust."[11] These annual meetings conducted by the CBIL not only helped maintain enthusiasm for the project, they also signaled local and national government officials and legislators that the gravity plan had appreciable support.

President Warren Harding visited Spokane in July 1923 and spoke favorably about the project, but died a month later. Calvin Coolidge, who did not view reclamation so favorably, replaced him. The Interior Department then fired Reclamation Service Director Arthur P. Davis, changed the name of the organization to the Bureau of Reclamation, and installed businessman David W. Davis as its first commissioner.[12] For some time the Bureau of Reclamation had not been a healthy agency. The agricultural depression of the 1920s and the Bureau's failure to produce numerous self-sufficient projects hurt it and the change in leadership to the businessman, David Davis, did not help. The Bureau increasingly angered the Department of Agriculture—which saw no need for new farm land—as it looked for a large, spectacular project somewhere in the West to boost its fortunes.[13] The two best prospects were the Columbia Basin Project or something on the Colorado River.[14] Eventually the Bureau settled on the Boulder Project in the Southwest and put all of its energy there. Consequently, it lost whatever interest it might have had in the Columbia Basin Project as it could not manage two enormous ventures at the same time. It did, however, keep Homer Gault in the field reinforcing the impression that it still took the project seriously, and maintaining its claim on the Columbia. Private correspondence between Gault and the Bureau's Denver office shows the Bureau most interested in information about the proposed Grand Coulee Dam, despite the public posture that the Bureau leaned toward the state's gravity plan.[15] The Bureau continued the game of secretly dismissing the Columbia Basin Project while making an outward show of backing it, to hang on to it just in case Congress should eventually authorize construction work there.

In March 1924, Gault submitted his report. He estimated the gravity plan at $241.38 an acre and the pumping plan at $246.58 an acre.[16] Startling were Gault's findings when compared with those of the state's Columbia Basin Survey Commission Report which estimated costs of $171.40 per acre, or the Goethals study that predicted $144.99 an acre. Further, Gault showed much of the land would be impossible to irrigate and he recommended more detailed studies.[17] Concerned over Gault's accuracy, considering the differences with the previous work, the Bureau of Reclamation appointed a review board.[18] Pressured by Spokane, the board allowed that the gravity plan was preferable to any other, but then it cautioned, "the high construction cost makes the project financially infeasible at the present time."[19] Suspicious, Interior Secretary Work appointed a second review board later that August. It finished its investigation on February 21, 1925, lowered the cost per acre estimate to $158, advised against any work in the immediate future, and sent its conclusions along to the new Reclamation Commissioner, Elwood Mead.[20]

The Coolidge administration and the Bureau of Reclamation, in addition to reluctance about the project for technical and financial reasons, were sensitive to other opposition. Throughout the Midwest, farmers, then in a financial depression, condemned Western reclamation efforts, believing crops grown there would further depress prices. They named the Columbia Basin Project specifically as extraneous. Closer to home, both the *Yakima Daily Republic* and the *Yakima Morning Herald* rejected the Columbia Basin Project. Washington Governor Roland Hartley deemed irrigation unnecessary and in November he bluntly advised the state to get out of the reclamation business.[21] The Washington State Grange restated that organization's opposition to irrigating the Columbia Basin. The State of Idaho took an even stronger position against using its water to irrigate land in Washington. Idaho's colorful Republican Senator, William Edgar Borah, announced that he did not want to see others make a "duck pond" out of Idaho's domain.[22] On February 3, 1927, Idaho legislators passed a bill prohibiting its Commissioner of Reclamation, or any other administrative officer in the state, from granting permits appropriating Idaho water for any purpose in the state of Washington. The *San Poil Eagle* wrote that the legislature had shoved the Pend Oreille project "under the guillotine and chopped off its head."[23] Backers of the gravity plan downplayed the law that they hoped to neutralize through negotiations. But in the end, they never overcame Idaho and Montana concerns.[24] Had the government ever seriously considered the gravity plan this opposition would have been a formidable and perhaps insurmountable hurdle, something Spokane people must have realized, but always denied.

Although Idaho forces, nor anyone else for that matter, did not realize it in 1927, two years earlier a seemingly unrelated event initiated a process that would have the greatest impact on the future of the Columbia Basin Project. On March 3, 1925, Congress passed its annual Rivers and Harbors Act with features that lead directly toward resolution of the conflict between the gravity and pumping plans. The act included authorization for the Secretary of War to investigate navigable streams with potential power sites, using the Army's Corps of Engineers and the Federal Power Commission to undertake the surveys.[25] This resulted in a study published on April 12, 1926, listed as House Document No. 308, 69th Congress, 1st Session—the original so-called "308 Report." It called for more extensive surveys of various rivers. On January 21, 1927, Congress again passed a Rivers and Harbors Bill. Senator Wesley L. Jones of Washington (known as "Yakima Jones"), who had promoted the 1925 law, worked quietly to insure that the new legislation included the Columbia River.[26] As before, the detailed surveys fell under the direction of the Army Corps of Engineers. The Corps divided the Columbia River into two sections and in 1928 assigned Major John Soule Butler, District Engineer in the Corps' Seattle office, investigation of the Columbia above its confluence with the Snake.[27] By insisting that the government include the Columbia among the rivers investigated, Wesley Jones deserves appreciable credit for Grand Coulee Dam. Although Nat Washington, James O'Sullivan, Rufus Woods, Billy Clapp, and the rest of the pumpers worked for the dam, the government would never have built it without detailed research and the strong endorsement provided by the Army. In that sense, Wesley Jones is one unsung hero whose actions precipitated its realization and construction.

In 1925, however, none of the principles in the debate over the Columbia Basin Project yet realized that those studies would resolve the debate between the gravity and pumping plans. Even after he began work in 1927, the full significance of Major Butler's work was not immediately obvious. Through 1925 and into 1926 the gravity plan received most of the public attention directed toward the Columbia Basin Project. The organizations that backed the pumping plan existed mostly in the heads of a few diehards; the *Wenatchee Daily World* had not printed any stories about a dam at the Grand Coulee since 1921. The only support for the dam, weak though it was, rested secretly with the Bureau of Reclamation, whose engineers saw it as preferable to the gravity plan. Then, in the summer of 1926, the few locals who kept the dam idea alive suffered a tragedy. On Saturday, July 10, 1926, Nathaniel W. Washington, the main driving force behind the pumpers, his brother James, and his sister Peachy, drowned while swimming in the Columbia River. Ironically, the accident occurred very near the place where Grand Coulee Dam stands today.[28] It was the low point for the pumping plan. James O'Sullivan, who wrote the stirring articles for the *Wenatchee Daily World*, was in Michigan tending to his family's business. Billy Clapp and Gale Matthews were busy with their legal practices. Rufus Woods, as usual, ran in many directions pursuing whatever caught his attention, turning out articles for his newspaper which then actually seemed to embrace, and certainly did not contest, the gravity plan.[29] With Nat Washington dead, pumping plan forces lost their leader and they fell apart. This left the field entirely to the Columbia Basin Irrigation League, which continued to push the gravity plan.

In the summer of 1927 the CBIL and the Spokane Chamber of Commerce arranged for two groups of congressmen to tour eastern Washington. The CBIL paid all the expenses.[30] With their favorable response, CBIL leaders decided to approach Calvin Coolidge directly. The President spent part of his summer vacation that year in the Black Hills near Rapid City, South Dakota. Through Senator Jones and other connections in Washington, Roy Gill arranged an appointment there. At the meeting, Gill briefly described the project. Then others, including Senator Jones, spoke. The President listened and promised to give the project, "careful study."[31] Apparently the group made an impression on Coolidge as some months later he told a meeting of the Union League of Philadelphia that development of "the Columbia basin is not far distant."[32] That, however, was the only tangible result of the meeting.

The CBIL returned to Congress. On December 12, 1927, Senators Jones and Dill, and Representative John Summers (also of Washington) simultaneously introduced what became the Columbia Basin bill. They worded it so that it did not commit the government to either the gravity or the pumping plan but asked the Congress only to authorize the project, leaving funding for later.[33] A Senate committee on reclamation and irrigation held hearings on January 11 and 13, 1928, followed by a similar House of Representatives committee that did the same on January 16 and 17. In February the Senate committee, and in March the House committee, forwarded the bill with positive endorsements. The Department of the Interior refused to approve the bill and the Department of Agriculture opposed it vehemently. Nevertheless, the Columbia Basin bill moved forward toward consideration before both houses of Congress. In Spokane the CBIL organized a letter-writing campaign that flooded congressmen with mail from businessmen in Spokane and other cities around the Columbia Basin.[34] But when the Coolidge administration failed to support the bill, Congress ignored it and adjourned without taking action. Political opponents of Wesley Jones accused him of secretly killing the bill to deny Senator Dill credit for it and to help the Republican administration's relations with farmers. Jones vigorously denied the charge and stated that Dill himself possibly worked to kill the bill so that the Republicans would not get credit for it.[35]

The CBIL next proposed that Congress conduct further surveys of the Columbia Basin. Then President Coolidge bluntly declared that Columbia Basin investigations might be appropriate eventually but the time was not then right for the government to pursue the project.[36] Despite the President's comment, the House of Representatives Committee on Irrigation reopened hearings on the Columbia Basin. It finished early in January 1929, approved the bill, and sent it on to the full House where it failed.[37] The unfinished Butler study by the Army Corps of Engineers may have influenced congressmen to wait before putting more money into the Columbia Basin. The administration's lack of support, Eastern disenchantment with reclamation, and the preoccupation of the Bureau of Reclamation were also significant impediments.

But increasing awareness of Major Butler and the possible ramifications of his work certainly influenced Rufus Woods and brought him back into the battle. Late in 1927 and into 1928, Woods's interest in the pumping plan rekindled. He always acted as a booster for his region, an area he dubbed North Central Washington, which he defined as the middle of the state along the big bend of the Columbia River and outside the domain of powerful Spokane. He sought economic independence from that influential city. Woods increasingly focused on power projects as the key to that end. "Great Power Development would Pay the Cost of Reclamation of Columbia Basin," read one headline early in 1927 that repeated the idea of free irrigation provided through sales of cheap power. Speaking before the Wenatchee Chamber of Commerce in November 1927, Woods played on local rivalries as he pointed out that Spokane's gravity plan would develop last the best land around Quincy. First would come water for the eastern sections of the project, which Spokane people dominated. Urging North Central Washington supporters to organize and fight on behalf of the Grand Coulee Dam, he slowly reinvigorated interest in the pumping plan. In August 1927 a group he organized at Soap Lake created the Columbia Basin Landowners League that vowed to keep people advised about the proposed dam, still sometimes called the Columbia Dam, but increasingly referred to as the Grand Coulee Dam. Frank McCann of Coulee City headed the organization with Billy Clapp and Gale Matthews serving as vice presidents.[38] This was a new beginning, but only a weak effort at first and for the next eighteen months.

Then, in February 1929, James O'Sullivan returned to Ephrata from Michigan. He still owned 800 acres in the area and hoped some day to live on them. He quickly realized the importance of the Army's survey and its close look at power sites. O'Sullivan took up the cause and made up for lost time. "This is the most wonderful project in the world, absolutely feasible from every standpoint and absolutely necessary," he wrote in an article that Rufus Woods printed.[39] Over the next few months the *World* featured more new articles by O'Sullivan, along with excerpts from his earlier efforts.

James O'Sullivan had an ability to focus on a cause, and once back in the fray he dedicated himself to the construction of Grand Coulee Dam. He left his wife and children behind in Michigan and willingly exposed them and himself to poverty while he waged what he saw as a sacred battle against the selfish, profit-oriented, dishonest Spokane private power interests that cared little about farmers. Generally argumentative and cantankerous, he saw people as either supporters or enemies. Unwilling to bend on any point, he irritated his adversaries while just as frequently annoying and quarreling with his friends. O'Sullivan used every opportunity to make his case for Grand Coulee Dam and, invited or not, took the floor at meetings and delivered long, arm-waving, fact-filled harangues that exasperated the gravity people and often embarrassed the pumpers. Indefatigable at his typewriter, he churned out letters, articles, and broadsides supporting the cause. He also shadowed Major Butler, much to the latter's annoyance, trying to look over his shoulder, influence his thinking, and discover what he might recommend.

On June 3, 1929, twenty-nine people from Soap Lake, Coulee City, Almira, Ephrata, Wilson Creek, Wenatchee, Trinidad, and Warden met at the Rock Cafe in Ephrata. There they formed what they called the Columbia River Development League (CRDL), a more impressive organization than its predecessor (the Columbia Basin Landowners League), dedicated to promote the construction of Grand Coulee Dam.[40] They named David R. McGinnis of Wenatchee as their president, and James O'Sullivan as executive secretary with a small salary. Later that summer McGinnis resigned and Rufus Woods took the leadership position.[41] O'Sullivan declared that the dam would produce power and that, "This power will pay for the entire Columbia Basin Project exactly as it is paying for the Salt river project in Arizona."[42] Ephrata provided the natural place for the group to meet and organize. Noting the pro-Grand Coulee sentiment and activity among its community leaders, the *Wenatchee Daily World* had already commented that "a dam university exists at Ephrata."[43]

Creating the CRDL was the first significant act by the pumpers in the three years since Nat Washington's death. It again provided them with leadership and direction. Rufus Woods was the well-known local figure who could draw the different factions together as had Washington, and O'Sullivan the dedicated worker who disseminated information and rallied forces to the cause rushing around from town to town, gathering contributions, meeting people, and promoting the dam. A worried friend wrote to Rufus Woods,

James O'Sullivan was pretty well knocked out when he got back from Coulee, and he had piles, a bad heart, and generally run down condition. He is going to have to be kept in the office, as he cannot stand the rough and rugged end of the game.[44]

O'Sullivan had other ideas and never confined himself to the office. Although his health suffered, he accepted low pay and hardship as the cost of realizing the dream of the dam. He eventually sold his land to support himself, leaving him without any personal stake in the outcome of the debate.

From its creation in 1929 until formation of the next Columbia Basin Commission in 1933, the Columbia River Development League, held together largely by O'Sullivan, fought Spokane's Columbia Basin Irrigation League's gravity plan. Always short of funds, the CRDL existed, over its four-year life, on around $13,000, mostly in the form of small donations. This contrasts with the wealth of the CBIL which spent well over $50,000 in 1927 alone.[45] Rufus Woods, however, reveled in playing David to Spokane's Goliath.

Despite all the activity of the CRDL and CBIL, the developments that really lead to Grand Coulee Dam proceeded apace elsewhere. From 1927 until 1929 Major Butler and the Army Engineers continued with their detailed inquiry into irrigation, flood control, power development, and navigation on the Columbia above the Snake.[46] Butler conducted the work somewhat secretly to avoid interference from partisans of the two irrigation plans, people like O'Sullivan and Roy Gill, both of whom tried to influence him. But when the Engineers began core drilling around the Grand Coulee site in August 1929, he could hardly shelter the activity from the public, and the *Wenatchee Daily World* heralded the event as if it signaled the start of work on the dam itself.

If not before, by early 1929 it became obvious to everyone concerned that Major Butler held the key to victory in the contest between the pumpers and advocates of the gravity plan. Born in Tennessee and educated at Vanderbilt University, he was a proud career military man then fifty-seven years old. He spoke with a mild Southern accent and, like most engineers, was precise in everything he said. In Yakima that November 1929, in one of his few public appearances, he outlined plans for development of the Columbia River that clearly emphasized hydroelectric power development.[47] In so doing, he presaged the death of the gravity plan. At

Grand Coulee Dam construction, 1936. *Courtesy Bureau of Reclamation.*

Roy Gill. *Courtesy Eastern Washington State Historical Society, no. 83-32.*

Willis T. Batcheller. *Courtesy Wenatchee World.*

Nathanial Willis "Nat" Washington. *Courtesy Nat Washington.*

James A. Ford. *Courtesy Eastern Washington State Historical Society, no. L93-18.64.*

Harold S. Ickes. Harris and Ewing photograph. *Courtesy Bureau of Reclamation.*

Bureau of Reclamation Commissioner Elwood Mead, Grand Coulee Chief Engineer Raymond Walter, Hydrographic Engineer E. B. Debler, Electrical Engineer E. N. McClellan, and Roy Gill of the Columbia Basin League, 1931. G. A. Beyer photograph. *Courtesy Bureau of Reclamation.*

Columbia Basin Commission consulting board, 1934. Upper row: Horace Smith, Alvin F. Darland. Lower row: David C. Henny, James O'Sullivan, W. C. Morse. *Courtesy Wenatchee World.*

Albert S. Goss. *Courtesy Wenatchee World.*

Chief Jim James and Governor Clarence Martin drive the first stake for Grand Coulee Dam, marking the official start of construction, July 16, 1933. *Courtesy Wenatchee World.*

Frank A. Banks. *Courtesy Bureau of Reclamation.*

John L. Savage. *Courtesy Wenatchee World.*

Senator Clarence Dill and President Franklin Roosevelt on train trip to the dam in 1934. *Courtesy Eastern Washington State Historical Society, no. L85-83.*

Opening the bids for Grand Coulee Dam construction, Davenport Hotel, Spokane, June 1934. *Courtesy Eastern Washington State Historical Society, no. L87-1.4142-34.*

Constructing the blocks forming the foundation of the dam, July 1936. *Courtesy Bureau of Reclamation.*

Dam and east-side slide showing refrigerating pipes for freezing, April 1937. *Courtesy Bureau of Reclamation.*

Driving steel piles in west cofferdam, August 1936. *Courtesy Bureau of Reclamation.*

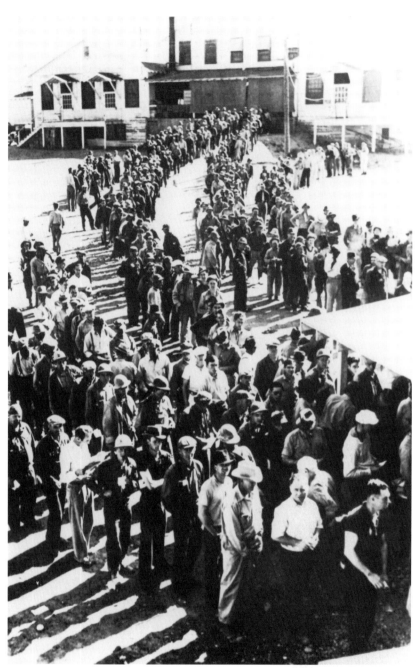

Dam workers stand in pay line, 1937. *Courtesy Bureau of Reclamation.*

Building a cable suspension support for one of the conveyor belts that eventually crossed the Columbia, 1935. *Courtesy Eastern Washington State Historical Society, no. L87-1.5319-35.*

Night view of construction, 1939. *Courtesy Bureau of Reclamation.*

Trestles and cranes with concrete mix plant and huge "Safety Pays" sign in background, 1938. *Courtesy Bureau of Reclamation.*

Ice forms during winter construction at the dam. Tunnels like this one penetrate the entire structure. *Courtesy Bureau of Reclamation, Columbia Basin Project Office.*

Pouring concrete at one of the blocks in the spillway portion of the dam, 1939.
*Courtesy Bureau of Reclamation.*

the same time he heightened the fears of the Washington Water Power Company and its supporters with the specter of a large block of cheap power, financed with public money, competing with them—something they had worked so hard to oppose.

The 1920s and 1930s were the fiercest years of the bitter public/private power fight throughout the West. It is fair to say that private ownership forces and hidebound capitalists gave ground only grudgingly and never without a struggle. They lost the battle in the case of the Boulder Project, and they vowed not to lose it again at Grand Coulee. In Washington state, especially in its lightly populated eastern half, they adamantly took the position that no market existed for any huge power source, something essentially true in the 1920s and even in the 1930s. In rebuttal, O'Sullivan tried to prove that a market did exist, or that it would in the future. Butler and his study would attempt to settle that critical issue.

The prospect of cheap power changed the mind of the Grange. In a surprising reversal, as so far the Grange had opposed the Columbia Basin Project, Albert Goss, its state leader and a man of considerable influence, wrote an article in the *Grange News* of August 1929. In it he revealed that the state Grange would now support the Columbia Basin Project, using the pumping plan, because of the power it would provide, as long as the government postponed the reclamation aspect until the country needed the land.[48] The Grange also positioned itself squarely against what it called power-trust holding companies that it feared would gain control of the Columbia River if the government did not develop it first. By late 1930 the stock market crash of the previous fall caused many of those companies to collapse and revealed the unethical practices and manipulations of utility holding company giants Samuel Insull, the House of Morgan, and others, discrediting private power, and increasing acceptance of public power and public utility districts in Washington.[49] At its annual convention in 1931, the Grange formally endorsed Grand Coulee Dam and added a call for its power to sell at one rate for all, with preference for public utilities.[50] In other words, the general political climate moved toward increasing acceptance of public power, another significant blow against the gravity plan.

At its eighth annual meeting, held in 1929, the CBIL announced that it was "ready for the final push" toward getting irrigation for the Columbia Basin.[51] Noticeably missing were comments specifically about the gravity or the pumping plans. Spokane people, now also realizing that Major Butler's work was stressing power, took on the loftier goal of backing reclamation generally. Mark Woodruff surprised everyone when he said that the CBIL did not endorse any particular plan to reclaim the 2,000,000 acres of the basin.[52] Incredulous, James O'Sullivan countered that the CBIL had always opposed the pumping plan. In reply, the League invited O'Sullivan, not once, but at least twice, to address first its executive committee and then its total membership, and it urged other supporters of the Columbia River Development League to come along and listen. Hardly the diplomat, O'Sullivan used the occasions to harangue the CBIL for its past practices. He blasted them for quoting inaccurate engineering reports, for putting the cost of Grand Coulee Dam at least $30 million too high, for covering up investigations favorable to the dam, and for allowing the Washington Water Power Company to improperly influence state studies.[53] At the end of one meeting Roy Gill smiled indulgently at O'Sullivan, probably through gritted teeth, and announced that the CBIL had no quarrel with anyone supporting a particular plan for irrigating desert lands, and that these differences of opinion should not detract from the main goal of getting the government to adopt and build the project. "The Columbia Basin Irrigation League will not now give its support to any particular plan—either gravity or pumping plans—until the government has recommended the best plan from all viewpoints," Gill concluded.[54] O'Sullivan had trouble believing his ears as he watched the enemy not only change its stand but rewrite history in the process.[55] For years the CBIL had only grudgingly admitted that the pumping plan even existed. Now they recognized the ground lost by the gravity plan and moved to salvage whatever they could.

Recognition of the Butler study's significance did more than re-animate Rufus Woods and force the CBIL to alter its position. It brought the Bureau of Reclamation to attention. For years the Army Corps of Engineers and the Bureau had jealously competed for turf, and neither wanted to yield any ground to the other.[56] With the Boulder Project now secure, and more certain of its abilities, the Bureau decided not to let the Army snatch away a potentially huge project in the Columbia Basin. To reassert its claim, Reclamation

Commissioner Elwood Mead visited the basin. "Big Dam Across Columbia River at Grand Coulee Inevitable," ran one headline in the *Wenatchee Daily World* that quoted Mead. Two days later the *World* proclaimed, "Mead Favors Boulder Dam Plan for Construction Grand Coulee Project." The article quoted Mead as saying, "I do not care what the cost of the dam will be so long as power revenues will retire the cost of construction within a period of fifty years and pay four per cent on the investment."[57] Then Mead stationed construction engineer Harry W. Bashore in North Central Washington to hold down the Bureau's claim and watch Major Butler.[58]

Rufus Woods enjoyed the more favorable position how held by the pumpers. Moreover, he delighted in exposing the activities and motives of his Spokane adversaries. A progressive Republican of the Theodore Roosevelt school and something of a populist in his tactics, he favored government activity on behalf of its citizens, and it irked him when his fellow Republicans opposed his causes. He made no secret that he believed the Washington Water Power Company orchestrated the push for the gravity plan from the start, and that it deliberately manipulated and even falsified studies to produce favorable results. He chafed when some of his fellows in Wenatchee gave him only minimal support or even sided with Spokane.[59]

From October 30, 1930 through November 4, the *Wenatchee Daily World* ran a box on its front page with a large headline inside that proclaimed, "Grand Coulee Dam," followed by a short piece, usually written by O'Sullivan. "Those interested should be urged to keep in touch with its [the dam's] progress through the columns of the *Wenatchee Daily World*," concluded each article. Repeatedly O'Sullivan argued that the power from Grand Coulee Dam would be easily absorbed by 1955 and that it would return to the government a revenue of $863,940,000 in fifty years. "The Grand Coulee Project," he wrote, "is [the] key to the future development of the entire Northwest."[60]

Toward the end of 1930, if not before, Frank Post, President of the Washington Water Power Company, members of the Spokane Chamber of Commerce, and its CBIL realized that their gravity plan was dead. Despite the secrecy surrounding Major Butler's work, word leaked out convincing them that the Army would stress hydroelectric power development.

At this point, an unusual natural phenomenon provided ammunition for the backers of reclamation and supporters of Grand Coulee Dam. On Tuesday, April 21, 1931, warm and exceptionally dry east winds suddenly swept the Western United States from Everett, Washington, in the north to as far south as Sacramento, California. The humidity dropped dramatically and, starting from eastern Washington, huge clouds of dust filled the air. In pillars that reached 5,000 feet into the sky and higher, the dust moved west, across the landscape, over the Columbia Basin and down the Columbia Gorge, through Portland, into the Willamette Valley, and finally as far as 500 miles out into the Pacific Ocean.[61] Writing to O'Sullivan the next day a county agent declared: "The greatest reason for pushing the development and irrigating of the Columbia Basin has been almost entirely overlooked, and that is the conservation of this rich, fertile top soil which is being scattered from the Rocky Mountains to the depths of the Pacific Ocean."[62] The incident and the comment reinforces one of the themes that backers used to promote the dam and irrigation project—that they were conservation measures as much as anything else.[63]

On June 28, 1931 the Columbia River Development League held a picnic at Park Lake, a green spot near the foot of Dry Falls in the Grand Coulee. Thousands attended.[64] James O'Sullivan, who wore a large button proclaiming "Columbia Basin Dam Will Pay Uncle Sam," addressed the crowd that came from as far as Walla Walla, Yakima, Pasco, Lind, and Wenatchee. They had a beautiful day as they listened to Representatives Sam Hill, John Summers, Ralph A. Horr of Seattle, and Senator Dill. Then the assemblage voted to urge the Washington congressional delegation to push the project. Afterwards they all drove home and waited for something to happen.

Something happened in July. Reclamation Commissioner Elwood Mead revisited the Columbia Basin and brought with him members of a House of Representatives appropriations subcommittee.[65] In Spokane the Commissioner indirectly admitted that the Bureau of Reclamation had not really been excited about the Columbia Basin in the past but he told supporters it was no longer afraid of such a large job. He predicted that the Pacific Northwest would someday become a "vast industrial development," something Rufus Woods had said for years and the *World* had reported with considerable enthusiasm.[66] In Spokane Mead declared, "Power is more and more becoming a predominant factor in the success of reclamation projects."[67]

Despite Major Butler's increased secrecy, bits and pieces of his findings continued to dribble out, giving more indications about his specific recommendations. As the significance of his work became apparent, everyone watched him more closely. With so many people ready to broadcast his every word, he never had any real chance of his verdict remaining entirely confidential. Exactly how much anyone knew in advance is debatable, but those who kept close watch on the major's study recognized his inclination toward power and the pumping plan. Butler finished his work in the summer of 1931 and although the government did not publish it until 1933, by the fall of 1931 everyone who cared had a very good idea what it said.[68]

The Associated Press broke the story on September 21, 1931, and the *Wenatchee Daily World* broadcast it that day under a large headline. "Army Engineers' Report Favors Grand Coulee Dam Says Washington Report to *Spokesman-Review*."[69] Over the following days more stories about the Butler report emerged in the *World*, which declared on September 26, "It will say gravity plan not feasible— project must be linked to power."[70] The *Spokesman-Review* held the story until September 24, and then ran it on page two. The *World* proclaimed on October 2, "Mead's Engineers Favor Grand Coulee Dam—Reclamation Survey Concurs with Army." "Army engineers agree with reclamation bureau engineers that the cheapest most feasible plan for developing [the] Columbia basin is by erection of a great dam at Grand Coulee," said a story in the *Seattle Post-Intelligencer* on October 24.[71]

Congress received the 1,845-page report on March 29, 1932. It carried an endorsement by Elwood Mead and by Colonel William J. Barden, senior member of the Board of Engineers for Rivers and Harbors. In all the report gave the details of seven different variations of the two plans proposed for irrigating the Columbia Basin. Butler and his engineers calculated the average cost of the pumping plan at $284 an acre against the average cost of the gravity options at around $493 an acre. And the report made clear that power revenues must offset the cost of irrigation, something impossible under the gravity plan.[72] On the question of demand for the power that a large dam would generate, the report argued that every two years the Pacific Northwest consumed 9.5 percent more electricity; its need doubled every 7.6 years. Peak demand would come in 1990 and in the interim there would be a market for power generated at Grand Coulee Dam.[73] As if still necessary, this sounded the final death knell for the gravity plan. Not everyone, however, accepted that opinion.[74]

On January 7, 1932 the Bureau of Reclamation's already well-publicized findings went to the Congress. The Bureau endorsed the Army plan that called for a large dam at the Grand Coulee, along with a pump plant, power generation station, equalizing reservoir in the Grand Coulee, canals, power stations along those canals, and drainage as needed. Less conservative than the Army study, the Bureau of Reclamation engineers estimated that the Grand Coulee Dam would accommodate only one-half of the region's additional power needs by 1955.[75] The report declared that the Northwest would absorb all Grand Coulee power within fifteen years. Both the Army and the Bureau, whose officials met in October 1931 to iron out differences in their reports, agreed that the entire project would completely pay for itself in sixty to ninety years.[76]

Major Butler emerges from the story as a hero in his own right. The Army Corps of Engineers felt considerable pressure from proponents of the gravity plan and opponents of public power, and the agency debated and divided within itself over endorsing Grand Coulee Dam. Butler remained firm during a week-long meeting and close questioning as his superiors considered coming out for the dam. Butler even held out for the dam when the Army changed its mind and its review board overruled him and his findings. In so doing, Butler hurt his career, although he stayed with the Army until he retired.[77] In maintaining his position and advocating not only Grand Coulee Dam but also the entire chain of dams on the Columbia River, along with the need to develop power, the Butler study, together with all of the "308 Reports," shaped the destiny of the Pacific Northwest and was among the most significant influences on its development in the twentieth century.

The long-awaited Army report finally settled the question about how to irrigate the Columbia Basin. The gravity plan gasped its death rattle and disappeared. By the end of 1931 few in Washington state doubted that when irrigation came to the Columbia Basin, it would include a dam at the Grand Coulee. To that end the forces of the Columbia Basin Irrigation League grudgingly moved into an uneasy alliance with the Columbia River Development League.[78] As the *Seattle Post-Intelligencer* put it, "Now it is up to all of us to induce congress at the coming session to agree to endorse the project, adopt it, and make initial appropriations for its start."[79]

An article in the *Wenatchee Daily World* on September 21, 1931, the same day that news broke about the Butler study, presaged how difficult achieving that goal would be. "Hoover Calls for Government Thrift," read the ominous headline that told how the depression had hurt the national economy and explained the President's desire to balance the federal budget. Convincing the government to spend money on reclamation in distant Washington state would have to wait for the next general election and some changes in Washington, D.C. At least there was now a decision. Grand Coulee Dam won over the gravity plan.

It would be wrong to see the fight as one entirely between the "good guys" and the "bad guys," where the "good guys" won.[80] Each side had its reasons for wanting either the pumping or gravity plans. The Spokane/gravity people, moved by the concerns of the Washington Water Power Company, sincerely feared public power and competition from the federal government.[81] In an era when capitalists saw the specter of socialism as a threat to civilization itself, let alone their pocketbooks, their concern is understandable. Like the railroads a few decades before, they wanted public money and resources, not interference and competition, and they firmly held their trust in *laissez faire* capitalism. They wanted an irrigation project they could control for their own benefit and profit—a position that had built nineteenth-century America. They wanted to see their influence and economic power grow, based on a strong agricultural foundation, just as it had for large urban centers in the East and Midwest. To that end, they promoted the gravity project by means fair and sometimes not so fair.

To its credit, the Spokane Chamber of Commerce kept the idea of Columbia Basin reclamation alive before Congress and the country over a period when no other agency or group could do that job. As James Ford, managing secretary of the chamber, correctly wrote in 1933,

> The Spokane Chamber of Commerce, through its Columbia
> basin committee, not only took the lead, but has been the inspi-
> ration for the 15 years of tireless effort which today brings it to
> the verge of development almost to the turning of the first sod.[82]

That organization publicized the project both throughout the
Northwest and nationally. It cleverly included among its trustees
prominent Northwest personalities who gave the effort the appearance of a regional and even a national movement rather than a local
plum for the benefit of a few.[83]

The Columbia Basin Irrigation League, the creation of the
Spokane Chamber of Commerce, was always an organization of
urban businessmen, bankers, and politicians. Professionals like
James A. Ford, Fred A. Adams, and Roy Gill provided the motivat-
ing force and inspiration and did most of the work, sometimes representing substantial expenditures of time and money. Railroads
and chambers of commerce from throughout Oregon and Washington contributed generously and frequently to support the CBIL.
At least $124,231.05 came from Spokane alone with $52,000 from
railroads, $43,000 from Seattle, $10,399 from Pasco, $11,710 from
Tacoma, and $6,500 from Portland. In all, the League raised nearly
$300,000 during its existence.[84] Between them, the Columbia Ba-
sin Irrigation League, the State of Washington, and the federal government spent well over a million dollars investigating and
promoting the Columbia Basin Project from 1918 through 1931.
This was hardly an effort unworthy of admiration, and while the
push for the gravity plan failed, one should not discount or dismiss
the contributions and influence of the CBIL. Had they not kept the
idea alive during the lean years, there might never have been a
study by the Army that in turn led to Grand Coulee Dam.[85]

The pumpers, for their part, dreamt of cheap water transforming their region into an agricultural/industrial empire free from
control by Spokane. The rural forces around Wenatchee and
Ephrata wanted government assistance to pay all the bills and they
would sacrifice some of their individualism to accommodate that
public assistance. They banded together to fight what they saw as
the evil power trust that either refused to serve them or charged
them exorbitant rates.

In contrast with the affluent Columbia Basin Irrigation League,
the Columbia River Development League, and those pumpers in

various organizations that preceded it, represented, with a few exceptions like Rufus Woods, a true grassroots movement. Their professional leaders were not wealthy, but were country lawyers, like Billy Clapp and Nat Washington. By the end of 1931 they had raised only about $4,500 with some of it in-kind rather than cash.[86] The CRDL, which existed from 1929 until 1931, had only one employee, James O'Sullivan, and it was never out of the red nor did it fully compensate O'Sullivan for his work. It thrived on being the underdog fighting the evil rich. Throughout the area little people like bakery owner Tillie Abelson scratched around and, despite the depression, contributed a dollar here or some food there to support "the cause." If ever an organization with visions of glory sustained itself on a shoestring, it was the Columbia River Development League.

The efforts of Woods, O'Sullivan, Clapp, and the others, however, hardly countered the CBIL's gravity plan. The Spokane people had too much clout to be stopped if everything else had gone in their favor. The economics of the 1920s, the Republican administrations' disdain for government spending on speculative projects, and the Bureau of Reclamation's reluctance to adopt an incredibly expensive and complicated venture made the gravity plan unlikely even had no pumpers existed. The suggestion for the great dam on the Columbia River had been around for a long time and it is possible that the same result would have emerged without the pumpers. The great accomplishment of Rufus Woods, James O'Sullivan, and the others was yet to come.

Through the entire struggle, neither the pumpers nor the gravity people really knew which plan would best irrigate the Columbia Basin. O'Sullivan had little training that qualified him as a competent engineer and nothing that prepared him to evaluate anything on the scale of the dam and irrigation project. Although the pumpers accused the gravity forces of falsifying facts and figures, they did not do much better themselves. Both sides played freely with the limited data available. O'Sullivan presented his information as if he was an expert and Rufus Woods abetted him in that charade. Their often-repeated claim that power would pay all the bills, for example, rested on speculation and hope, and in the 1950s and 1960s it returned to haunt the Bureau of Reclamation.

Until the Butler Report emerged, everyone operated on limited information based on economic conditions that existed before World War I. This led to excessively optimistic predictions on both sides. Even Major Butler conceded the technical feasibility of both projects. His hard look at economics alone pushed him toward endorsing the pumping plan. Even then, changes already coming, and more that came in the 1930s, with World War II, and later, invalidated many of Butler's figures. As Grand Coulee Dam and the Columbia Basin Project showed repeatedly, it is almost impossible to plan a huge project, build it through periods of economic change, and successfully use it as originally envisioned in another era. In the end, however, Butler's refusal to buckle under pressure from Spokane and his honest, professional, detailed assessment laid the foundation for Grand Coulee Dam. Butler, neither a "good guy" or a "bad guy," was the determining factor in the battle between the gravity people and the pumpers.

It is fortunate that nobody built the Columbia Basin Project during the 1920s. Had the government or private sources financed the gravity plan, the depression of the 1930s and the farm surpluses of those years would have brought devastating financial failure. If, by some miracle, the pumping plan had prevailed before 1933, it certainly would have involved only a low or small dam at the Grand Coulee, making the great dam of today unlikely. In the 1920s, the time was not right for the Columbia Basin Project.

The Columbia Basin Irrigation League staged the last act in the debate between the two plans. On October 25, 1932 it held its eleventh and final convention. Reclamation Commissioner Elwood Mead attended and supported the dam. President Hoover sent a telegram expressing his hope for the future success of the project. Rufus Woods moved that reclamation backers meet in Salt Lake City at a national reclamation conference then planned for December 5. There Marshall Dana, editor of the Portland *Daily Journal*, became the first president of the National Reclamation Association.[87] The remaining members of the CBIL became the core of the new organization and the moving force behind its future activity. With the National Reclamation Association in place, the Columbia Basin Irrigation League slowly deteriorated and after a few years disappeared.

The victory of the pumpers did not end their problems. They had to find somebody, either the federal government or the State of Washington, to finance their dam and project, and they met continued opposition, some of it from their old enemies in Spokane. But with the Butler Report as their banner, they moved forward to irrigate the Columbia Basin.

Work has just started on Grand Coulee Dam in this photo from the fall of 1933. View looks south. *Courtesy Wenatchee World.*

# Chapter Five

# Washington State's Dam

Now all in the world we want, gentlemen, is a slab of concrete right across here!

*Rufus Woods*[1]

The government had all but shunned the Columbia Basin Project through the 1920s. Then the Army Corps of Engineers' Butler report determined how best to irrigate the Columbia Basin but not who should do it, when, or how to pay for it. Suddenly the Bureau of Reclamation looked at the job with renewed interest. So did the Army—off and on—as factions within that organization debated their preference for navigation or reclamation projects. Washington state also pondered building the project. Through late 1931 those agencies considered when and how to build the dam and where to find the money. They knew that the answer depended on the federal government. But the Hoover administration, mired in the financial depression and determined to balance the federal budget, fearing increasing farm surpluses, and already committed to the Boulder/Hoover project, saw Columbia Basin irrigation coming only far in the future.[2] The President appeared to back the project, but only at some distant date, and he offered no money for the present.

When the 1932 election carried Franklin Delano Roosevelt and his New Deal into the White House liberal programs suddenly offered prospects of immediate funding for public works ventures offering relief, especially for the unemployed. But people in the Northwest squabbled over who should benefit from those dollars. In the background, private power forces worked tirelessly to keep any dam built at the Grand Coulee as small as possible to lessen its competitive threat. And as often as not, even those who backed and fought for the dam and the irrigation project disagreed, distrusted, and openly argued with each other over the best way to proceed.

In that fluid and contentious setting government officials decided who would build Grand Coulee Dam. Instrumental in the process were Franklin Roosevelt, his Secretary of the Interior Harold L. Ickes, Reclamation Commissioner Elwood Mead, representatives of the Corps of Engineers, Rufus Woods, State Grange Master Albert Goss, James O'Sullivan, Senator Homer Bone, Representative Samuel B. Hill, Washington Governor Clarence Martin, and Senator C. C. Dill. Most of all, the Bureau and the Army, each jealous of the other, positioned themselves to clash over which would build the Grand Coulee Dam. As events unfolded, however, that clash never occurred.

Well before the advent of the New Deal both the Bureau and the Army had developed preliminary plans for a dam at the Grand Coulee and a reclamation project for the Columbia Basin. Anticipating eventual authorization and funding by Congress, both agencies sent more investigators into the field. Local workers, under contract with the Corps of Engineers, dug test pits at the Grand Coulee site.[3] Bureau Commissioner Elwood Mead wrote to engineer Harry Bashore, "The most important reason [for having engineers look at the site] is that when this matter comes before Congress, representatives of the Bureau will be asked, 'Does your recommendation represent judgments formed on the ground or are they based on reading reports in Washington?'" Mead wanted his men at Grand Coulee to provide him with accurate information to bolster the Bureau's claim on the project.[4]

In September 1931 Rufus Woods and Roy Gill went to Washington, D. C., where they visited with Mead to discover if the Bureau now really backed the dam. Woods distrusted Gill despite his appearance of support for the dam. He feared that Gill might still try to talk Commissioner Mead, with whom he had a personal friendship, into ignoring the Butler report and building the gravity project. Woods also had misgivings about Mead, who the pumpers thought had drug his feet all the way through the late 1920s while playing a cozy game with the CBIL. As Woods wrote that September,

> It is a tragic thing to have Roy Gill go east to see Mead at this time. Mead is getting along in years and is regarded as somewhat shifty. He some times [*sic*] takes one side and then

another. And yet Roy Gill goes down as the representative of the Columbia Basin League to subvert the army report. Already he has given out that all they want is to have the project "adopted" and call for a small appropriation. What I think we should go after is enough appropriation to get the dam well on its way. There ought to be an appropriation for at least $50,000,000.[5]

When the Bureau of Reclamation's final report committed it to the dam, Woods and Gill switched their emphasis to President Hoover.    The Wenatchee publisher met with the President on October 8 and assured him that the Columbia Basin Project would pay for itself. Senator Wesley Jones, who went with Woods, pointed out that the project would also provide much-needed employment.[6] Roy Gill met with Bureau officials in Denver, then with Hoover and Elwood Mead. Hoover remained uncommitted and Mead released a few vague stories to the press indicating that the Bureau anticipated starting the Columbia Basin Project shortly, although he knew better. Late that October he wrote confidentially to his chief engineer in Denver,

> We must make a recommendation to the Secretary [of the Interior] as to legislation on Columbia Basin. The Secretary and the President do not favor legislation authorizing and approving the project. They do approve of an appropriation large enough to enable the Bureau to prepare plans, specifications and estimates of cost, and to make a showing as to financial feasibility.[7]

Hoover said as much to Senator Dill in December. When Dill met with the President and urged him to appropriate $350,000 for borings and other preliminary work on Grand Coulee Dam, Hoover refused and explained that it would amount to starting the dam itself. This Hoover said he refused to do without congressional authorization, which he would not recommend.[8]

At a meeting in Seattle, sponsored by the Washington State Chamber of Commerce, organizations that backed Grand Coulee Dam decided, despite Hoover's reaction, to push for the dam and send Roy Gill, James O'Sullivan, and J. J. Underwood of the chamber to Washington, D. C., to lobby Congress to introduce legislation authorizing the project and appropriate enough money to start it.[9] In addition, the Columbia River Development League also sent

Billy Clapp. Aware of O'Sullivan's habit of going off on his own, Woods sent him a letter and warned,

> One of the most important things on this end as we see it is for a regular team-work job there in Washington. There is some fear that some of you boys may be trying to work single-handed. If you do, it may turn out to be not for the best.[10]

Woods and O'Sullivan distrusted Gill, and Woods feared that unchecked, O'Sullivan might alienate not only his fellows, but also people important to the cause. They undoubtedly sent Clapp as much to watch O'Sullivan as anything else.[11]

Senators Dill and Jones studied the legislation that authorized the Boulder Project and then drafted a similar bill for the Columbia Basin Project and Grand Coulee Dam.[12] It called for a special Columbia Basin fund of $260,000,000 and, before any construction, contracts sufficient to return the government's investment within fifty years for both power and irrigation, plus 4 percent interest on the power component.[13] The Bureau of Reclamation developed plans for a dam 350 feet high, a lake 150 miles long ending at the Canadian border, and irrigation; total cost—$400,000,000.[14] The bill mandated the highest possible dam producing maximum power, something Spokane people opposed.

Despite the demise of its gravity plan, the Spokane Chamber of Commerce remained determined to lead and direct the drive to irrigate the Columbia Basin, and to do so with a low dam. As events unfolded in the national capital, the men of the CRDL became even more suspicious of the Spokane group, convinced that its members would never sanction or support the high dam at Grand Coulee. In this they were correct. The Butler report and the possibility that the government might build the dam strengthened the resolve of the Washington Water Power Company to keep such a dam as small as possible, limiting its power-production capacity, and restricting the venture to irrigation.

Secretary of Agriculture Arthur M. Hyde also emerged as a formidable roadblock. He wrote to the Army Engineers Review Board,

> We already have a productive agricultural capacity far beyond our needs. . . . The market is glutted with farm lands at depressed prices. It is plainly and indisputably against the interests of the farmers of Washington and of the adjoining states to

undertake a project that would bring into production 12,000 more farms. . . . We fail to see how the project can possibly constitute a sound opportunity for any prospective settler. The proponents of this project are not farmers. Farmers have not besieged the Congress with demands that the Columbia Basin irrigation project be undertaken.[15]

President Hoover agreed. Bowing to administration pressure, on February 23, 1932 the Army Corps of Engineers, in a move that eventually cost it the project, reversed itself and rejected the idea of the Columbia Basin Project.

O'Sullivan fumed at the army and called Hyde "obstreperous."[16] Elwood Mead took the whole thing more philosophically. "I am in agreement with your conclusions," Mead wrote to the War Department. "This development to be solvent must be based on the revenues from power and that these revenues must contribute to the cost of the irrigation works to avoid injurious burdens on irrigation farmers."[17] Exactly what Mead thought privately remained unrecorded, but the wily commissioner had his men continue exploring potential markets for Grand Coulee power.

On May 25, 1932 the House of Representatives Committee on Irrigation and Reclamation began hearings on the already moribund authorization bill. A month later they voted to table it. At the same time, the Senate Irrigation and Reclamation Committee held hearings for one day and then also tabled the bill.[18]

Discouraged, Rufus Woods explored a new tack. He wrote to friends, "I am wondering if the state of Washington itself is not about ready for a state program."[19] Woods had decided that the state should build the dam as a self-liquidating development with private companies buying the power. He reasoned that abundant cheap electricity would lure industry to the region and bring about his "agricultural-industrial empire." James O'Sullivan backed Woods, and in July he urged the State of Washington to apply for $7 million through the Reconstruction Finance Corporation (RFC), which made money available to states for large, self-liquidating projects.[20] When the RFC put up $62 million for construction of the Oakland-to-San Francisco bridge, Woods and O'Sullivan vowed to go after their own RFC money.

The approaching 1932 general election became increasingly significant. In three crucial races Franklin Roosevelt challenged

Herbert Hoover for the presidency, Republican John Gallaty opposed Democrat Clarence Martin of Cheney for the position of Washington governor, and incumbent Senator Jones ran against public power proponent Homer Truett Bone of Tacoma. Martin declared that the state should back Grand Coulee Dam but Rufus Woods swallowed hard and backed his opponent. Jones ran against the popular Homer Bone, who made his reputation changing political parties and fighting for public utility districts and public power. Roosevelt openly favored public power development and the construction of large dams by the federal government.

In September, Roosevelt took his presidential campaign to the Northwest. Dill joined the Roosevelt party in Butte, Montana, and helped Roosevelt prepare the speech the candidate gave in Portland on September 21. In the city's municipal auditorium Roosevelt announced that the next great power development would be on the Columbia River. He added that water power belonged to the people, and he advocated public power as a "yardstick" against which consumers might measure the high cost of private power. The statement received prolonged applause.[21]

In announcing his backing for Columbia River development, Roosevelt courted the Northwest. Washington people interpreted him as meaning Grand Coulee Dam and the Columbia Basin Project, while Oregon people knew he preferred the Warrendale/Bonneville site just east of Portland. Hoover, on the other hand, despite urging from Rufus Woods, refused to support the dam or even mention it. Nevertheless Woods and the *World* stayed with Hoover.

How much his statement about developing Columbia River power contributed to Roosevelt's election is debatable but he carried Oregon, Washington, and Idaho.[22] In Roosevelt's wake Clarence Martin became Washington's new governor and Homer Bone unseated Wesley Jones.[23] Ten days later, his health broken by the campaign, inconsolable in his loss, Jones suddenly died.[24] For the Republicans of Washington state it was indeed "The World Turned Upside Down."

The *World* played up the Roosevelt victory and acted almost as if it had backed the President-elect all along. Democrat James O'Sullivan reveled in the election results but warned that the Warrendale/Bonneville proposal posed a threat to power development in Washington.[25] A change in the Oregon constitution provided a way for that state to finance hydroelectric power plants on interstate streams.[26] If Oregon built Warrendale, O'Sullivan wrote, it could delay Grand Coulee twenty-five to thirty years. He urged Washington to build the big dam. "The state could make $15,000,000 yearly," he declared.[27] Governor-elect Martin took O'Sullivan and Woods seriously, suggesting an independent Columbia Basin Commission to oversee construction of the dam and secure federal funding.

Meanwhile, Albert S. Goss of the Washington State Grange commented in a letter to Rufus Woods,

> We have been behind the dam for many years with the proviso that the irrigation features must be so handled that land will be brought in only as there is a sound economic demand for it. My understanding is that this is now the attitude of official Washington, D. C. If the matter could be handled purely as a power project, I know it would be possible to get the support of farm organizations instead of their opposition.[28]

Woods and O'Sullivan took the cue, changed strategy, minimized irrigation, and emphasized creation of jobs, power, and improved navigation. O'Sullivan argued that 100,000 men would find employment connected with the project, but he never lost sight of the irrigation goal and he never forgave the Corps of Engineers for changing its position on the dam.[29] He argued that the Bureau of Reclamation should do the work, hired by the state, under direction of the new commission proposed by the governor.

The Twenty-third session of the Washington state legislature opened on Monday, January 9, 1933. At his inauguration Martin formally proposed his commission and stated his support for Grand Coulee Dam.[30] Elmer Fisk Banker (sometimes called 'Ed,' but generally referred to as 'E. F.'), from Winthrop in the Okanogan, introduced House Bill Number 11 which became the "Banker Bill" authorizing the Columbia Basin Commission. On February 1, 1933 the Washington House of Representatives passed the bill by a vote of 89 to 6. In another measure the state Senate allocated $27,000 for preliminary work at Grand Coulee.[31] The full state legislature eventually passed the bills, which also give the new Columbia Basin Commission the authority to seek RFC funds. This completed the legal framework then thought necessary for the state to proceed with the project. These bills were the significant acts by which the state gave itself the authority to build the dam. With the governor's approval, they marked the true start of Grand Coulee Dam and were a major landmark in its story.

Governor Martin cheerfully signed the measures and on March 26 named members of the Columbia Basin Commission. He appointed Democrat Elmer F. Banker as the new director of the Department of Conservation and Development which, under the law, made him the commission chairman. Democrat James E. McGovern of Spokane, Republican Rufus Woods of Wenatchee, Republican Albert S. Goss of Seattle, and Democrat Harvey Smith of Neppel (now Moses Lake) filled the other four positions. And Martin appointed James O'Sullivan to the only paid position as secretary. The state directed that the commission be housed in Spokane and, in an ironical accident, O'Sullivan found himself working in the very office that once held his adversary, the Columbia Basin Irrigation League. In fact, the CBIL left its files behind, and O'Sullivan delighted going through them and discovering what the group did through the 1920s.[32] For the Spokane forces, the composition of the Columbia Basin Commission was a crushing blow they loudly lamented. Roy Gill complained to Elwood Mead, "Well, the Democrats are not letting jobs get away from them. We, in Spokane, realized that we were licked when the Governor named the Columbia Basin Commission. It is a Wenatchee Commission."[33]

Suddenly, in light of the changed administrative stance and the improved climate for project funding, the army reversed itself again and asked the state to hire it, not the Bureau of Reclamation, to construct the dam. It argued that its people would be more suited to oversee what it maintained would be mostly a power development, with reclamation coming much later.[34]

Elwood Mead continued his correspondence with Roy Gill, and in March the commissioner wrote,

> It is reported here from those who have talked with President Roosevelt that he includes [the] Columbia Basin in his emergency works program, but I have seen no printed statement to that effect. There are plans to introduce a bill, with the President's backing, to provide for the creation of small farm colonies to take care of the jobless workers in cities. We were consulted about the bill and hope it may have a trial.[35]

Representative Charles Henry Martin of Oregon introduced just such a bill, but in the mountain of work that overwhelmed legislators during the so-called "hundred days" of the New Deal, it sank from view. The *Spokane Chronicle* reported that the Columbia Basin Project would be the largest single employment project with the largest hydroelectric potential in the nation and that it would employ 100,000 men.[36] Not everyone agreed. Republican Representative Francis Dugan Culkin of New York, representative of Eastern forces unfriendly to reclamation and the New Deal's costly ventures, attacked federal projects such as the proposed Tennessee Valley Authority and the Columbia Basin. President Roosevelt had to contend not only with those who wanted to build the Columbia Basin Project but disagreed among themselves on how best to do it, but also with others opposed the idea in any form.

A great deal depended on Roosevelt. Realizing this, Grange Master Albert Goss resolved to approach him directly. He worked with Representative Sam Hill and occasionally with Senator Dill, but he found the latter unresponsive. Goss could not discover why Dill seemed indifferent toward the project, did "his best to discourage it," and advised waiting before going to the President.[37] In any event, Goss and Hill pushed Dill, and together they secured an appointment with Roosevelt on April 17. Unfortunately, Hill became snarled in a traffic jam and missed the meeting.

Goss and Dill outlined the Columbia Basin Project to Roosevelt, described it as a way to end unemployment in Washington state, and urged the President to spend $400,000,000 on the development. Startled by the price, Roosevelt suggested a more modest dam or the foundation for a large dam, and stressed the power aspect of the project, although he feared overloading the market. Roosevelt doubted that Congress would approve the project and he recommended the state do it in stages with RFC money. He told Dill to approach the Bureau of Reclamation[38] and see what it could design in the way of a smaller dam with a larger dam coming later.[39]

"Apparently the Oregon people have been working on him [Roosevelt] with reference to improving the lower Columbia, where plants can be put in at lower cost," Goss wrote to Governor Martin immediately after the meeting.[40] "I was disappointed in the

President's attitude, although Senator Dill apparently felt quite elated," Goss wrote to O'Sullivan the same day. "The president was chiefly concerned over our ability to market the power. He had the matter of gradual development in his head so strongly that nothing could be done until we got the figures before him."[41] Dill wrote to O'Sullivan,

> I can say that Mr. Goss and I had a conference with the President. He is studying the Columbia Basin project for the first time. He told us he really had never had any definite or official information about it. I left a copy of the hearings with him and I am going to try to carry out his suggestions as to what we may do toward getting something started this summer if possible.[42]

Dill immediately saw Elwood Mead and on that same day Mead wrote to engineer Raymond F. Walter,

> Senator Dill has talked with the President about Columbia Basin. The President wants to know if we have given any thought to a lower dam, with less initial outlay and less power to be absorbed in the market at the start. Have you given this matter any thought recently and do you think the plan of a lower dam, with a higher one later on, is financially feasible?[43]

The Bureau of Reclamation's engineers looked into lower and smaller dams.

Four days later Dill wrote to Mead that Roosevelt would consider a low dam costing about $50 to $60 million.[44] Mead wrote to Dill that the Bureau could build a dam about 145 feet high for $60 million. Dill announced the news to the press, adding that the low dam had the President's blessing, that it would not require congressional approval, and that money for the project would come from the RFC and Washington state unemployment relief funds.[45]

Mead had sent his chief design engineer, John Lucian "Jack" Savage, to look over the Grand Coulee.[46] Savage thought he had capped his career when he worked on Hoover Dam. "[It] meant a wonderful climax—the biggest dam ever built by anyone anywhere," he said.[47] But he had a much bigger dam ahead, the quintessential dam, and he realized it when he looked at the Grand Coulee site. Always well in the background, Savage oversaw design work on the various models proposed for the dam, including the form it ultimately took, and he was responsible for the configuration of the successful structure the government finally built.

Under Savage's direction, Bureau engineers first designed a multiple arch structure 3,400 feet long containing two million cubic yards of concrete and speculated that it could be raised later.[48] Gill and the Spokane people still wanted only a small dam, not one that would grow into a giant generating prodigious amounts of power and flooding their Kettle Falls site, and they wanted it tied to reclamation which would consume the electricity. Concerned, Gill wrote to Mae Schnurr, Commissioner Elwood Mead's assistant,

> There is much apprehension here over the situation since the Governor has ignored the OLD CROWD entirely and turned the State Commission over to the Wenatchee radicals. None of the Leaders who carried on for 15 years are being consulted in anything that is being done. Mr. Goss, who fought the Project for 12 years is the strongest man on the Commission, with Rufus Wood [*sic*] the next. Banker, Smith and O'Sullivan are "Yes, men" to Goss & Wood. Our man McGovern is ineffective. I will be glad to get any further information you have on this "Low Dam." It would seem that the cost per kilowatt hour must be much increased over the High Dam. The question in my mind is, with a higher cost of power and no irrigation, would the power companies be interested???? Unless the Country is rehabilitated by Irrigation, what will the Power Companies do with additional power??? I will be glad to hear further from you about the LOW dam.[49]

Gill had cause to worry. O'Sullivan, acting for the Columbia Basin Commission, applied to the Federal Power Commission (FPC) for permits on the Grand Coulee site, and he asked that agency to revoke the Washington Water Power Company's Kettle Falls permit.[50] The company objected because the government did not then contemplate a high dam. The FPC held the matter in abeyance until the government settled on the configuration of Grand Coulee Dam.

Increasingly, Woods and Goss felt that the state should build the dam and they wondered about involving the Bureau of Reclamation at all, fearing that the Bureau would dominate the project and overshadow local control or, based on its performance so far, would go slow and never complete the task. They increasingly favored the Army Corps of Engineers. Woods telegraphed O'Sullivan, then in Washington, D. C.,

> Lay off pushing reclamation department in this project  Keep all avenues open  Sentiment growing rapidly favorable [to] state project  This unanimous opinion in our group  Reclamation service too dilatory and has always had to be pushed into action in past  State wants revenue as state must guarantee repayment.[51]

O'Sullivan knew better. Happy with the support he received from Elwood Mead, and irked at the Army Engineers, their reversals and review boards, he decided that the Bureau of Reclamation should build the project because it offered the best hope of following through with the irrigation. O'Sullivan initiated negotiations on a contract between the state and the Bureau. Shaken at the speed with which O'Sullivan moved, Goss wrote to Woods that he only hoped the state could control the Bureau. "It would be a serious mistake to turn the dam over to the Reclamation department and have the people of the State of Washington pay to perpetually keep this Department operating," he added.[52] Despite their apprehension and disapproval, members of the Columbia Basin Commission acquiesced in O'Sullivan's decision. Thus, James O'Sullivan determined the builder of Grand Coulee Dam. In doing so, he guaranteed that sooner or later it would be a reclamation project irrigating the Columbia Basin.

Elwood Mead and his people worked on the contract between the Bureau and the State of Washington. They estimated that preliminary work, preparing plans, and doing surveys would cost $377,000. At this point the state still had no guarantees of federal money and the Columbia Basin Commission debated between approaching the RFC or waiting for Congress to pass the pending National Industrial Recovery Act (NIRA) with its Public Works Administration (PWA) which held more promise for large projects. Meanwhile members of the Columbia Basin Commission and Governor Martin decided that the $377,000 needed to start work could come from Washington state's $10 million relief fund.

Congress approved the NIRA and Roosevelt signed it on June 13, 1933. The Columbia Basin Commission immediately applied for funds and quickly met competition. James Delmage Ross, Seattle City Light superintendent, also actively petitioned for money to complete construction of the Diablo Dam and start the Ruby Project (now Ross Dam), both on the Skagit River in the state's northwest corner.[53] In his application to the PWA, Ross boldly stated that surplus power from the Skagit could easily be transmitted over the Cascade Mountains to the Columbia Basin where it might pump water from the Columbia River to irrigate the basin project.[54] Dill wrote to Ross that he and Bone would recommend government money for the Skagit as a flood control measure, but not for power generation because they did not want it to compete with Grand Coulee Dam.[55]

Ross may have talked directly with Roosevelt about his Skagit application.[56] O'Sullivan, Woods, and the Columbia Basin Commission knew of Ross's friendship with the President and they saw him as a serious threat. Although Ross argued later that he never opposed Grand Coulee Dam, in the summer of 1933 he did everything he could to deny it PWA support. Ross even claimed that Senator Dill owned land that would benefit from Columbia Basin irrigation and that Dill was part of a "gigantic land swindle" along with his former secretary, Frank Bell.[57] Ross ordered his aides to scour land records for evidence proving the charge.[58] Although they found none, Ross continued debating proponents of Grand Coulee, who increasingly saw him as an enemy.

The *Bellingham Herald* joined the fray with an editorial labeling Grand Coulee Dam a "folly" and insisting that there was "no call for more agricultural land," or "no demand for hydro-electric power." It ended, "Washington has started into an enterprise that is almost certain to be disastrous. It should stop it NOW!"[59] The *Yakima Daily Republic* also attacked the dam, claiming a power surplus already existed in the Northwest. *Outlook Magazine*, in a reference to O'Sullivan and Woods, described Grand Coulee as "the impulsive scheme of a country lawyer and a country editor."[60]

Further opposition came from Oregon. Early in May Senators Charles McNary and Frederick Steiwer and Representative Martin called on President Roosevelt and urged him to sidetrack the Columbia Basin Project until the government developed the lower Columbia River for hydroelectric power, navigation, and flood control.[61] Roosevelt, who also liked the Bonneville proposal, faced the prospect of justifying two large projects close together in states with small populations or of choosing between them and making political enemies. As was typical of Roosevelt in such cases, he postponed the decision.

In Washington state, on June 30, 1933, an inordinately large headline in the *Wenatchee Daily World* announced that the State Relief Commission had allocated of $377,000 to finance work by the Bureau of Reclamation on the Grand Coulee Dam.[62] Governor Martin and Senator Dill both described the event as something "outstanding in the economic history of Washington."[63] Not to be out done, Rufus Woods wrote, "As it is the beginning of the most monumental program of its kind ever attempted in the United States, the story will be not only of local interest, but also of state-wide and national interest."[64] On July 6 members of the Columbia Basin Commission signed three copies of the contract between Washington state and the Bureau of Reclamation and sent them to Denver where Mead signed on July 11.[65] The contract committed the state to build the dam and the money assured that work would at least start.

On Saturday morning, July 15, 1933, a special train, provided by the Northern Pacific Railroad, left Spokane and headed for the Grand Coulee. As no track yet extended to the site, the reporters and spectators left the train the next day and drove the last few miles in busses. That Sunday morning, July 16, over 200 cars left Wenatchee in a caravan heading to Almira and then over a twenty-mile stretch of dirt and gravel road to the dam site. Fox Movietown News, Pathé, and Universal movie companies sent cameras to record the event. Undaunted by uncertainty over funding and rumors that the federal government might take the project away from the state and build the dam itself, the Columbia Basin Commission went ahead with a gala ground-breaking ceremony.

Perhaps 5,000 people attended the event that Senator Dill had suggested. Fifteen years and one day after the Billy Clapp article appeared in the *Wenatchee Daily World*, at 1:30 p. m., Governor Martin hit an engineering stake with a sledge hammer and Senator Dill turned a shovel of dirt. O'Sullivan, in his role as executive secretary of the Columbia Basin Commission, acted as master of ceremonies with Martin, Dill, and Bureau of Reclamation engineers standing around him. A few Native Americans, residents of the nearby Colville Indian Reservation, showed up in full regalia, then stood by and watched as the temperature reached 105 degrees.[66]

In Spokane the *Spokesman-Review* noted,

> A new chapter in the history of the Pacific northwest, less colorful perhaps than some of the pages out of a thrilling past, but more important from the viewpoint of the development of an industrial and agricultural empire, was started yesterday afternoon when the first actual preliminary work on construction of Grand Coulee power project was put under way.[67]

A few days later the balloon seemed to burst. It started when the Portland *Oregonian* printed a story, based on a statement by Representative Charles Martin of Oregon, that President Roosevelt had granted $31,000,000 for Bonneville Dam. On July 23 the *Spokesman-Review* carried the news and interpreted it as meaning that suddenly Grand Coulee Dam was dead.[68] But Senator Dill found the Public Works Board still considering the Grand Coulee proposal despite Oregon's propaganda to the contrary.[69] The next day, July 26, Dill managed another appointment with Roosevelt, hoping to persuade the President to back the project. When he left the White House Dill said only that he was "encouraged" by his talk.[70]

The next day, July 27, 1933, Secretary of the Interior Harold L. Ickes met with the Public Works Board and that body appropriated the $63,000,000 needed to build a low dam at the Grand Coulee.[71] Apparently Roosevelt decided to support both projects. Dill wired O'Sullivan, "President has big plans for this development which I will explain later."[72] Roosevelt himself made a more modest announcement at a press conference the next day.[73]

Catching his breath, Dill fairly bubbled in front of reporters. "President Roosevelt will use the Grand Coulee dam at Seaton's ferry in the Columbia river to 'bust' the power trust," he told a correspondent from the *Spokane Daily Chronicle*.[74] The *Wenatchee Daily World* announced 1,500 jobs for the coming winter and cited Secretary Ickes as the source for that estimate. Elwood Mead said that the government would employ 12,000 men at the dam. Ecstatic at the prospect of prosperity just around the corner, the Spokane American Legion invited the President to visit, and in honor of Dill, a Spokane cafe created what it called its "Grand Coulee Sandwich complete with a Dill pickle."

Dill returned to Spokane on August 1 to what the *Spokane Daily Chronicle* called a "hero's welcome." "Senator C. C. Dill came home last night from Washington where, in the face of strong opposition, he won a glorious victory in assuring the Grand Coulee Dam," reported the *Spokesman-Review*. The *Wenatchee Daily World* reported that 10,000 people greeted Dill at the Spokane railroad station where the Senator gave a dynamic speech broadcast by local radio stations.[75] Dill and O'Sullivan toured the region, taking bows and reveling in their success. Only Representative Sam Hill grumbled a bit about Dill. Hill justifiably criticized Dill for grabbing the Grand Coulee spotlight and ignoring the ten years that he put in working for the dam.[76] Nobody paid much attention.

O'Sullivan and the five-member Columbia Basin Commission celebrated repeatedly during those halcyon July and August days. On July 31 Governor Martin turned over a check for $50,000, the first installment on the $377,000, so that the commission could begin work. The next day O'Sullivan awarded a core drilling contract to Lynch Brothers of Seattle and a second to the Rumsey Company for test pit and trench work.[77] On August 31 the Federal Power Commission issued a preliminary permit for Grand Coulee Dam. In that action, the FPC noted a better utilization of the water power potential at the Grand Coulee site than at Kettle Falls. This did not, however, settle the Kettle Falls issue.[78]

An unexpected problem quickly developed. Washington state officials discovered that they could not simply take the land needed to build the dam and let the courts determine compensation later. And Washington state law required that it gain title before development.[79] The lawyers guessed that resolving this issue could stall work for months or perhaps years.[80] In a hurry, having promised to provide jobs immediately, the government began hinting about taking the project from the state and building the dam itself.[81]

This led to animated discussions between Washington state, the Bureau of Reclamation, and the Department of the Interior. Beneath the surface, Reclamation Bureau legal counsel Bernard E. Stoutmeyer, in a letter to Elwood Mead, put his finger on an underlying consideration that undoubtedly moved the Bureau as much or more than the apparent difficulties. Stoutmeyer pointed out that the Army Corps of Engineers, smarting at the loss of the Grand Coulee project, would attempt to build Bonneville Dam in record time to demonstrate that the Army worked more efficiently than the Bureau. The comparison would be clear as both projects, then about the same size, began at the same time:

This may be a friendly competition but if the Grand coulee [*sic*] project, which was started first, should be delayed so that the Bonneville project gets under way with a large force of men engaged in actual construction work while the Reclamation Bureau is still fiddling around with only a handful of men employed on surveys and investigations, it will lead to some rather severe criticism of the Reclamation bureau and the public will feel that the Army Engineers have made good on their claim that they are the people who should handle work of this kind.[82]

Moreover, Columbia Basin Commission members had a tendency to argue with each other, and at times the infighting destroyed their ability to make decisions. E. F. Banker and James McGovern, both of whom had backed the gravity plan and who wanted a relatively small Grand Coulee Dam, clashed with Rufus Woods, and especially with James O'Sullivan. This lack of harmony troubled Mead and Interior Secretary Ickes who knew only too well the long history of contention among Washington factions. All of this eventually led Ickes to take control of the project.

When Senator Dill returned to Washington, D. C., he told reporters that the Grand Coulee Dam would be built but, as was typical of Dill who always carefully looked before he jumped, he suggested that it might end up as either a state or a federal project.[83] Woods urged Dill to help retain the project under state control but Dill already knew that the Department of the Interior saw federal responsibility as the best course. As early as mid-October, Elwood Mead concluded, "I am convinced that Grand Coulee will have to be built as a Federal reclamation project, along the lines of Boulder Canyon."[84] On October 9, 1933, Secretary Ickes officially released the first million dollars for work at the dam.[85] By then it was obvious that they would not use the entire $63 million immediately, and on October 24 the Public Works Administration reduced the allocation for fiscal 1934 from $63 million to $15 million.[86] Then Secretary Ickes strongly recommended that the Grand Coulee Dam become a federal project.[87] Republican Rufus Woods never reconciled himself to the federal government taking the project. Reluctantly, however, after largely secret negotiations with the Department of the Interior, the Columbia Basin Commission agreed to a contract giving the project to Ickes with Washington state scrambling to salvage consultation rights and permission to keep its representatives on the construction site but, in the long run, losing the project completely.

On November 1, 1933 two related events altered the fate of Grand Coulee Dam. In Washington, D. C., Harold L. Ickes, in his capacity as Secretary of the Interior, signed the memorandum to himself as Administrator of Public Works, stating that the Grand Coulee project would be "a federal project to be constructed, operated and maintained by the Bureau of Reclamation." The memo went on to say that the "cost of the project should be repaid from the net revenues obtained by the sale or other disposal of hydroelectric energy," and that the allotment of $63,000,000 be considered "an allotment for a Federal project."[88] Grand Coulee Dam then officially became Public Works Project No. 9.

Also on November 1, the Columbia Basin Commission met in Olympia with Governor Martin, Senator Dill, and Bureau representatives, and tentatively approved a new contract between the state and the Bureau of Reclamation. It stipulated that project costs would be paid with power revenues and that the federal government would build the dam with an option for the state to lease or acquire it at some future date.[89] Government lawyers checked the contract through November and into December. Meanwhile Mead, on a visit to Seattle, announced that although the government had not finalized the contract, the Bureau of Reclamation would go ahead with the dam.

In Spokane, Mead made another announcement. On December 1, 1933 he revealed that the Bureau of Reclamation had reconsidered plans for the dam. After looking over nearly a dozen options, it scrapped the multiple-arch dam and chose instead a somewhat lower and considerably thicker, solid, gravity-type structure. The massive concrete block could later serve as part of a high dam. Mead said it would be cheaper for the government in the long run. If the Bureau built the multiple arch dam first, it would have to remove it, wasting its $63 million cost, before placing another dam, because Bureau designers determined they could not raise the multiple arch dam after all.[90] The Reclamation Commissioner pointed out that the new plan required more than $63 million but that his engineers now felt sure that a high dam could be imposed on this structure.[91] O'Sullivan and Woods considered this a partial victory, but they intended to press the government to build the high dam immediately or at least the foundation of the high dam, rather than this smaller structure.

In Washington, D. C., Secretary Ickes complained that Washington state had already received more than its share of PWA money, and that the PWA would not fund anything else there. Dill argued that Ickes had promised that large projects would not count against the state's general share of available relief money. Dill further pointed out that Ickes had yet to spend any of the Grand Coulee money and that so far everything had come from the state.[92] On November 17 the *Spokesman-Review* announced that Dill won his skirmish when President Roosevelt ordered Ickes to rush federal civil works projects and not count "Grand Coulee when considering applications from eastern Washington cities and towns."[93] The Bureau of Reclamation would build Grand Coulee Dam as a federal project. The dam would be low and for power, with the possibility of a higher structure and reclamation much later. At last, all the key decisions were in place.

Governor Clarence Martin's selection of the Columbia Basin Commission marked the direction that the project took. Martin clearly favored the position of Rufus Woods and the pumpers and he worked well with the Roosevelt administration's New Deal goals. He took chances and made enemies when he rebuffed the old guard in Spokane. He gambled on the success of the big dam, and he stood to lose if the project floundered or failed. But his gamble paid and Martin became one of the heroes in the Grand Coulee Dam story.

If the dam had a step-parent, it was the depression. The economic catastrophe that plagued the country through the first years of the 1930s brought Franklin Roosevelt to the White House. With Roosevelt came the goal of planning and the desire to tame private utilities with the "yardstick" of public projects generating cheap power against which to measure power rates.[94] The New Deal not only backed the multiple-purpose concept, started with the Boulder Project under the Hoover Administration, but it also added federal control and planning for development of whole river basins.[95] Although the low dam at the Grand Coulee, entirely for power generation, did not meet these goals, the potential for a larger project with irrigation, flood control, recreation, and navigation exactly fit Roosevelt's concept. The Grand Coulee Dam met Secretary Harold

Ickes's notion that public relief should go to projects that helped recovery and left the country with a valuable product able to pay its own way.[96] Toward this end, Roosevelt entered not only into the public-versus-private power fight, but also into the struggle between federal agencies themselves, and into disputes within states.[97] What happened to the Grand Coulee Project is an example of federal officials establishing authority over the waters of the West as a by-product of those struggles.[98]

Franklin Roosevelt and his New Deal started many projects through executive order during the first months of the New Deal. Roosevelt side-stepped Congress by providing the money out of the general fund. This left Congress to pay for them or abandon them, still unfinished, at some later date.[99] In that sense the Grand Coulee Dam followed a new model rather than the path established by Herbert Hoover. Where the Boulder Canyon Project Act of 1928 entirely funded Boulder Dam, Roosevelt began Grand Coulee Dam under what amounted to presidential fiat.[100] This later became an annoying hurdle for Grand Coulee backers. But that lay ahead. First the workers arrived and started building the dam.

The long conveyor belt on the west side of the river used to move aggregate. *Courtesy Eastern Washington State Historical Society, no. L87-1.5140-35.*

# Chapter Six
# Preliminary Work

Once I built a railroad, now it's done—
Brother, can you spare a dime?
Once I built a tower to the sun—
Brick and rivet and lime.
Once I built a tower, now it's done—
Brother, can you spare a dime?

*Popular Song, 1932[1]*

Through June and July 1933, bundles of mail containing up to 150 work applications daily inundated the Columbia Basin Commission's project office in Spokane. Tired, almost desperate men, some out of work for months or longer and anxious for employment—any employment—frequently showed up in person as the Great Depression of the 1930s left one-third of the nation's labor force with nothing to do. Afraid and looking for jobs, many in the West saw the great dam on the Columbia as a second chance. They were drawn by the often-repeated promise that Grand Coulee Dam would provide thousands, perhaps as many as 100,000, jobs.

The *Wenatchee Daily World* cautioned the jobless to write before traveling to Grand Coulee. One article begged people not to apply in person and explained that all applications went to the Bureau of Reclamation. The government, it said, currently needed only qualified engineers who lived inside Washington, and the contract between the Bureau and the state gave hiring priority to Washington residents. Besides, nobody knew yet exactly when or how many workers the task would require. There might be additional work if a contractor eventually built a railroad to the site, but it would be a while before many found work at Grand Coulee.[2] Despite the warnings a migration started. Single men, men who left their families behind, and men with their wives and children in tow appeared at a desolate spot along the Columbia River in the middle of nowhere. They moved toward a place where the hot sun dried up everything in the summer and cold winters made living uncomfortable.

On August 1, Frank Arthur Banks arrived at the dam site.[3] Raymond Fowler Walter, the Bureau of Reclamation's chief engineer, had named Banks the chief construction engineer for Grand Coulee Dam.[4] He would supervise the work and see that the various contractors accomplished their tasks correctly and on time, a formidable task that occupied Banks for much of the rest of his life. Frank Banks was the central figure who dominated construction of Grand Coulee Dam and, later, parts of the Columbia Basin Project. He and John Savage, the chief designer, together determined the dam's appearance and how it would be built.

Born on December 4, 1883 in Saco, Maine, Frank Banks always liked engineering and chose it as his major at the University of Maine. During his sophomore year the Reclamation Service, hoping to lure young men into a government career, sent a group of upperclassmen to look at projects in Montana. When they returned they gushed about the wonderful West and its engineering opportunities. Intrigued by their stories, Banks decided to move away from New England, where his family had lived for over two centuries, and when he graduated in 1906 he headed West.

In Montana young Banks secured a job with a survey gang as a rodman working on the Lower Yellowstone Project near Billings in eastern Montana. From 1909 until 1913 he worked as the division designing engineer at the Bureau's headquarters in Boise, Idaho. There he prepared the preliminary plans for Arrowrock Dam that, for a few years, was the highest in the world. About that time Banks married Theodora L. Drummond of Boise. She too thrived on life in the West and willingly followed her husband to every forbidding and isolated place he worked.

In 1913 the Reclamation Bureau named Banks as its construction engineer at a dam near isolated Jackson Lake, Wyoming. Neither Banks nor his wife avoided or complained about hardship.[5] She had to travel to a distant hospital for the birth of her first child, and then she and her baby daughter returned to Jackson Lake in a covered wagon that carried them over the mile-high Teton Pass during the middle of winter.[6] Later Theodora Banks, remembered by all who knew her as a gracious lady who offered warm hospitality to friends and acquaintances, had a son. Banks stayed at Jackson Lake until 1917.

From 1917 until 1920 Banks was the construction engineer on the Minidoka Project in southern Idaho, and from 1920 until 1927 he filled the same role at Idaho's American Falls Project. Banks worked six years on the Owyhee Reclamation Project in Ontario, Oregon, overseeing construction of the Owyhee Dam that then became the world's highest—until Boulder Dam surpassed it.[7] By 1933 he had spent twenty-seven years dealing with dams in one form or another. Round-faced, bespectacled, and soft spoken, Banks exhibited the precision of an engineer and the authority of a good administrator. He was calm with an even disposition and a knack for winning the good will and active cooperation of those around him. *Pacific Builder and Engineer* called him, "one of the most engaging personalities in the construction world of the Far West."[8] In the eight years it took him and his assistant, Alvin F. Darland, to supervise Grand Coulee construction, his black hair, which had just started to show a touch of gray, turned to white.[9]

Banks wasted little time when he arrived at Grand Coulee. He opened a construction office in an abandoned bank building in Almira, about thirty miles south of the dam site. With one clerk and a stenographer as his entire staff, he began. Although he worked for the Bureau of Reclamation, he also had to deal with Washington state's Columbia Basin Commission (CBC) that then still controlled the project, and listen to politicians like Senator Clarence Dill, who repeatedly insisted on influencing decisions and complicating the formidable technical chores. Despite years of investigations, no one had identified the best place to locate the dam. Banks needed an engineering staff and he needed it quickly.

The CBC announced it would hire thirty to forty engineers for field work and the *Wenatchee Daily World* printed the story emphasizing that the commission would consider only men residing inside the state before August 1, 1932.[10] It already held over 7,500 applications and more than 2,000 of those qualified. From them Banks selected and recommended thirty.[11] Over the next two months his crew increased, filled the old bank building, and then expanded into a nearby lodge hall.[12] Banks used the buildings in

Almira until March 1935, when he moved his headquarters to the new town at Grand Coulee.

No railroad passed near the dam site and the Bureau of Reclamation stressed the importance of immediately laying track into the area. Both the Great Northern and the Northern Pacific showed interest and the NP started surveys for a spur originating in Hartline, while the GN investigated an eighteen-mile-long route starting in Mansfield. The GN estimated the job at $750,000 and, in a spirit of friendly competition, it boasted it would have its track in place well before the NP could complete its anticipated extension.[13] Later the two railroads joined forces and applied to the Interstate Commerce Commission stating that they would do the job together. Then suddenly they demanded that the government guarantee shipment of all material to the dam by rail.[14]

Truck drivers in Spokane declared that they would protest to the Interstate Commerce Commission any contract with an exclusivity clause. They argued as unfair the request for a monopoly on all hauling, and insisted that the railroads stood to realize over a $3 million profit. Representatives of the railroad countered that they would not lay any track if the government forced them to share the traffic with trucks.[15] On March 3 the Northern Pacific told the press it would not build any rail line to the dam site since no government agency would guarantee it the exclusive right to all business. The railroad did agree to give its surveys and engineering records to the government. Consequently, the Bureau of Reclamation decided to build the railroad itself. In the town of Odair on May 17, 1934, it opened bids for the thirty-mile-long branch line. On June 12 the Bureau awarded the job to the David H. Ryan Company of San Francisco.

On September 5, 1933 the CBC received preliminary permission from the Federal Power Commission to proceed with engineering and initial construction at the dam.[17] Acquiring title to the land proved more difficult. James O'Sullivan managed only to buy a few parcels including eighty-seven acres, previously the Charles Osborne ranch, on the west side of the river about one-half mile below the dam site—a place that seemed a good spot for the engineers to locate their houses and offices. Banks looked at a much larger parcel on the other side of the river, still held by its owner, as the best place for a construction camp. Both the state and federal government withdrew public land around the project.[18] Speculators hovered about, and the people who owned acreage on either side of the river held onto it, hoping its value would increase and that the government would pay them inflated prices.

Time was important as Banks and the CBC attempted to create jobs quickly. Proponents had touted the dam as a way to put unemployed laborers to work, and the state's $377,000 came entirely from money earmarked to create the thousands of expected jobs. Still, it looked as if it would be months before serious large-scale activity began. Pressured by Senator Dill and the Columbia Basin Commission to employ more men immediately, on September 1, Banks awarded a core drilling contract to Lynch Brothers of Seattle and one for test pit work to Rumsey and Company, also of Seattle.[19] The skilled and unskilled laborers used on the job would receive eighty cents and sixty cents an hour respectively, put in a thirty-six hour work week, and pay $1.50 daily for their meals and lodging. They were excellent wages—something that did little to discourage people from seeking jobs at the dam.[20]

The contractors arrived on September 5, and Saturday, September 9 marked the official start. Setting out in a steady, cold, driving rain, Frank Banks and five of his engineers[21] took one full day to set the stakes that marked the axis or center line indicating the length of the dam. Twenty-seven other engineers began different tasks assigned by Banks. Early the following week the core drilling and test pit digging started in earnest.[22]

The Rumsey Company built a temporary mess hall and a bunkhouse for forty men. The buildings sat close to the axis of the dam so the workers lost no time going to and from the job. The total number on the job, including engineers, increased to over 150 by the end of September. A few small frame buildings clustered near the mess hall and bunkhouse, while many men lived in tents.[23] The *Spokesman-Review* reported,

> Grand Coulee dam site has changed within the last month from a sleepy spot where the Grant county ferry crosses the Columbia, to an area where machines whir day and night. From the top of the Columbia river gorge as one starts the descent to the dam site, a little town of tents and unpainted board buildings is seen. The west side is dotted with drill rig structures.[24]

Every few days the local newspapers reported construction milestones, some momentous and others dubious. On October 23, 1933, for example, the engineers poured the first bucket of concrete at the dam site. It was just that too—one bucket—dumped as a survey marker. Despite the event's insignificance, the *Wenatchee Daily World* gave it front-page coverage. Happenings at Grand Coulee held the attention of the region and, for a while anyway, even tiny incidents took on considerable importance.

Senator Dill increased pressure on the Bureau of Reclamation to employ more laborers and announced to the press that over 800 jobs would exist by that fall.[25] Disturbed by Dill's optimistic and premature predictions, Frank Banks called the report "erroneous." And he added, "There are now 75 idle men at the dam, some of them with families, and all living in squalor, under unsanitary conditions." He anticipated that conditions would worsen as winter approached and he added that the "influx of squatters . . . is aggravating and interfering with the work of those employed."[26]

A few entrepreneurs opened stores and other businesses around the construction site. Neither the Columbia Basin Commission nor any other state agency had anticipated the problem of unregulated growth and no mechanism existed for granting or denying permits of any kind. The CBC commissioners pondered what to do should things get out of hand. Finally they instructed Joe Mehan, their representative there, to chase away anyone not employed at the dam. Mehan wrote back to James O'Sullivan,

> As instructed I have notified all squatters and others not connected with the work that they must be off the damsite not later then [*sic*] midnight of November 19, this also conforms with Mr. Banks [*sic*] wishes. It has been necessary that I spend a lot of time at [the] damsite at night lately [on] account of petty fights, gambling, etc, not having a car works quite a hardship.[27]

Mehan struggled to keep away drifters, job seekers, and would-be shopkeepers. Once having traveled that far in search of work, however, the men hardly wanted to leave, and the problem continued.[28]

Joe Mehan himself irritated Banks. In a confidential letter to Elwood Mead, Banks wrote:

On September 9th a typical "ward heeler" by the name of Joe Meehan [*sic*] called at the office, stated that he had been brought over from Olympia by Mr. O'Sullivan to act as the personal representative of the Columbia Basin Commission and requested a desk, an automobile and the use of a stenographer. I didn't like his looks, his attitude or the general idea but learned from Mr. O'Sullivan that Mr. Meehan had to be taken care of on the job and he did not know what to do with him. I explained that I had no objection to his representing the commission on the job but that there was no place for him in the engineering organization, no room for him in their office, and no equipment available for his use.[29]

Although Mehan remained a few months, Banks ultimately managed to rid himself of the man. Mehan was one of many who hoped that Grand Coulee would provide him with a job. In his case, he found work, but like many who followed, he stayed only a short time. He was more fortunate than many who never found a job and was typical of the labor force at Grand Coulee which remained transient throughout the construction period. He also demonstrated the friction between the Bureau and the CBC, both of which wanted to direct the project. Had the federal government not taken over the job, it is doubtful that the two would have worked harmoniously for very long.

News of jobless men accumulating at Grand Coulee increasingly appeared in newspapers around the state and beyond, hardly presenting the image of an expensive project easing unemployment. The Public Works Administration released the second million dollars for use at Grand Coulee late in October and Reclamation Commissioner Elwood Mead ordered Frank Banks to spend it immediately on preliminary excavation. It was work needed eventually and, since engineering studies now showed fairly clearly where the dam site should be, it would not waste money. Above all, it would create jobs.[30] On November 2, Banks called for bids on the earth removal job.[31]

In the meantime, in November, Ickes transferred control of Grand Coulee from the State of Washington to the federal government. Frank Banks and the other government engineers welcomed elimination of the state, since dealing simultaneously with both their own bureaucracy and the CBC was a nuisance. Surprisingly, the Bureau of Reclamation continued using, and the state continued supplying, money out of the state's $377,000 allocation. Washington taxpayers provided $77,000 for preliminary engineering work, the state paid for a few cars used by Bureau engineers, and Banks did not transfer state employees or some workmen to the federal payroll until late spring 1934.[32] By then he had spent most of the state's money, leaving none for road work.

On November 20, 1933 over 300 contractors showed up at the Adair engineering office to watch the bidding for the excavation contract.[33] The specifications detailed removal of 2,040,000 cubic yards of earth and rock called overburden, half of the work to take place on each side of the river.[34] Of the twenty-two bids submitted, the David H. Ryan Company won the award with an estimate of $534,500. Ryan revealed he would build a small camp on each side of the Columbia and employ up to fifty men through the winter and spring.[35] Ten days later the Bureau verified Ryan's bid and formally awarded the contract.

The David H. Ryan Company issued two subcontracts. Goodfellow Brothers of Wenatchee began work on the west side of the river early in December while the Roland Construction Company of Seattle started digging on the east or Okanogan County side about two weeks later.[36] On December 5, in a heavy snow storm, one of the large machines took its first bite out of the earth. The contractor moved in more equipment, including larger shovels. On Christmas Eve 1933 drillers working for Lynch Brothers got drunk and damaged the camp, completely wrecking one small building. Such incidents occasionally livened things, but otherwise the work proceeded uneventfully. Early in January the Roland people started a fifteen-kilowatt diesel power plant that turned a generator supplying power to large floodlights allowing night work.[37] Although Goodfellow workers met veins of hard clay that required blasting, both subcontractors made satisfactory headway.

Despite the contract for overburden removal, critics repeated complaints that the Bureau did not move fast enough or employ the thousands of men promised. The Columbia Basin Commission credited this to "pressure from other work" that faced the Bureau, and the Commission's own failure to expedite activity.[38] Some argued that the Bureau still lacked the experience to handle such a large project which should have gone to the Army Engineers, then showing considerable skill and speed at Bonneville. Bureau of Reclamation engineers chafed hearing such comments, but determined to continue in an orderly manner and not rush into heavy construction before ready. There remained considerable preliminary work before anything could start on a large scale.

At the dam, different complaints came from the unemployed. Despite its remote location, and many warnings, men continued to scramble for jobs at Grand Coulee. Rumors spread that some of those who secured positions, either as laborers or engineers, failed to qualify as Washington state residents. When those charges reached the Grant County public prosecutor he asked the CBC to investigate.[39]

Others contended that the contractors ignored the spirit of the National Recovery Act and showed favoritism in handing out work. Something approaching 150 men put together a quasi-organization they called the Grand Coulee Unemployed Workers' League. The group held "indignation" meetings at different locations around the construction site and appealed to Frank Banks for "fair treatment" in hiring. Leaders argued that a few men worked 80 or even 100 hours weekly while others sat idle. They pointed to carpenters, recently arrived from California, who successfully found jobs while experienced locals waited. An announcement in April that the Boulder project, now past its construction peak, would soon lay off 2,500 men did nothing to allay rising fears at Grand Coulee.

Privately, the men told reporters that they really wanted the government to fire C. C. Beery. Since the dam received funds under the National Industrial Recovery Act, the National Reemployment Service (NRS), organized under the United States Department of Labor, had to screen and clear everyone hired. The NRS set up an office at the dam site and ordered first preference for ex-servicemen, especially those with dependents. Next came residents of the three counties near the dam with the rest of the state following. Sometimes arrogant and dictatorial, Beery ran the NRS office, and those who failed to find work increasingly blamed him for their troubles. Nothing much came of this initial flap but Beery remained a source of aggravation, and problems with him and the NRS recurred throughout the construction period.[40]

Frank Banks oversaw and dealt with everything from major decisions to minor irritants. He recommended that the government officially name the project Grand Coulee Dam rather than the Columbia River Dam, the designation so far used in official documents and communications. Banks argued, successfully, that everyone called it Grand Coulee and that the government might just as well join in and avoid confusion.[41]

On the other hand, the federal government did not agree to build roads to the dam.[42] Senator Dill, and just about everyone else, concluded that the state had to finance the roads, and Dill argued that the money should come out of the state's $377,000 allocation. But Banks had spent all of that.[43] Early in 1934, forty Civilian Conservation Corps (CCC) boys began work on a gravel road between Nespelem and the dam, twenty miles to the south.[44] It was the first CCC activity in the area, with more to follow. Then warm weather and heavy rain turned everything to mud, and creeks, ordinarily dry, washed-out sections of the road running southeast to Wilbur. Two stretches of highway near Broadax, each over 100 yards long, sat underwater. This hindered contractors as they struggled to move heavy equipment and dramatized the immediate need for a good highway. On January 16, 1934, Governor Clarence Martin signed a bill authorizing the state highway department to spend over $600,000 to build and maintain the routes to Grand Coulee. Officials, still holding the illusion that the dam would immediately create thousands of jobs, reasoned that road construction would thus generate enough business, and then taxes, to repay the cost. The *Wenatchee Daily World* announced that over 1,000 men would build the new state highways. Officials investigated a number of proposals and in February finally agreed to pave one main highway from Coulee City, through the Grand Coulee, to the dam, plus three lower-grade roads linking other nearby towns. Work continued through the spring and, ironically, dust became a major problem for the contractors when they started laying the hard surface around the middle of April. They put gravel on the road to Nespelem and for that job the state hired only Indians from the Colville Reservation. By the summer of 1934, getting to and from the dam became much easier.[45]

No bridge crossed the Columbia River anywhere near the Grand Coulee and everyone depended on two small ferries run by locals. The state argued that it could help with design but did not have enough money to build a bridge, which it estimated at over $650,000.[46] When the Bureau of Reclamation finally agreed to build a suspension bridge it encountered pressure from businesses and government officials around Seattle, one of whom telegraphed Commissioner Elwood Mead:

> Understand they are proposing suspension type bridge for Coulee Dam Strongly urge substitution of cantilever type bridge which will give Washington industry and labor opportunity to do the entire work. No work has been allocated to western Washington. With this type of bridge they will at least have opportunity to quote.[47]

After debate in Denver, the Bureau requested bids for both a suspension and a cantilever-type bridge.[48] On February 28, 1934, citing geologic problems that eliminated the suspension bridge, the Bureau opened only bids for a cantilever structure and awarded the contract for its two piers to the Western Construction Company.[49]

Western Construction immediately placed cofferdams (temporary structures meant to divert water away from an area, allowing work there) at each of the selected pier locations. The cofferdams kept the sites dry until men could remove all earth and other material from the underlying bedrock, erect the piers, and secure them in place. On May 3 the company abandoned one of the two cofferdams because the Columbia rose dramatically in the annual spring flood—from approximately 100,000 cubic feet of water per second to over 325,000 cubic feet. The high water eventually forced the company to leave both sites and suspend all work until the river subsided. It was only the first problem to plague the bridge.

Near the end of December 1933 the Rumsey Company completed its test pits. Bureau engineers labored to establish permanent markers at various elevations around the dam, including concrete bench marks indicating the low and high levels of the dam itself. All the work indicated a solid rock base of granite and that the valley behind the dam would effectively hold water.[50]

Downstream from the dam Bureau workers began grading and leveling the site of its "model city" where engineers would live.[51] Late that spring the Bureau had a contractor busy putting up seventy-five houses and an administration building. On the east side of the river a few miles away, twenty-four men completed a primitive runway that became an airport. Built with funds provided by the Civil Works Administration, the small field, finished that summer, helped link the construction site more quickly with the outside world.

Things seemed to be going well. Then on January 30, 1934 a small article on a back page of the *Wenatchee Daily World* first reported a problem that quickly intensified and plagued the entire eight-year-long construction period. On the previous night, 75,000 cubic yards of earth on the east side of the river shifted and slid into the excavation area.[52] Geologists working for the Bureau explained the cause. More than 15,000 years before, when ice dammed the Columbia River and formed lakes that filled with water melted from the glaciers, an extremely fine silt, called glacial flour, settled out, accumulated, and formed much of the overburden that the contractors now labored to remove. Unstable, especially when wet, this material showed little ability to maintain its shape without the support of surrounding earth or rock and it tended to slide, sometimes suddenly and without warning. The formation of the slope on the east side of the river itself appeared to be the result of some huge ancient slide.[53]

Despite the threat of slides, excavation continued. In mid-March the original job was over 60 percent complete and the Bureau extended the Ryan contract to include an additional 500,000 cubic yards of material.[54] Then a crack appeared in the west hillside, widened dramatically, and three days later, on March 27, 1934, the earth moved again. In all, 1.5 million cubic yards of earth moved that day and inundated the area where the west abutment of the dam would eventually stand. The sliding ground buried two tractors, but the men scrambled to safety in time. It took the Goodfellow crew until April to clear away the mess.[55]

A second slide hit the east early in April when 35,000 cubic yards of material fell into the excavation. Nearly 125 men waited while the Bureau's engineers decided what to do next and while the west side continued to rumble and shift. Finally the government chanced putting the men back to work, hoping they could finish by early summer without another serious incident. Although nobody died, a small slide early in June hurt two men. Without further mishap, Ryan and his subcontractors completed their tasks by the middle of that month.

The unexpected slides had both short- and long-term ramifications. The Bureau had to alter the main roadway and the railroad route into the construction site and preliminary work on the railroad slowed while the engineers pondered where exactly to put it. To guarantee the railroad's stability, the Bureau had to remove nearly one million cubic yards of earth it had not anticipated. This added expense concerned the engineers, but the potential for future and larger slides caused even greater anxiety. Early on it became obvious that the unstable ground and resulting landslides would be one of the biggest headaches connected with building Grand Coulee Dam. The first contractors, however, managed to complete their work and by early June released their crews. Some activity continued at the town for a few engineers, while the rest looked for other means to support themselves and their families as they waited.

At last, however, construction had started. But the frequent occurrence of slides indicated that building the dam would be more difficult than anticipated. And the project failed to employ thousands as promised. The desperation with which men sought employment resulted in the labor squabbles and fierce competition that continued almost through completion of the dam. The number of people who came to the site, the distance they traveled, and the desperation they exhibited constantly amazed Frank Banks and other officials. When Banks announced a temporary lull until sometime in July when the first major contractor would really begin the dam, the hiatus temporarily worsened already bad conditions around the construction site. Everyone eagerly waited for the long-anticipated heavy work finally to start.

Constructing steel cells connecting the downstream end of block 40 with the west cofferdam, September 1935. *Courtesy Bureau of Reclamation.*

# Chapter Seven
# MWAK's Giant Cofferdam

Carl said he had taken a job working on the new dam that was being built east of the mountains, at Grand Coulee.

I said something about that being as good a way as any to earn money to come back to school.

"It's not the money," he said. "I just want to have worked on that dam. I'd like to be a part of it."

*As told by Murray Morgan*[1]

Early in the summer of 1934 Helen Thomas, an attractive young brunette from Soap Lake, won the title of "Miss Coulee Dam." Carrying that honor she was part of the ongoing hoopla with which locals surrounded everything related to the dam. On the morning of June 19, 1934 she occupied a prominent place next to Washington Governor Clarence Martin, Reclamation Commissioner Dr. Elwood Mead, representatives from the Nez Perce Indian Nation, just about everybody who was anybody in eastern Washington, and 10,000 excited spectators. All of them assembled in Spokane to watch the government open the bids for the construction of Grand Coulee Dam.[2]

On March 3 the Bureau of Reclamation had called for bids. Preliminary excavation was well along and Bureau engineers wanted work on the dam itself underway that summer. The slides prolonged overburden removal and design complications postponed bidding until June 18. The final specifications estimated the job at about four-and-one-half years, resulting in a dam 290 feet high with powerhouses but no irrigation facilities.[3] Thirteen companies purchased copies of the plans but most lost interest when they read that whoever won the job assumed all risks while diverting the Columbia and pouring the concrete. Although the government took on that liability at Bonneville, it declined the responsibility at Grand

Coulee, where moving the river out of its bed posed a formidable and complicated problem.[4] There were some who openly declared that the Columbia could neither be diverted or dammed. Whoever won the bid also won the risk.

Representatives of the Six Companies, the half-dozen units that banded together under the leadership of Henry John Kaiser and Warren A. Bechtel and built Boulder Dam, made no secret that they wanted Grand Coulee next. By 1934 the giant dam at Black Canyon was nearly done.[5] The Kaiser people doubted that many competing outfits could produce the required $2 million bidding bond and James E. McGovern of the Columbia Basin Commission privately assured Six Companies representatives that the job was theirs.[6] Confident of success, each of the firms within the Six Companies established an office in Spokane's Davenport Hotel and, on June 12, six days before the bidding, opened a combined headquarters in Almira.[7] Everybody in the know assumed that the Six Companies all but owned Grand Coulee Dam.

But nobody figured on Silas Boxley Mason. A Virginian by birth but a resident of Kentucky by adoption, Mason played the part of a Southern gentleman and in 1934 he headed one of the oldest contracting firms in the United States. He loved breeding, training, and racing horses and his Kentucky bluegrass farm had the reputation as one of the nation's finest. He eventually named a favorite horse Grand Coulee. Silas Mason did not like to lose a race—any race.[8]

The problem for all bidders was with the Columbia River. Its volume fluctuates from a low over the winter to a high sometime in the spring, declining again during the summer and fall. Heavy rains upstream or an unusually quick snow melt can rapidly double or triple the river's size.[9] Mason's people studied the river diversion problem and came up with a novel, money-saving plan. Rather than dig a new channel for the river, they would block parts of it with cofferdams and construct the main structure a piece at a time. The river would then flow over, around, or through the finished parts while work continued on the remaining sections.[10] The engineers estimated that diverting the river by even this plan would cost between $6 and $9 million.[11] And misjudging the height of the spring flood could mean physical and financial disaster.

Besides his diversion plan, Mason held a second trump card that he saved as a surprise for the bid opening. With his secret well kept, he and his associates put together a conglomerate and calculated their eleventh-hour estimate. June 18 would see which group guessed correctly and submitted the lowest cost.

People started gathering in Spokane on the Saturday and Sunday before the big day. They filled the city's hotel rooms and latecomers scrambled for someplace to spend the night. At 10:00 a. m. in the auditorium of the city's Civic Building, with full radio coverage, the drama unfolded. At the front of the room stood a long table before a curtained stage, under a large American flag. Frank Banks, Elwood Mead, Raymond Walter, Bernard Stoutmeyer (chief legal council for the Bureau), Governor Martin, other Bureau engineers, E. F. Banker, and the six members of the Columbia Basin Commission took their places. A second table, facing the first, held clerks, reporters, more officials, and Bureau engineers. Behind them sat the 1,000 people lucky enough to secure a place in the small auditorium. The rest waited outside.

Frank Banks sat at the center of the head table with four large envelopes in front of him. After formalities the auditorium quieted and Banks opened the first envelope. "It is from three firms which have linked together," read Banks. "They are the Silas Mason Company of New York, the Atkinson-Kier Company of San Francisco, and the Walsh Construction Company of Davenport, Iowa and they have combined and incorporated under the name MWAK Company. Their bid is for $29,339,301.50." Some in the auditorium laughed at the fifty cent finale concluding the long string of numbers, but others gasped and the audience buzzed with excited whispering. No one expected such a bid since no bond company had reported Silas Mason acquiring the money needed to support his entry. That was Mason's secret, which he now revealed. He posted a cash bond. Such an unusual maneuver meant that nobody knew his plans; the strategy stunned the Six Companies' people.

The second bid, for just under $36 million dollars, came from Albert C. Wiltse of Ritzville. Wiltse worked there as an attorney and while his entry into the competition showed considerable thought, it lacked supporting figures, documents, or a bond so Banks set it aside as unacceptable. The third envelope, Banks announced after he opened it, came from "Mae West" who apparently had no address. The bid appeared in the form of a poem which Banks read to the amused crowd. Miss West, later revealed as Edwin F. James, a veteran Seattle real estate promoter, offered, among other things, to "span the river any old day," and to "offer plenty of diversion, with adequate foundation therefore," and "she" signed off with the request that during construction the engineers "come up and see me sometime." With a wry smile Frank Banks called the offering "informal," and set it aside.

The last thick envelope held the Six Companies' bid of $34,555,582, over $5 million higher than the MWAK total. Everyone cheered, and Frank Banks announced the obvious: the MWAK group won the job. The whole affair took about thirty minutes and then the crowd went outside where the festivities continued.[12] Henry Kaiser put his best face forward and announced, "The United States government is fortunate in having received such a wonderful bid from such a capable group of contractors."[13] Army airplanes roared over the city and a band played. Both Elwood Mead and Governor Martin spoke to the crowd, praising the MWAK bid and the dam that promised "a new day" for Washington and the Northwest. Then a parade wound through downtown Spokane. The *Spokesman-Review* reported,

> With construction of the Grand Coulee Dam an assured fact, residents of Spokane and the Inland Empire who swarmed the downtown streets yesterday afternoon saw a parade seldom equaled for color, length, and trueness of theme. Symbolic of the huge project, enormous trucks, tractors, steam shovels and other heavy equipment that will be used in building the dam vied with pretty floats, colorfully garbed Indians, bathing girls, blaring bands and snappy drill teams. It was a parade worthy of the great event it commemorated.[14]

That evening the dignitaries held a banquet while fireworks played over the city.[15] They celebrated the start of construction—perhaps not the high dam that some wanted and had worked for over many years but, nevertheless, a dam that might eventually support Columbia Basin irrigation.

Silas Mason named himself chairman of the MWAK Company board and said he would live at the dam some of the time. Manley Harvey Slocum, known to everyone as Harvey, became the general construction superintendent. Likable and popular with working men, and trusted by Mason, Slocum had little more than an eighth grade education and a troublesome penchant for heavy drinking, but he knew how to build big things and Mason counted on him to divert the river and oversee placement of the concrete.

Interior Secretary Harold Ickes officially awarded the job to the MWAK Company on July 13, 1934. The agreement limited the contractor to 1,650 days, or about four-and-one-half years. The government also maintained the option to switch, with thirty days notice, to the high dam.[16] The MWAK Company agreed to build a straight gravity dam 350 feet high, 3,400 feet long from one side of the river to the other, holding 3,100,000 cubic yards of concrete.[17] When complete the dam would be the second largest in the world, coming in after Boulder Dam.

Not long after MWAK began work, President Roosevelt visited the West. In a special edition of the *Seattle Star*, Governor Martin spread the news that Roosevelt would view the construction site late that summer.[18] The President stopped first at Bonneville Dam on August 3, 1934, and that night went by train through Spokane to Ephrata. From there the party, including the governor and Secretary of the Interior Harold Ickes, continued in cars. At 11:05 a. m. Roosevelt arrived at Grand Coulee, where he spoke for twenty minutes.[19] In his diary Ickes noted surprise at the crowd of over 20,000: "It was perfectly astounding to see so many people in a desert country. Some of them must have driven two or three hundred miles to see and hear the President."[20] Most telling, perhaps, was a remark attributed to the President's wife, Eleanor Roosevelt. After looking around she said, "It was a good salesman who sold this to Franklin!"[21]

Proud officials took Roosevelt across the river on a ferry, then up the hillside to view the project. On that hot day, people used handkerchiefs or shirt sleeves to wipe away sweat, found whatever they could to drink, and cheered the President, who moved quickly from one vantage point to the next, talked to Banks and other officials, and then left. Newspapers around the state marked the occasion with special editions filled with hyperbole. "Dam will Boost State's Wealth!" declared Seattle's *Post-Intelligencer* in a gush similar to the others. Senator Clarence Dill wrote an article praising the President and the cheap electricity that would help the state and the nation.[22] The only thing missing was the finished dam.

MWAK had to remove the remaining overburden from both sides of the river, fully exposing the bedrock on which the dam would rest. That job alone involved relocating over 12,000,000 cubic yards of material. The volume of earth eventually moved exceeded the size of the high dam the government ultimately built, and was considerably larger than the low dam then contemplated.[23] The preliminary contractors had taken away only the first of the overburden. Hauling off the rest was a mammoth task and slides that already plagued the project made it complicated and dangerous. On August 15 at 3:15 in the afternoon, in temperatures exceeding 100 degrees, a mechanical scraper began pushing earth around on the west bank of the Columbia River. Without the fanfare usually accompanying every beginning at Grand Coulee, general superintendent Slocum began the MWAK effort.[24]

Downriver, the Western Construction Company resumed work on the highway bridge. With nearly half of its allotted time consumed, Bureau officials wondered if the contractor could successfully place the two piers that would support the superstructure. Meanwhile, to bypass the slow ferries, the MWAK Company put up a temporary bridge. Company employees floated logs down the Columbia from Kettle Falls and, over a three-week period, drove them into the river. The first traffic crossed the structure near the end of October.[25] MWAK announced that its bridge would remain until the spring high water period, but it came down when an ice jam lodged against it in mid-January 1935.

MWAK then built a cat-walk, suspension, pedestrian bridge about fifty feet upstream from the damaged structure, providing a safe place for foot traffic.[26] While crews put up the suspension bridge a platform, carrying fifteen men at a time and held by a cable, swung back and forth across the river slowly shuttling workers. Some waited up to three hours for their turn at what became a significant bottleneck.[27] Finally the suspension bridge opened, but lasted only a few weeks. The company tore it down late in February because it was in the way. In April 1935 a new suspension foot bridge stood near the dam site. About one-half mile downstream the company added a combination railroad/highway bridge and removed the ice-damaged log structure.[28]

All these costly temporary bridges filled in for the troubled highway bridge that eventually carried all traffic except the railroad. Their necessity demonstrated the difficulty the river presented, not only in diverting it, but just in crossing it. The J. H. Pomeroy Company of San Francisco won the bid for the highway bridge superstructure, but could not begin work until the first company completed the two piers. By January 1935 the Western Construction Company finished the east pier and began the concrete caisson for the west one. Seven months behind schedule and already paying late penalties, the hapless contractor struggled to finish. Then the Pomeroy company began the superstructure. Pomeroy workers moved quickly and by late summer the two halves of the span nearly met over the center of the river. But in August a small landslide along the east side of the Columbia pushed overburden against the base of the bridge pier, causing it to lean.[29] A drag line and rocks dumped around it stabilized the pier and held it in place.[30] Finally, in September, the bridge that the *Wenatchee Daily World* called "a jinx from the start," finally spanned the river.[31] Six months late and without fanfare, Frank Banks ordered the highway bridge open for traffic on January 24, 1936.[32]

The Bureau of Reclamation continued enlarging what everyone called "Engineers' Town." It already had a few long streets running north and south parallel to the river, short connecting east-west avenues, a few houses bracketing the future site of the administration building, and a school—all on the west side of the Columbia just below the dam on a narrow flat piece of land. Fifty more houses went up during the summer of 1934. The Bureau left Almira and moved into the new school on March 22, 1935. When its large white administration center opened, it moved again. In the spring of 1935 the government ordered twenty-five more houses and later doubled the number.[33]

The Bureau received hundreds of applications from small businessmen wanting to open stores, but decided to create a "model town," strictly controlling its atmosphere by limiting the size and appearance of all commercial ventures. At the Boulder Project the government created a "federal zone" around the construction site and regulated activity inside it.[34] Grand Coulee had no federal zone. The only areas entirely under Bureau jurisdiction were Engineers' Town and the dam itself. The rest came under State of Washington or county authority.[35]

On a larger piece of flat land on the east side, a few yards below Engineer's Town, the MWAK Company established Mason City. From the start the contractors vowed to use only electric power to make the place the world's first "all electric city."[36] The fear that no market could absorb Grand Coulee power continually plagued its construction and MWAK hoped this exposure would sell people on the convenience of electricity and promote its use statewide and beyond. Using only electricity also eliminated the need to haul coal or some other heating source during the winter. In September, MWAK contracted with the Washington Water Power Company for power lines to the construction site. They stretched from Coulee City thirty miles away. Six weeks later the first lights went on in Mason City.[37] Ironically, the Washington Water Power Company, which for so long through the 1920s and 1930s fought against Grand Coulee Dam, provided much of the power that facilitated its construction.

On October 2, 1934, MWAK awarded a $200,000 subcontract to the White Pine Sash Company of Spokane for 360 houses. In addition MWAK built dormitories, a two-section cookhouse to feed 1,000 men, and a thirty-three bed hospital.[38] Workers lucky enough to live in the million-dollar camp paid a modest $32 monthly for a three-bedroom house, or $38 for one with four bedrooms.[39] In the center of Mason City a large grassy circle became Mead Park. Ickes Avenue ran north and south, bisected by Roosevelt Avenue which crossed the highway bridge. The latter separated the houses on its north from the dormitories to the south.[40] The few Republicans in the community chuckled when they noted that eventually Roosevelt Avenue became a "dead end street."

On high ground above the main part of town, 200 yards from the new airport, stood two special houses—one for Silas Mason and the other for George Atkinson. Mason moved into his that November and, until he died, lived much of the time in the city that bore his name.[41] A tall man who gave the appearance of a gray-haired gentleman, and who always carried a cane, he liked to walk around and visit all parts of the project.

Thirty miles south of Mason City at Odair, seventy men with the David H. Ryan Company and its subcontractor, Crick and Kuney, worked on the million-dollar government railroad. On September 8, 1934 they laid the first rail for what then was the longest railroad under construction anywhere in the United States.[42] On December 5 the rails reached Grand Coulee, and on Saturday, December 8, Bureau of Reclamation officials held a ceremony to mark completion of the track. Governor Martin showed up along with Senator Dill and together they drove a "golden spike" to mark the occasion.[43]

By July 1935 the Ryan Company finished the project, including a 200-foot-long tunnel, installation of switches and other safety features, acquisition of engines and equipment, and the rest of the necessary paraphernalia for a railroad. The first train, complete with a load of steel, arrived on Wednesday, July 24, 1935. Five days later the government put on a show to mark its completion. Governor Martin, dressed in the garb of a traditional engineer, stood at the throttle when the "official first train" pulled out of Odair. After the customary speeches, and as the train moved away, dignitaries christened the engine by shattering a bottle filled with Columbia River water against it. After more hoopla at the dam, Frank Banks, acting for the Bureau of Reclamation, presented the railroad to the MWAK Company. Then everyone adjourned to the cookhouse for a big meal and more speeches.[44] Another contractor paved the road between Coulee City and the dam—a route that became known among the locals as the Speedball Highway.

Immediately after the final go-ahead on its contract, MWAK began removing the remaining overburden. Rather than use trucks to haul away the earth and rock, MWAK gambled on an innovation untested on such a scale. It built large conveyor belts to carry the material to a new location more quickly and on a continuous basis. One spectacular assembly on the west side of the river was billed as "the largest ever to be built in the history of the construction world."[45] Sixty inches wide and three-quarters of an inch thick, the belt stretched over a mile upriver and, in time, the company doubled its length. It cost $750,000 and carried 52,000 cubic yards of earth and rock daily. In short order the belt and large electric shovels, each capable of biting into between four and five cubic yards of earth, repeatedly moved record amounts.[46] By the end of 1934 the system had already relocated over a million cubic yards of material.

Conveyors had been used elsewhere long before the work at Grand Coulee, and unprecedented here was only the size of the machinery. The scale of the operation drew considerable attention, with the national trade journal *Engineering News-Record* noting, "In using conveyors for the major transportation and disposal of excavated dirt at Grand Coulee, the contractors have shown that the pioneering spirit of the West is still a vital American characteristic."[47] Articles in other newspapers and magazines complemented the conveyor system and played up its "world's largest" claim, the first of many such "biggest" records that followed throughout the nine years it took to build the dam.[48]

The conveyor did not operate without some difficulties. Large rocks damaged the belt and the operating mechanisms. Installation of a grizzly—a framework which holds bars that allow dirt and small pebbles to pass through but which shorts out larger pieces—helped.[49] On January 22, 1935 thirty-eight-year-old B. M. Layport of Soap Lake, working near the machinery, stumbled and became caught between the heavy conveyor and the rollers over which it ran. Layport died as the equipment savagely tore him to pieces.[50] A second, similar death followed some days later. Two other workers, while dealing with the belts, suffered heart attacks and died but did not qualify for the count of work-related fatalities which was already rising. With large construction work came risk, and Grand Coulee had its share of accidents. Indeed, accidents—particularly fatalities—constantly troubled workers. Whenever someone died, the news spread rapidly and played on the emotions of the others. But the numbers of injuries and deaths did not equal those that plagued the Boulder Dam construction. Considering the size of the undertaking, the number of people involved, and the speed of the work, Grand Coulee Dam proved to be a comparatively safe work place.

The long conveyor belt carried the overburden upstream, about 3,000 feet south of the dam, to Rattlesnake Canyon. There the material formed a pile that grew daily. On March 30, 1935 the heap of earthen debris shifted and 750,000 cubic yards of material cascaded down on itself, hitting and damaging the end of the conveyor belt. Although everyone managed to scramble out of harm's way, the slide briefly stopped work and idled over 700 men.[51] Eventually

MWAK abandoned the original dump at Rattlesnake Canyon and moved to a new location closer to the Columbia.[52]

Slides harassed MWAK just as they had its predecessors. The hillsides above the dam gave way on November 15 and around two million cubic yards of material slid into the excavation.[53] Ralph J. Kugelman, the Columbia Basin Commission's contact man at the dam, wrote to James O'Sullivan, "For the past week the slide area has been causing the engineers of the Reclamation Department a great deal of concern and perhaps sleepless nights."[54] They had reason to lose sleep. On November 19 an additional 750,000 cubic yards broke lose and this time carried fifteen men with it, hurling them 100 feet down into the excavation on a ride that left them shaken but, amazingly, unhurt.[55] Heavy spring rains aggravated the problem.[56] Saturated with water, the unstable earth tumbled into any opening. The contractors struggled, not always successfully, to remove it or keep it from sliding. "In the slide area more trouble has been experienced in the past week," wrote Kugelman. "About noon on Saturday, April 27, the lower side of the main road started slipping and by four o'clock it had dropped about eight or ten feet."[57] By June it was obvious that the route of both the road and railroad needed changing. Cracks that opened up in July on the hillside above the excavation added to the concern about the safety of the men working below. Three weeks later a slide on the east side of the river mimicked the activity across the Columbia.

To remove the accumulating water, which softened the earth, made it slippery, and exacerbated the slides, the contractor dug drainage shafts into the slide area, hoping this would stabilize the earth and diminish its tendency to move. This eased the problem somewhat, and by mid-summer the contractor announced that the excavation job was well over half complete with nearly six million cubic yards of material moved. Nevertheless slides continued to plague both the west and east sides of the project.

In addition to the slides, MWAK faced a deadline in diverting the river. It had to build its first temporary cofferdam during the low-water period. If the annual flood began before its completion the company would lose work time during the spring and summer when the swiftly moving, treacherous river rose and invaded the site. The Bureau of Reclamation built models in its experimental laboratory at the University of Colorado in Boulder, testing different plans for diverting the river, while the MWAK Company conferred with its consulting engineers. All the investigations confirmed that the soil around the dam site, the same material that had a tendency to slide, would not hold water. So MWAK could not put a cofferdam on top of it with any success, as seepage would be an insurmountable problem.[58]

The engineers finally chose a long cofferdam, made of steel sheet pilings, running north and south, parallel to the river on the west side.[59] Large drivers would pound the pilings deep into the overburden and down to bedrock, forming a solid barrier confining the river. With its ends turned toward the river bank, it would resemble a very long, flat, letter U. Once in place, it would keep the river away during excavation and later while the men poured the dam's concrete foundation. The trick would be to build the cofferdam before the river swelled in the spring of 1935.

Despite winter temperatures occasionally dropping to eighteen degrees below zero, a pile driver placed the first of the long steel sheets on January 1, 1935.[60] A chronicler for the Bureau of Reclamation called it the "start of the greatest river control structure ever undertaken in construction history."[61] Difficulties quickly developed when the piles hit obstacles. Ralph Kugelman reported to James O'Sullivan,

> Many of the piles in Southwest Cell Block D have struck large boulders and have become badly bent or twisted. Most of the time this week has been spent in extracting these piles. Probably twenty-five or thirty have been pulled out and more will be extracted. A few of the boulders are said to be from ten to fifteen feet in diameter.[62]

The drivers learned to stop when they hit something solid, and in some cases the piles never reached bedrock, due to the rocks.[63]

The rush to start pile driving brought with it a spate of accidents. Thirty-six-year-old Ransom Burke of Wenatchee, a welder's helper, died one afternoon when the 200-pound bottom plate of a steam pile-driving hammer broke loose and dropped on his head.[64] One week later a load of steel sheet piling, being hoisted by a crane, struck and killed another man.[65] Forty-five-year-old Thomas Newton, sitting in the bos'n's chair of his stiffleg crane, maneuvering some steel sheet pilings, had the load swing back and knock him into the river when a boom on the crane snapped. They recovered his body days later.[66] Through the entire month of February the company recorded 157 accidents, with more in March.

The State of Washington paid the expenses of men hurt or killed on the job. The widow of any married worker collected $35 monthly for the rest of her life, unless she remarried. She also received $12.50 for her first child and $7.50 for each additional youngster. Startled by the sudden increase in accidents and especially the deaths, the safety department of the MWAK Company published its first issue of the *MWAK Columbian* on March 1. The weekly in-house magazine carried local news and promoted safety.[67] In addition, MWAK placed a huge sign, nearly thirty feet long, over the east side of the project. On it, in letters six feet high, everyone in the area could easily see the words, "Safety Pays."

Work on the cofferdam continued in three shifts with eighteen drivers banging away around the clock. As many as 1,200 men labored on the cofferdam. Many of the men hired by the company had few skills and trained on the job. The terrible cold of January relented, leaving knee-deep mud. By mid-February the steam-driven hammers had placed about 80,000 of the ultimate 290,000 linear feet of pilings. The contractor added a dozen more steam hammers. The crews rushed the work, watching the river to see if it started to rise, noting that they would lose the race if the last steel was not in place before high water came.

On Saturday, March 23, 1935, fewer than ninety days after starting, the last piece of steel-sheet piling reached bedrock, finishing the main work on the cofferdam.[68] Local trade journals gave the MWAK Company high marks for quickly and efficiently completing what was then the "world's largest cellular cofferdam."[69]

As if to confirm that the struggle to build the cofferdam had indeed been a race that the MWAK people could have lost, the Columbia began its seasonal rise about mid-April.[70] A month later the river had doubled its average volume, reaching flows of nearly 200,000 cubic feet of water per second (cfs). MWAK people added a five-foot addition to the top of the dam, just in case. On June 5 the river hit 308,000 cfs and trucks dumped more earth and rock behind the dam. On June 19 the river peaked at 344,000 cfs, over three times its average size and thirty-two feet higher than normal.

From 200 to 600 gallons of water seeped into the area behind the dam every minute, but pumps easily removed it, and the engineers complemented themselves on a job well done.[71] Behind the coffer-dam, 105 feet below the surface of the river, crews safely removed overburden, digging down and exposing the bedrock foundation.[72]

On the east side of the Columbia the plan called for a coffer-dam similar to the one on the west side, but the MWAK Company abandoned the idea as too difficult. Instead, the new strategy called for removing as much overburden as possible during the low-water period. On May 1 the contractor started a very small earthen cofferdam that extended digging time on the east side.[73] When the water went down during the summer MWAK men built a dike around the east side excavation and used pumps to remove the remaining water. Then, fighting the continuing slides, they carted away the unwanted dirt.[74]

In September work began on a timber crib cofferdam on the east side. Twenty feet wide, it ran along the river for 1,130 feet and consisted of piles driven approximately ten to fifteen feet into the ground. "On the east side work has started over a week ago on what is usually referred to as a cofferdam but it is hardly more than a long crib to keep the river out of the east side until about May of next year," commented Kugelman.[75] About 20,000 cubic yards of earth and stones filled each of the cribs, giving them enough weight to hold against the rising water and keep it away from the excavation for six to eight weeks the next spring.[76] In all, except for the constant slides, the excavation went well. By the end of October 1935, the contractor had removed almost 12,000,000 cubic yards of material. But it had been costly: a dozen men had died so far.

With the MWAK contract, Grand Coulee Dam finally began to employ men in significant numbers. The limitations that gave hiring preference first to those living near the dam, and next to those in Washington state, caused resentment. Senator Dill and Governor Martin recommended easing the limit and giving equal preference state-wide.[77] Those employed all had to go through the National Reemployment Service, whose officials struggled to keep increasing numbers of destitute men from simply showing up at the dam expecting work. The regulations now required that each man apply for a job in the county of his residence and then wait for a call from the government. Many refused to wait, making their way to the growing community that sat at the edge of the Grand Coulee, perched on the bluff above the construction site. There they settled in and hoped. Some found jobs, eventually; others did not.

Voicing another complaint, the largely Republican *Spokesman-Review* wrote, "Only Democrats and supporters of the New Deal are being accepted for minor positions in the engineering department at the Coulee Dam."[78] Republican Rufus Woods had evidence that the *Spokesman-Review* was correct. A frustrated job applicant wrote him and complained that in seeking work he was informed that, "only Democrats who had worked for the party and had taken active part in its campaigns were entitled to receive work from this development."[79] Okanogan, Grant, and Douglas county commissioners, who had large Republican constituencies, criticized officials with the National Reemployment Service.[80] Woods agreed that the problem lay with a preference rule that gave party hacks influence over hiring. Senators Dill and Sam Hill pressured the Columbia Basin Commission and the Department of the Interior to change the regulation and loosen its limitations, but the issue became moot when the number of jobs increased and the problem eased.

"The name of Walter D. Muntz, carpenter, will go down in the history books written about Grand Coulee Dam," declared the *Wenatchee Daily World* when Muntz became "the first laborer to be hired for the actual construction of the main contract for Coulee dam."[81] Muntz, a resident of Grant County and a World War I veteran, had worked at the site earlier when he helped build the wooden platform from which Franklin Roosevelt spoke. Like others who secured employment, Muntz considered himself fortunate, and he took the job happily. The first, and one of the few, women hired at the dam was Helen Thomas, the "Miss Coulee Dam" from Soap Lake who helped open the bids in Spokane. She had been working for the Reemployment Service when MWAK hired her as an office secretary.[82]

When the mess hall opened in Mason City women also found jobs there. Ralph Kugelman wrote,

> Feeling that the refining influence of the feminine presence will have a salutary effect upon the boys in camp and pep them up somewhat the MWAK Company are [sic] going to give the women a break in securing employment here at the damsite.

No "Mess punks" [young boys] are to be employed in Mason City—no indeed—only feminine pulchritude will adorn the festive boards in this most favored construction camp. Here's hoping that whoever does the picking of the pulchritudinous damsels will have more of a penchant for bright eyes and red cheeks than for bright red fingernails. (If the last three sentences are too facetious for a report kindly delete same in copies sent to members of the Commission.)[83]

By October 1, 1934, 900 men worked at various jobs on the project, a number that rose to almost 1,400 by November 1. Some of them slept in their cars, moved in with friends, or just settled wherever they could find some place to stay because none of the camps had enough housing. And even when Mason City was complete, it hardly held all of the laborers. The overload had to fend for themselves. Despite the hardships and difficulties, the growing surplus of men competed for the scarce jobs.

Ignoring the cold weather, in early January 1935 over 2,500 men worked at the dam, with new arrivals coming daily to fill the places of others quitting. The older men proved most stable, with recent high school graduates more likely to stay only briefly. A few of the engineers hired that winter came directly from positions at Boulder. By February 1935 a total of 3,300 men and women had jobs of some sort. They worked to remove overburden, build the railroad, put up houses in Engineers' Town or across the river in Mason City, bridge the Columbia, pave the highway, and maintain the cofferdam. At the end of October all of them together had put in over six million hours working on the dam, and so far no one had poured any concrete.

A few labor organizers, and a couple of others labeled as communists by the *Wenatchee Daily World*, appeared near the construction sites and handed out literature, but nobody paid much attention to them.[84] At this stage of the construction, few had complaints, and if they did, they kept them to themselves, happy just to have a job. In time that changed.

Through the summer of 1935 the various crews removed overburden from the west side of the river and, considering the high water, wherever possible on the east side. With the cofferdam finished they prepared for the concrete pouring that would initiate work on the dam itself. That activity began MWAK's second phase.

The satisfactory completion of the cofferdam proved that MWAK could handle the job, at least so far. Despite the skeptics, the river had been diverted. It was the first milestone in construction of the big dam.

Francis D. Culkin. *Courtesy Oswego County Historical Society, New York, no. 90.1.1195.*

# Chapter Eight
# Making it Bigger and Making it Legal

The population of the four Pacific states of Nevada, California,
Washington and Oregon may increase so greatly that it can
sop up all that extra light and power. . . . It is not probable,
perhaps, but it may be possible. . . . There might be a war which
would crowd the Pacific Northwest with eager workers and their
Saturday nights with wassail.

*Public Utilities Fortnightly, 1934*[1]

Since the Grand Coulee dam is a federal project, its completion
is subject to congressional appropriations. Not only the state of
Washington, but the entire Pacific Northwest should be prepared to go to bat for the essential appropriations when the
time comes.

*Pacific Builder and Engineer, 1934*[2]

The start of construction at Grand Coulee did not end the political
wrangling that had preceded the dam. And with the New Deal came
more detractors who fixed on the dam as evidence of the mistakes
being made by the deficit-spending, wasteful Democrats. When
Franklin Roosevelt approved Grand Coulee Dam, he bypassed
Congress and financed the work through the Public Works Administration. In 1935 the conservative Supreme Court returned the
project to Congress. That year became a crucial one for the dam,
with its future tentative at best until the legislators finally legiti-
mized it. Just before the Supreme Court ruling, the Roosevelt administration decided to build the high dam, rather than the less
imposing barrier started in 1933. In looking at the proud concrete
monolith blocking the Columbia River today, few realize how close
it came to being much smaller and less impressive—or perhaps
not even existing at all.

As they had through the 1920s, critics of Grand Coulee Dam and its attendant reclamation centered on the region's ability to absorb power and the overabundance of land around the country already producing surplus crops. One author in *Public Utilities Fortnightly* wrote,

> If the statements which are to follow are accurate, and I believe they are, it [Grand Coulee and the Columbia Basin Project] is the most colossally humorous stunt of our notably funny generation. Secretary of Agriculture Wallace has said that if one third of the land now under cultivation in the United States could be thrown out of production every one would be better off. But if the Columbia river is dammed to produce power that no one needs it is too much to expect the engineers to refrain from reclaiming the 2,000,000 acres that no one wants. . . . If the 1,500,000 potential kilowatts of the Columbia Basin project are forced on the power market the privately owned power plants in this area will be crippled and their bond and stock holders and widows and orphans will be unable to pay as much tax as they pay now.[3]

Undoubtedly also thinking of the widows and orphans who owned stock in private power companies, Republican Representative Roy Orchard Woodruff of Michigan called the project "idiocy," and "tragedy."[4] *Barron's* complained,

> Unlike the Tennessee Valley development, this little-known Pacific Northwest venture has not received Congressional endorsement, yet it plans to build in three successive stages the greatest coordinated water power development in the United States, and at an ultimate expenditure of approximately one billion dollars in federal funds.[5]

The *Chicago Sunday Tribune* attacked the project for its remoteness[6] and *Collier's* ran a lengthy article titled "Spare that Desert!" that criticized the government for paying $300 million to take 43,000,000 acres out of production while at the same time planning to spend another $227 million for "dams, dikes and ditches" to irrigate two or three million new acres. "Only the land speculators will benefit," the article concluded.[7]

The Bureau of Reclamation brooded over the power market problem[8] and when the government awarded the Bonneville Project to the Army Corps of Engineers, that worry became serious. If one dam on the Columbia River would oversupply the region, what would happen with two? And this in addition to the Rock Island Dam near Wenatchee—a private power project then nearing completion. John W. Kelly of the Portland *Oregonian* best described the dilemma when he explained that the current plan limited the low dam to generating power for which there might be no market. Grand Coulee Dam would be successful only if the government built the high dam with extensive reclamation. But that would increase by at least $123 million the $63 million so far allocated, and this before the project's irrigation component.[9] No matter how the doubtful looked at it, the whole thing seemed an enormous mistake.

Roosevelt and his Secretary of the Interior, Harold L. Ickes, knew the arguments against building Grand Coulee Dam but Roosevelt hoped that, for every acre added through reclamation, the government would withdraw five acres of sub-marginal land elsewhere. The idea was to transplant people onto land where planning guaranteed their success.[10] "We have reached the end of the pioneering period of go ahead and take. We are in an age of planning for the best use of everything for all," Ickes declared.[11] Only the high dam and the reclamation venture it promised would allow the planning that Ickes advocated.

Rufus Woods, James O'Sullivan, and Columbia Basin Commission members welcomed the Bureau of Reclamation's decision to build a small gravity dam, but they also understood it would not support irrigation and they urged the Bureau to build the high dam.[12] But Senator Clarence Dill cautioned O'Sullivan to go slowly and not annoy either the Bureau, Congress, or the President. "It is unthinkable to reopen this question for more money at this time," he wrote O'Sullivan. "Once we get the low dam actually started and have the foundation actually worked upon, you will find me just as aggressive for the high dam as you have ever been."[13] O'Sullivan, however, distrusted Dill, and he went ahead with his efforts on behalf of the high dam.[14]

Tired of O'Sullivan's activism, James E. McGovern, Columbia Basin Commission representative from Spokane, attacked him and asked the other members to fire their outspoken secretary. Senator Dill also approached Governor Clarence Martin and recommended removing O'Sullivan.[15] McGovern forced the issue but lost, and the Commission retained O'Sullivan.[16] Undoubtedly McGovern reflected Spokane's and, more specifically, the Washington Water

Power Company's resentment of O'Sullivan's success so far and their fear of power from the high dam.

When Roosevelt visited Grand Coulee in August 1934, Dill insisted that nobody mention the high dam, arguing that the whole problem represented something of an embarrassment for the President. O'Sullivan felt that, in deference to Dill, the Columbia Basin Commission maneuvered his schedule to keep him away from the President, but he ignored them and positioned himself in Soap Lake near a large display of agricultural products put out for Roosevelt's inspection.[17] When Roosevelt stopped, O'Sullivan approached the chief executive's car and introduced himself to the President, who acknowledged knowing who he was.[18] O'Sullivan had also arranged for Adrian Grange members and others to hang banners and signs along Roosevelt's route so that everywhere the President looked, he saw demands for the high dam. How much the lobbying swayed either Roosevelt or Ickes is debatable.[19] A few days later the Secretary announced that he looked forward to building Coulee High Dam and irrigating 1,200,000 acres, but that the present plans called only for the low dam.[20] Nevertheless, he added, his views concerning Grand Coulee and the entire Columbia Basin Project had changed and he was ready to back the high dam.[21]

At the same time, Grand Coulee itself came under renewed assault from Seattle City Light Superintendent J. D. Ross. Ross repeated his claim that Grand Coulee would rob Seattle of a power market it deserved and that Skagit Project power (from Seattle City Light's series of dams then under construction on the Skagit River in Northwestern Washington) would be considerably cheaper than anything generated at Grand Coulee.[22] Ashley Holden of Seattle wrote to Rufus Woods,

> Without exception everyone agrees that he [Ross] is honest and sincere but so drunk with power that he is worse than a fanatic. He really intends to put over his deal to purchase the properties of the Puget Sound Power & Light Company and absurd though it may seem, he has the backing of the City Council and almost enough fanatical voters to put the deal over. Ross dreams of an electrical empire greater than anything Insull ever created, with himself as the head of it. He really believes that the coulee [*sic*] development is responsible for Skagit not getting a Federal Loan.[23]

Rufus Woods wrote to Ross, "You do not seem to realize this is not merely a dam with a power plant but that it is a program for the whole Northwest, particularly for arid sections of Eastern Washington."[24] Throughout the debate J. D. Ross remained a thorn in the side of Grand Coulee supporters. Years later, when President Roosevelt needed a leader for the Bonneville Power Administration, he chose Ross, but over the strong objections from eastern Washington, where people remembered the Seattle superintendent's efforts at what they interpreted as attempts to scuttle Grand Coulee Dam.

In the summer of 1934 the *Wenatchee Daily World* pointed out that if the government were to change to the high dam, it would need to do so during the next year. The work to be completed by mid-1935 would make alterations thereafter expensive and perhaps technically impossible. But the complicated political and economic situation inhibited Roosevelt's and the Bureau's ability to embrace the high dam openly despite the fact that both had undoubtedly wanted it from the start.

O'Sullivan, lobbying in Washington, D. C., found strong opposition to the high dam. He wrote to Rufus Woods, "The east an[d] middle west are trying to get at us both from the standpoint of POWER as well as of IRRIGATION. Hardly a day goes by without some of the newspapers whacking away on these subjects."[25] Power and reclamation were hardly the only things Grand Coulee opponents attacked. Using strong language, columnist James Rorty wrote an article in *Nation* accusing Roosevelt of sounding like Huey Long of Louisiana when he spoke about "socialized national planning" during his visit to Grand Coulee in 1934. Rorty also charged that the "Inland Empire Crowd" blocked funding for Ross's Skagit project while favoring spending that would help land speculators in the Columbia Basin. Rorty turned what had been whispered rumors into blunt allegations when he specifically mentioned ex-Senator Dill as the leader of the speculators and added that Interior Secretary Ickes knew all about the scheme.[26] Dill denied the claim and although it resurfaced from time to time, nobody ever proved it.

Despite these controversies, the first hint that the government might switch to the high dam came on January 17, 1935. Senators Homer Bone and Lewis Schwellenbach revealed to the press that

Bureau of Reclamation designers had started redrawing their diagrams.[27] Representative Sam Hill sent Rufus Woods a confidential letter on April 9 telling him that Elwood Mead would soon sign the order to modify the dam's design. Bureau of Reclamation engineers began running high-level survey lines but, still wary of criticism, assured everyone that the activity signified nothing. Then, after months of speculation, work undertaken somewhat in secret, and denials, on June 5, 1935, Secretary Harold Ickes signed Interior Department Change Order No. 1 providing for construction of the base or foundation of Grand Coulee High Dam.[28] At last the government committed itself to the high dam rather than the low dam then under construction. This meant that the MWAK Company, then at work, would not build the entire Grand Coulee Dam but only its foundation. The bulk of the dam would come at a later date under a second contract. The change order indicated that the government would build the high dam but it did not specify when.

Dramatic though it seemed, the switch to the high dam was more of an anti-climax than many of the principals realized. Evidence in Bureau of Reclamation files indicates that its officials had made the decision by the previous December and perhaps much earlier. For example, on December 4, 1934, Bureau engineers held a conference in Denver and discussed the prospects for successfully imposing a high dam on the lower dam. John Savage, chief design engineer, urged avoiding a joint of new and old concrete if possible. The report added that with the construction of Bonneville, a low Grand Coulee Dam existed only as a relief measure since the low dam would produce power not immediately needed. On the contrary, the high dam would provide both reclamation and relief.[29] On December 22 the engineers sent their report directly to Secretary Ickes.[30] Five days later, on December 27, Mead wrote Ickes, "The Bureau of Reclamation recommends that the Grand Coulee power project be changed to the Columbia Basin power and irrigation project. This change contemplates a high dam in place of the low dam."[31] Although many factors contributed to the decision to build the high dam, the technical problems involved with later raising the low dam were among the more significant. In any event, by the end of December 1934, Roosevelt, Ickes, and Mead had committed themselves to the high dam at Grand Coulee and to the Columbia Basin Project.

Without doubt Roosevelt himself wanted the high dam and reclamation. The project epitomized his goal of cheap public power and a planned project where the government could relocate farmers from other parts of the country. With an election coming in 1936, Roosevelt needed to move on these plans or risk not seeing them materialize at all. And the high dam was the key. How much Elwood Mead and Harold Ickes played in the decision is uncertain but clearly, as with the agreement to build the low dam, the Bureau of Reclamation could not switch to the high dam without the President's personal approval. As with the decision to build the dam in the first place, Roosevelt was the central figure who both faced the opposition and eventually authorized Grand Coulee High Dam.

In making these decisions, Roosevelt attempted to have the PWA pay for the project and avoid including Congress directly. Congress would provide blanket PWA funding, but the administration would determine how to spend the money. Congress would never vote directly to authorize the dam as such. In the debates over the PWA bill, however, legislators discussed Grand Coulee Dam, and its detractors decried it as being typical of the New Deal's ability to waste public money. Republican Representative Francis Dugan Culkin of New York led the assault against Grand Coulee, a role he filled from 1935 through the outbreak of World War II. Culkin charged that Elwood Mead, who pioneered social planning in Australia and California, "wrecked Australian agriculture and had practically wrecked California agriculture."[32] Culkin eloquently expressed conservative complaints that now Mead, along with Roosevelt and the New Deal, were off on a suspicious, almost socialist, exercise with visions of planning that departed radically from the American Way. Grand Coulee Dam and the Columbia Basin Project became the focal point of Culkin's attack and he assailed the dam with considerable vigor.

Rufus Woods correctly pictured Washington Congressman Sam Hill as the high dam's ablest supporter.[33] Hill worked hard for the project, put up with James O'Sullivan, and kept his friends back home informed about developments in Congress. In the spring of 1935 he wrote Woods that the Public Works Administration considered allocating only $25 million for work on the dam during 1936 rather than the $70 million that would cover the entire dam and some preliminary work on irrigation. "The reason assigned for postponing work on the reclamation project is that the President is determined to curb the operation of land speculators," he wrote to Woods. Hill guessed that the government might try to buy as much as 500,000 acres of basin land and he went on at length about Roosevelt's dislike for land speculators.[34] If the price of the land rose beyond the reach of farmers, it would ruin the dream of transplanting agricultural refugees to the planned development.

The *Spokesman-Review* reported that Roosevelt considered having the government buy what it did not already own of the entire 1,200,000 acres the project would irrigate for about $6 million, or around five dollars an acre.[35] Secretary Ickes announced that there would be no irrigation project at all if speculation continued. There were more rumors in Washington, D. C., like those published by James Rorty, that some people in the national capital bought land for as much as $20 an acre with the hope of enormous profits. Woods countered that the land speculation rumors were nothing more than a tactic of project opponents like J. D. Ross and he urged the Columbia Basin Commission and the Columbia River Development League to counter such propaganda with the truth.[36] The flap over land speculation would not end until 1937.

During the spring of 1935 Congress considered a $4,800,000,000 bill for work relief, including PWA funding. An amendment to the measure, proposed by Democratic Senator Patrick Anthony McCarran of Nevada, required that the government pay prevailing wages at all federal projects. Grand Coulee backers saw this as an attempt to make the dam so expensive that the government would abandon it. On March 15 the Senate voted 50 to 38 against the proposal and a few days later it passed the appropriation, appearing to end Grand Coulee funding worries.[37] Undoubtedly the Roosevelt administration had waited until Congress approved this bill and guaranteed PWA funding before announcing the high dam decision. In doing that, it avoided, at least temporarily, a battle over the high dam issue among the legislators, many of whom opposed the dam and especially its link to reclamation.

The relief was temporary. Roosevelt's clever maneuver that paid for Grand Coulee through the PWA suddenly backfired and put the project in jeopardy. On April 29, 1935, in the case of the *United States v. Arizona*, the Supreme Court held unanimously that Parker Dam construction on the Colorado River had been illegally authorized. The Rivers and Harbors Bill of 1899, and subsequent legislation in 1910, prohibited construction of barriers across navigable streams except with the consent of Congress.[38] The Court argued that the National Industrial Recovery Act, under which Roosevelt authorized Grand Coulee and a good many other dams, had not repealed the former legislation.[39] In blunt terms the Court ruled that either Congress had to approve Parker Dam or the administration must stop building it. Concerned about the legitimacy of its other projects, including Grand Coulee, the Roosevelt administration immediately rushed legislation to sanction the twenty-some federal dams then under construction. The Senate held hearings on May 22, 1935 and passed the bill without debate on May 27.[40]

Five days later the Supreme Court found unconstitutional parts of the National Industrial Recovery Act (NIRA) itself—specifically the National Recovery Administration (NRA). The implications for Grand Coulee, if any, were unclear at first. Some felt it made passage of that year's Rivers and Harbors Bill, with authorization of Grand Coulee and the other dams, all the more imperative. They reasoned that Congress would eventually have to fund those projects directly, although the Public Works Administration, also part of the NIRA, appeared in no immediate peril. Senator Schwellenbach announced that the NRA decision had "endangered completion of the Grand Coulee project," and went on to say that the Parker Dam decision, among others,[41] all had a direct bearing on the future of Grand Coulee. As Schwellenbach pointed out, the Court cut deeply into New Deal power and indicated that it would hold the administration to strict constitutional limits.[42]

Elwood Mead and the Bureau of Reclamation appeared uncertain about exactly how to continue, since nobody could predict what Congress would do. The Columbia Basin Commission considered seeking money through other relief programs, fearing that Congress might only authorize the dam but not fund it and that the Court would eventually shut down the PWA. Grand Coulee Dam supporters greeted with considerable relief the news that on June 17 the Public Works Administration allotted $23 million specifically for Grand Coulee Dam and that Roosevelt had approved the amount himself.[43] That would carry construction through June 1936, but after that nobody knew what would happen.

In the House of Representatives the Senate bill approving the dams went to the Committee of the Whole House on the State of the Union, which never called it up. Senator Roy Samuel Copeland, Democrat of New York, then proposed adding the contents of the bill, along with a number of other measures, as amendments to the Rivers and Harbors Bill that the House had just passed and sent to the Senate. The Senate adopted all the amendments and sent the measure back to the House. In conference the House agreed to all the amendments (one of which approved Bonneville Dam) except the one mentioning Grand Coulee. From there the amendment containing Grand Coulee authorization, along with the bill, went to the full House where together they met considerable resistance and appreciable maneuvering.

House Republicans insisted on voting separately on Parker and Grand Coulee dams.[44] Representative Culkin again led those opposing Grand Coulee. On June 15 the New York Republican went on the radio, under the auspices of the National Grange, and declared that reclamation projects in the West would destroy existing farmers. "Let me illustrate this," he declared:

> If the two million acres of the Grand Coulee in Washington is put in work it will unhorse the fruit farmers of the West Coast. ... This project was condemned by the U.S. Engineers in 1932 as uneconomical either as a reclamation or a power project. It has never had the approval of Congress. The Grand Coulee is a vast area of gloomy table lands interspersed with deep gullies located in northern Washington.

In addition, Culkin charged, the dam would kill the entire $300 million Columbia River salmon industry.[45]

Rufus Woods wrote Culkin a blunt letter in which he said, among other things,

> As one citizen of the state of Washington, I wish to say that your statement is so full of errors that you ought to be ashamed of yourself. We have more squawking and howling from some of you fellows about our irrigation—and yet this is one of the very few Government activities which pays itself out over the years. ... Why can't you fellows down east let our western country alone—or at least come west and find out what the facts are. Your talk aggravates me so much that I can scarcely write you a decent letter; but if you will come west, I will be glad to give you some much-needed information.[46]

Not content with that, Woods's *Wenatchee Daily World* printed an eight-page special edition that carried the banner headline "Two Million Wild Horses" above a collection of articles about the power and irrigation that Grand Coulee would provide. Woods declared that the paper told legislators the truth about the Columbia Basin Project and he saw to it that every member of Congress received a copy.

It was hard for some opponents and, for that matter, some of the backers of Grand Coulee to follow the switches in tactics by project proponents. For years they advertised the dam as a reclamation project and then, suddenly, it became a power project. Now the Roosevelt administration again returned the emphasis to reclamation. Backers deepened the confusion by promoting whichever aspect of the project appealed to, or least annoyed, the group they addressed.

In the Rivers and Harbors bill debate the project's nature changed again. As opponents tore into reclamation, Representative Hill stressed power and argued that it might be fifteen to twenty-five years before the dam delivered irrigation water. Hill reasoned the dam had a better chance if he played to recently quickened pro-federal power sentiment while softening the emphasis on irrigation that irritated anti-reclamation forces throughout the East and Midwest.[47] In rebuttal Representative Culkin charged, "There is no market in the Pacific Northwest for power except rattlesnakes, coyotes and rabbits."[48]

On August 19, the House passed the Grand Coulee measure with 61 percent of those present favoring it. Senator Schwellenbach telegraphed James O'Sullivan,

> SAM HILL TODAY ALONE AND SINGLE HANDED SAVED THE GRAND COULEE ON THE FLOOR OF THE HOUSE STOP YOU HAVE NO IDEA THE VICIOUS EFFORT OF EASTERN CONGRESSMEN TO ELIMINATE COULEE FROM RIVERS AND HARBORS BILL STOP HILL TALKED THEM ALL DOWN AND IS ENTITLED TO FULL CREDIT.[49]

Hill himself telegraphed Rufus Woods,

> IN HARDEST FIGHT OF MY LIFE WON GRAND COULEE VALIDATION TODAY IN HOUSE AGAINST MOST VICIOUS AND DETERMINED OPPOSITION STOP THE VOTE WAS ONE HUNDRED NINETY NINE TO ONE HUNDRED TWENTY SIX STOP I HAVE BEEN ORGANIZING MY

FORCES FOR DAYS I RESORTED TO EVERY POLITICAL STRATAGEM TO BUILD UP THIS ORGANIZATION.[50]

Perhaps exaggerated a bit, the otherwise modest Hill later refused the honor of a gala homecoming. The House then approved the entire Rivers and Harbors Bill as amended. Two days later the Senate passed the bill and the President signed it on August 30, 1935.

With the Rivers and Harbors Act (Public Law No. 409), the Seventy-fourth Congress authorized Grand Coulee Dam for the purposes of flood control, navigation, stream flow regulation, storage for and delivery of stored waters, reclamation of public lands and Indian reservations, and for the generation of electrical power as the means for financially assisting those undertakings.[51] In other words it sanctioned selling power to pay the bills. Although this was not the last legislative hurdle for Grand Coulee Dam, it was the most difficult and the most significant. Whatever else might happen, the dam finally received congressional blessing and needed legal underpinnings. No longer just a New Deal relief measure, it was finally a national project.

Governor Clarence Martin lays the first concrete at Grand Coulee, December 1935. *Courtesy Bureau of Reclamation.*

# Chapter Nine
# MWAK Pours Concrete

Concrete placing at Grand Coulee dam has settled down to a
well regulated, highly systematized procedure resembling the
operations in a straight-line production manufacturing plant.

*Western Construction News, 1936*[1]

Washington governor Clarence D. Martin enjoyed participating in
the ceremonies that marked various milestones in Grand Coulee
Dam construction. At 2:06 in the afternoon on December 6, 1935,
dressed as a concrete worker, he stood on a spot sixty feet below
the level of the Columbia River on its west side and there, before a
large crowd, pulled down the release handle on a four cubic-yard
bucket filled with concrete. The gray, heavy, soggy material plopped
down into the framework that marked one of the blocks that would
eventually form the dam's foundation. Looking pleased with himself, the governor placed eight more loads and massaged them into
place with a large electric vibrator.[2]

When the government switched from the low to the high dam,
it altered the MWAK Company contract. MWAK would now build
only the foundation for the huge structure. A second contract for
the bulk of the dam would come sometime later. Toward completing its part of the task, MWAK had accomplished a great deal in
fifteen months—enough to make possible Governor Martin's brief
adventure into concreting. Its electric shovels dealt with slides and
removed more than 16,000,000 cubic yards of earth and rock, with
a few million more remaining. Its crews worked nights under hundreds of flood lights that filled a twenty-five-square-mile construction zone on a project that so far cost well over $19 million or, at the
peak, about $1,750,000 monthly. Fifteen men had died, eleven work-
ing with MWAK.[3]

As overburden removal proceeded, attention shifted to preparations for the concrete pour that would lay the foundation. Through the spring and summer of 1935, Bureau of Reclamation engineers reviewed the latest advances in concrete formulation and investigated ways to enhance its quality and strength. Frank Banks stated the government's intent to lay a foundation as permanent as the existing technology could guarantee.[4]

Among its other ingredients, the concrete for the dam foundation required tons of small pebbles called aggregate. The Bureau found sufficient supplies on a plateau 700 feet above Mason City on the east side of the river, one-and-one-half miles northeast of the dam. But testing there revealed that the mix held more sand than needed, so the contractor built a large aggregate plant that eliminated the sand and sorted the pebbles into uniform-sized lots. Large pumps, installed near the river, moved thousands of gallons of water up the hill to the plant where it washed away unwanted sand, eventually forming a huge pile clearly visible today at the northeast edge of the present town of Coulee Dam. When in operation the aggregate plant produced about 12,000 cubic yards of rock daily.

More spectacular than the aggregate plant was the method of getting the rock to the dam. A total of forty-three conveyor belt operations moved aggregate from the spot where machines dug it out of the ground to the mixing plants where it became part of the concrete. While most were small, one was breathtaking. Unusually long, it crossed the valley from east to west on two suspension spans, each 1,500 feet long, one over the river and the second traversing the excavation before ending at the concrete plant on the west side. *Pacific Builder and Engineer* billed the completed structure as the "largest suspension bridge ever built for a conveyor belt."[5] The thirty-six-inch-wide belt delivered 700 tons of material to the concrete mixing plant every hour.[6]

In June 1935 construction began on the concrete foundation for Westmix, the large concrete batching plant that perched 200 feet above the dam's west abutment.[7] As crews put up the building on what looked like a framework of metal stilts, plans went ahead for a sister structure on the other side of the river. Together, the two mixing plants produced 640 cubic yards of concrete each hour, or eight cubic yards every forty-five seconds, the largest capacity of any plant built to that time.[8] In Denver the Bureau of Reclamation opened bids from seven companies eager to supply the necessary "modified Portland cement." Five firms—all located in Washington state—won contracts and made separate arrangements to transport their allotment to the construction site.[9] Most of the cement arrived by train and went into silos that sat near the mixing plants.

As soon as the machines and men exposed the bedrock foundation, other crews drilled holes down into it. First eleven holes, each thirty-six inches wide, reached anywhere from 30 to 200 feet into the granite foundation. They allowed engineers to descend and inspect the rock to guarantee its strength.[10] Next came 671 small holes about twenty feet apart and between twenty and thirty feet deep. Bedrock usually has thousands of cracks and seams running through it which, after construction, would allow water, under great pressure, to seep through the ground, eventually undermining the dam above. The holes gave access to the cracks and allowed pumps to force a special mortar, called grout, into the holes and cracks. Once in place, the grout solidified and formed a seal that prevented the troublesome leakage. In other words, this work built a solid dam below bedrock surface so that water could not wear away the granite under the concrete dam above, weakening its underpinnings. Grouting started on November 16, 1935 and continued through the winter.[11]

The MWAK Company settled on a combination of trestles and cranes to shuttle the mixed concrete from the batching plant to the waiting forms. Two trestles, at first only 1,300 feet long, extended over the west side excavation. The taller trestle, on the south or upstream side, preceded the low trestle, next to and downstream from it. Thirty feet wide, each of the trestles held three standard-gage tracks and, along the outer edges, rails for gantry cranes. Together they required 9,500 tons of steel and their supports, when buried in concrete, became part of the finished dam.[12] The concrete moved onto the trestle in four-cubic-yard capacity buckets, with five carried on a railroad flat car.

A large concrete dam is not one solid mass but rather a series of columns built very close to each other. In time grout fills the spaces between the columns, leaving the appearance of a monolith. Wood forms shaped the columns, called blocks. At each pour the buckets dumped a five-foot layer of new concrete onto a block, each of which had an identification number. After waiting seventy-two hours, workers removed the forms and reassembled them at a higher level, then repeated the procedure. In this manner the blocks rose side by side, forming the dam.

In August 1935 work began on the first box, or set of forms and, without fanfare, the Bureau and the contractor practiced placing concrete. On the afternoon of Wednesday, November 27 they set all the machinery in motion, from gravel pit to dam, mixed a batch of concrete, and sent it on its way. Rather than putting the material into one of the forms, however, they dumped it into a waiting truck. The *Wenatchee Daily World* called it a "tune-up pour, not to be confused with the first official pour."[13] For the most part, the trial-run succeeded although it revealed the need for some tinkering. The next day they restarted the equipment and did the job for real:

> The first actual test pouring of concrete was made on Thanksgiving Day, November 28, starting at 9:45 a.m. and ending at 2:45 p.m. About 250 cubic yards of material was [*sic*] poured and the pour was made under the personal supervision of Mr. Frank A. Banks, engineer in charge for [the] Bureau of Reclamation.[14]

The Bureau of Reclamation first scheduled the "official pour" for a day just before Thanksgiving, but unexpected cold weather and a conflict on Governor Martin's calendar prompted a postponement, first to December 4, and then to December 6.[15] By that date the contractor already had 1,500 cubic yards in place. The Columbia Basin Commission sent 800 invitations to politicians, celebrities, and regional dignitaries urging them to attend the ceremonial laying of the "first" concrete for the dam.[16]

Governor Martin took center stage at the festivities. As part of the speeches, extensive ceremonies, and hoopla, Martin formally enrolled as a concrete worker. Dressed in overalls and a workman's hat, he moved inside the forms that surrounded block 16G. Looking at an American flag nailed to a balustrade in front of him, he placed as a cornerstone a box filled with documents and other souvenirs. As more than 1,000 spectators watched from the surrounding scaffoldings and other vantage points, a crane on the trestle high above lowered a huge bucket and the governor poured concrete.[17]

As Martin pulled the handle on the second load of concrete, someone on the speaker's platform tossed one of the state's aluminum tax tokens so that it landed on top of the soggy gray mass. As

the Governor wrestled with one of the agitators, used to spread
and pack the concrete, he noted the token, often called "Martin
Money" by his detractors. He smiled, nodded, and acknowledged
the symbolic gesture.[18] Amusing at the time, the act was prophetic
considering the tons of aluminum that power from Grand Coulee
Dam produced during and since World War II.

When Martin stopped pouring concrete, a representative of
the MWAK Company handed him a check for seventy-five cents
covering his services for one hour as a common laborer.[19] Noting
the event, the Portland *Oregonian* reported, "actual construction
on Grand Coulee, which may make possible a huge industrial and
agricultural empire in the arid Columbia basin, is under way."[20]

With the ceremonies over and the dignitaries safely out of the
way, the concrete flowed in earnest. The men worked in three shifts,
each seven hours long allowing one-hour intervals for machinery
repairs and discussion of problems. After a few days, however, the
company changed to eight-hour shifts. By the end of the first week
MWAK had laid 6,000 cubic yards of concrete but it lamented delays installing the forms, preventing the batching plant from working full time.[21] On December 20 they put the first layer of concrete
into block No. 40, next to the cofferdam and the most easterly part
of the dam then available. Block No. 40 was crucial because it would
connect with the future cross-river coffers that would allow work
on the rest of the foundation, and eventually it linked with the eastern section of the dam.[22]

With the laying of the concrete, the pace accelerated and the
contractor hired more men. The Grand Coulee payroll became the
largest in the state, amounting to around $120,000 weekly. Work-
ers collected an average of eighty cents an hour and most worked
thirty-six hours every seven days—excellent wages and good hours
for those times. With a rapidly expanding crew, the living facilities
quickly filled to capacity and beyond. The MWAK Company turned
the dining hall into a bunkhouse each night and the men housed
there used cots or slept on and under the tables. More crowded
into the boom towns west of the construction site.

The winter of 1935-36 started out cold and grew steadily colder.
By mid-January temperatures plunged further and early in February the Columbia River froze, remaining that way for the rest of the
month.[23] For a while the concrete batchers used steam to heat the

water that they mixed with cement. More steam moved through pipes heating the granite bedrock, and electric heaters kept the working site warm. Thirty-six steel-framed canopies helped capture the heat and prevented the cold from spoiling concrete, which must not freeze during the first seventy-two hours while it sets.[24] Finally, on January 31, the foremen gave up and stopped most of the concrete pour. One thousand men immediately lost their jobs, and another thousand followed in a matter of days. The economy of the entire area suddenly collapsed as the weekly paychecks ended.[25] Most of the men lived well when there was work, but they and their families immediately suffered when the money disappeared. The layoff period began the boom-and-bust cycle that became part of life around the dam.

As a result of the cold, several of the pours froze and men later had to blast away the ruined concrete. On Valentine's Day the mercury dropped to eleven degrees below zero. Steam continued to heat the concrete laid so far and the canopies held out some of the cold. Everyone watched the river to see when the ice might begin breaking-up, worried that it would again damage the bridges. Through the end of February the remaining workmen placed only 500 cubic yards a day, or even less. Then the weather warmed. By early March the work force again exceeded 3,000 and it gained from 100 to 250 more weekly for the next few months. As the ice broke up in the river, a flow crushed one of the piers that supported the conveyor belt, but MWAK quickly replaced it. By the end of the month concrete covered 40 percent of the west foundation.[26]

The winter and spring of 1936 brought sad news to the Grand Coulee project. The American flag at the dam and at the administration building flew at half staff after news came on January 27, 1936 that Dr. Elwood Mead had died. Mead, who had turned seventy-eight just ten days earlier, had served as Commissioner of Reclamation since 1924.[27] Those connected with the Columbia Basin Project felt a sense of loss, as Mead had paid a number of visits to the site and was a friend who supported the effort.[28]

Almost three months later, at 8:30 in the morning on April 14, Silas Mason, then fifty-six, suffered a heart attack while working in his office near the dam. An ambulance rushed him to the hospital in Mason City where he died. The *MWAK Columbian* wrote,

Death removed not only one of the nation's great contractors but a friend to all he met, high or low. Numerous is the testimony by workmen, even those who met him but once, of the friendliness and interest manifested by the chairman of their employing company.[29]

Mason's wife, then in Kentucky, flew to Grand Coulee and accompanied his body back to their home in the East. On Saturday, April 18, at 2:00 p. m., as the family held funeral services in Lexington, a whistle sounded across the valley below the ancient Grand Coulee. It signaled the start of a five minute work stoppage held as a memorial for the man who had been one of the main forces behind construction of the great dam.[30]

Spring brought warmer weather and the annual flood. The river rose ten feet by April 21 and five days later it was up another eighteen inches, first seeping, then pouring into the east-side excavation. It rained hard in June and twice that month work stopped entirely because of the deluge. Surprisingly, despite the delays, MWAK was six months ahead of schedule. The summer continued uneventfully and by mid-July the Columbia receded enough that east-side overburden removal resumed. Employment increased, and by mid-summer the payroll included over 5,000, increasing the housing shortage with men again sleeping in the cookhouse while others found quarters in tents built in Mason City and elsewhere. For the most part, they worked around the clock placing concrete in the west-side foundation.

Construction of each block followed a set routine. After overburden removal, blasting and drilling dislodged loose material leaving only solid bedrock, which government inspectors checked to insure soundness. Next, drillers sunk holes deep into the rock and then pumps forced in grout to seal up the cracks. Government engineers laid out the corners of each block and carpenters built forms, shaped to fit snugly against the rock base. In each form workers set small pipes that later allowed cooling water to run through the concrete. After another inspection, sandblasters freed the bedrock surface of any remaining loose material which others carefully swept away, followed by one last check by government inspectors.

Immediately before pouring concrete, the men laid down a three-fourths-inch-thick layer of grout evenly over the surface of the rock. Then the buckets began descending from the cranes above and with a government inspector always watching, a crew of six to eight men dumped load after load, massaging and leveling with shovels and vibrators, compacting and maneuvering the gray masses into their final locations. Buckets continued coming as layers, each about one foot thick, built up in the block. After five layers, making the pour five feet deep and reaching the top of the forms, work in that block stopped. After twelve hours, high pressure air hoses and water blasted three-fourths of an inch of concrete from the top of the pour, leaving a surface that would bond with the next pour.[31] After twenty-four hours, crews removed the forms and raised them, preparing for the next pour. No longer concerned with bedrock, the additional layers of concrete simply rested on those below. Seventy-two hours separated each pour and the process continued until the column reached its required height.

The practice of pouring the concrete in five layers, each a foot deep, or five-foot lifts in the jargon of dam building, made it impossible to bury anyone under the concrete. There are no bodies encased in Grand Coulee Dam. Legends of men entombed behind walls, between decks, or, in this case, in the concrete, hover over most large projects. They come from the myth that the concrete poured rapidly and covered forever those who fell into what then became a monumental tomb. The legends at Grand Coulee claim that anywhere between one to perhaps hundreds of corpses remain in the dam. The legend is so dramatic it is almost a pity that nothing of the kind ever happened.[32]

As it sets, concrete generates heat. In a dam as large as Grand Coulee, normal cooling could have taken as long as 100 years or more. In addition, as the concrete cools it shrinks slightly and, if not controlled, often cracks. To solve this problem one-inch-thick pipes ran at five-foot, nine-inch intervals horizontally and five-foot intervals vertically through each block. Eventually totaling more than 2,000 miles in length, they carried cooling water that drew off the heat.

Through its interior, longitudinal galleries fifty feet apart, one above the other, ran the length of the dam. Traverse galleries laid out from those cave-like tunnels, and vertical shafts connecting the galleries, gave access to the cooling pipes through which pumps pushed Columbia River water to remove the heat. This activity began on

December 11, 1936, when the mass of concrete reached sufficient volume to require it.[33] The process reduced the cooling to three months and thermometers, left in the concrete, kept the engineers posted on internal temperatures.[34]

After the concrete properly aged, pumps forced grout into the contraction joints and the cooling pipes, closing them permanently. The sides of each block had sculpted concrete grooves called keyways that interlocked with neighboring columns.[35] Filling in between the columns or blocks came later when more grout solidified the structure into a single monolith, albeit one honeycombed with passageways and galleries. What may appear as a solid edifice on the outside is, in reality, a mass riddled with tunnels and shafts that gives someone walking inside the eerie feeling of being in a huge concrete catacomb.

A steady parade of whirley and hammer-head cranes moved along the trestles lifting buckets off railroad cars and lowering them to the waiting blocks below, then raising them and returning them to the batching house for more concrete. On the river bed itself the blocks slowly formed a checkerboard pattern. During the summer of 1936 the outline of the dam's foundation took recognizable shape, making the structure's eventual dimensions discernible, especially from vantage points above. At last, people could begin to imagine what the finished Grand Coulee Dam would look like. Its enormous size surprised the thousands of spectators who came to take a look.

Through the summer of 1936, MWAK rushed the west-side foundation and prepared to divert the river over it to expose the central and east sections. On June 6,500,000 cubic yards of concrete filled the various blocks, a figure that doubled by August 12.[36] As the concrete piled up, the government was spending $35 each minute on construction, and by the end of 1935 the dam had already cost over $27 million.

Workers finished block No. 40 first. Next they poured concrete for the pump station base where machinery would one day sit and push water up the hillside into the Grand Coulee and onto the irrigation project. Then the powerhouse foundations and the locations of the penstocks that would carry water from the reservoir to the turbines appeared. But MWAK concentrated most of its efforts on the dam itself because it needed to finish the west side before the end of the 1936-37 low water period. That would allow construction of the new cofferdams that would divert the river across the new foundation.

During September the contractor attempted to increase the average amount of concrete poured daily. More men found jobs, and the payroll increased to over 5,300. Some of those new employees worked on the east side where, in February, MWAK began the second concrete batching plant, which it finished in May when the plant produced its first concrete for a six-inch-deep layer that reinforced the top of the east-side cofferdam.[37]

An old headache resurfaced on the east side. The winter rains loosened the clay mass there making it prone to movement, and constant earth slides tumbled down into the work area, jeopardizing men and equipment. As if the east-side slides were not enough, just before the spring flood engineers discovered a new problem. Near the center of the excavation, along the axis of the dam, a long narrow trench, 100 to 120 feet deeper than the normal level of the bedrock, ran roughly parallel to the river. Removing the overburden from the area proved unusually difficult due to the size and depth of the hole, the increased threat of slides, and water that seeped into the resulting pit. At first MWAK engineers built a temporary arch dam to stop earth movement above the gulch, but that failed. Then they suspended work during the high water period. Weeks later, when the Columbia dropped and after pumps cleared the area, the earth began moving again, this time taking on the appearance of a slowly creeping "mud glacier."[38] The only solution seemed to be the expensive removal of all overburden for a considerable distance on each side of the pit.[39] Construction superintendent Harvey Slocum grumbled, "Buzzness is Terrible."[40]

Then the engineers hit on an innovative freezing technique used to stabilize soggy unstable earth in mine shafts.[41] No one had done it on a large scale, but the principal appeared sound. It required 102 pipes driven into the clay to freeze the earth for 100 feet across a stretch of embankment over forty feet high and twenty-five feet thick. Starting late in August, pumps forced super-cooled salt brine through the six miles of three-inch black pipes stationed at thirty-inch intervals.[42] The experiment worked and by late September the south slope of the excavation stopped slipping.[43]

The *Seattle Daily Journal of Commerce* reported,

> Government engineers, faced by a critical emergency at Grand Coulee, achieved an unprecedented solution that attracted national attention by clapping a giant artificial ice pack on a 200,000 cubic yard slide of earth that had been moving at two feet an hour.... This was hailed as one of the year's greatest engineering feats.[44]

The ice dam cost $30,000 but it saved the $100,000 expense of extra overburden removal.[45] It remained in place until April 1937.[46]

The main task ahead was the second diversion of the Columbia River. Early plans suggested building the dam in three sections, first the west side, followed by the east side, with the center coming last. On October 19, 1935, Frank Banks and C. Douglas Riddle, chief engineer for MWAK, visited the Bureau's hydraulic laboratory at Fort Collins, Colorado, and viewed models of alternative diversion plans. Based on those studies, MWAK engineers decided to build the foundation in two, rather than three, steps. After completion of the west side, two long cofferdams would stretch across the Columbia River from the east bank, meeting at the end of the new west-side structure. These would divert the flow of the river over the west-side foundation, through channels left in the concrete. MWAK could then drain the area between the two cofferdams, remove the overburden, and pour the rest of the concrete. But the new plan carried more risk than the three-step plan. Should an unusually high spring flood occur, it could destroy the temporary structures and undo much of the work. Placing the two long cofferdams would be a major accomplishment. Maintaining them would tax the ability of MWAK's people.

Work on what trade journals called "the largest timber crib in the world" began about the middle of February 1936.[47] The task of placing the ten-foot-square cells and cribs required 3,000,000 board feet of heavy timbers and a formidable amount of steel-sheet piling, rock, and gravel. In August, as the work on the west side neared its conclusion, the first sections of the upstream cofferdam slowly marched across the river.[48]

Over 50,000 soundings mapped the river bed to determine its configuration, and then carpenters built the timber cribs so that their bases matched those contours.[49] Using barges as guides, the individual sections floated to their ultimate destinations.[50] Divers descended into the cold water to confirm that the timber cribs rested properly, and then gravel, dumped into the cribs, held them in place.[51] Finally, interlocking 100-foot-long steel-sheet pilings, driven fifteen to twenty feet into the river bottom, faced each crib.[52] Steel-sheet piling, arranged in huge circles or cells, and filled with earth and rock, formed other sections of each cofferdam. When the diversion started, water amounting to only 26,000 cubic feet per second (cfs) ran past the dam—one of the lowest levels for the river in years.[53] This facilitated construction of the new 1,100-foot-long crossriver dams which the *Wenatchee Daily World* called "the greatest engineering battle of all time."[54]

Workers began dismantling the first cofferdam at the same time.[55] By August 17 the wings of the old dam had disappeared and on November 5, 1936, at 4:10 p. m., open gates in the remaining section of the old cofferdam allowed water into the diversion channel, filling the space in front of the new foundation.[56] It rose, and a few days later a trickle moved through the thirty-two-foot-wide slots left to accommodate more than fifty million gallons of water a minute.[57]

The long conveyor belt on the west side reappeared across the river where it carried earth and dirt away from the exposed river bottom.[58] Work proceeded rapidly and on November 28, 1936, without ceremony, MWAK poured the first concrete on the east side.[59] It happened quickly and with so little fanfare that almost nobody noticed.

People did notice when the entire river flowed over the new west-side foundation. On December 15, 1936 the *Wenatchee Daily World* declared, "Columbia is Diverted."[60] The story was somewhat premature as only the downstream cofferdam completely spanned the river. A 100-foot gap remained in the upstream length. Filling this, however, was easy as the downstream dam reduced the river above it to a huge pond and channeled the main flow westward into a quickened torrent that now rushed through the diversion slots in the west-side foundation. The *Grant County Journal* noted that this marked the first time the Columbia had left its bed since the ancient ice flows displaced it thousands of years before.[61]

With the cofferdams in place, pumps struggled for six days to remove water from the area between the structures. On January

11 the bed of the Columbia became visible to anyone who cared to look, and many did. The MWAK Company announced it was sixty days ahead of schedule and needed only to lay the rest of the foundation to complete its contract. Looking ahead, however, MWAK's paymaster included a warning from job manager George H. Atkinson with the 6,347 paychecks handed out just before Thanksgiving:

> Many now here are familiar with the severe weather experienced in January and February of the winter of 1936 and a repetition of such conditions would force us to suspend all work except cofferdam operations and absolutely necessary repairs.[62]

And the freezing weather proved to be worse than the year before. On December 31, 1936 the concrete mixing plants shut down and half the labor force went home, out of a job for the duration. The remnant lake between the cofferdams froze and the contractor predicted the resumption of work around February 15. By January 9 the whole Columbia froze. Around 2,000 men put the finishing touches on the cofferdams. Then, on January 29, eleven inches of snow fell and the contractor postponed new work until March.[63] Those who lost their jobs sat and waited.

When the weather warmed, the ice broke up and blocks flowed past the dam, but fortunately did no damage. Early in March workers began pouring concrete into the blocks on the east side of the river and MWAK not only rehired its former employees, but also sought 400 to 500 additional common laborers. Within three days the *Wenatchee Daily World* reported all jobs filled and the hiring offices swamped with applicants as men eagerly grabbed any available position.[64]

Thousands remained unemployed as the depression continued. Poor economic conditions hurt minorities especially hard. In March 1936, representatives of the Urban League approached the government and complained that black men had not been able to secure jobs at the dam. After a number of meetings both in Washington, D. C., and at the dam, Raymond F. Walter, then acting Commissioner of Reclamation, noted:

> Negro mechanics, skilled in construction work, are to be employed on the Grand Coulee Project at once. This decision was reached this morning by the Mason, Walsh, Atkinson and Kier

Company, general contractors, after a conference held in the office of Mr. F. A. Banks, Construction Engineer.

The meeting in Mr. Banks' office followed a week of intensive investigation in which every phase of the question pertaining to the employment of Negroes on the huge dam project was gone into. I found Negroes in Seattle in a resentful mood. They spoke of the Coulee Dam project as another evidence that the government is not interested in the welfare of its Negro citizens. They told of their efforts to secure work there, and several referred to a legend about one Negro who had walked all the way to Coulee Dam from Seattle seeking work, only to be turned down. (I was unable to find this Negro, or to find anyone who knew him personally, although several seemed to have heard about him. The incident [was] also unknown to [the] United States Reemployment officials at the Dam site.)

Negroes constitute less than two percent of the total population of the State of Washington. The terms under which the Coulee Dam project was accepted by the MWAK company included a specification that an area of first preference should be established, this area to include Douglas, Grant and Okanogan Counties, the three nearest and adjacent to the project. Veterans in this area were to get first preference, provided they met the other requirements as to physical fitness and to experience. As it became necessary to increase employment, other zones were set up. Only one Negro was found in the area of first preference who qualified for employment on the project, but he had not been employed. In the second Zone, which included Spokane, Mr. Sullivan declared that he could, without difficulty supply at least 15 or 20 men who could meet all the requirements, but had not done so for two reasons (MWAK said they would not hire Negroes. No Negroes had applied because they had heard that they would not be hired). . . .

It was at this stage that the conference was held in which all the persons listed above attended and took part. In answer to my question about the housing situation, it was brought out that the MWAK Company had built accommodations for 1200 persons. At present 1400 persons are living in them. There are at present over 4200 persons working on the dam, which means that 2800 white persons are living in surrounding towns. I suggested that if 2800 white persons could work on the dam and live in surrounding towns, I saw no reason why Negro workers should not do the same thing. Certainly, I saw no reason why the Company should be expected to evict white workers to house Negro workers, and if that were the only holdup, it was my opinion that Negroes should be hired and allowed to secure their own lodgings within walking distance of their jobs.[65]

Some blacks did then secure jobs at the dam as common laborers. Those hired often found themselves shunned by the other men as white workers even refused to drink water out of buckets, carried about the site on warm days, if any of the black men used them.

Banks only reluctantly hired blacks, as the case of Robert Flowers illustrates. On January 17, 1936, Frank Banks requested permission to hire a messenger and office worker for the engineering headquarters building. Five days later, the Denver headquarters approved.[66] Then, on February 14, Raymond Walter's office forwarded the application of Robert Flowers of Tacoma to Banks and asked him to give Flowers special consideration.[67] Banks replied,

> This application reveals that he has had no office experience along the lines outlined in our job description, which stated that 60% of the time would be required in operating office appliances such as mimeograph, blue print machine, etc. Mr. Flowers states that he was an assistant janitor for 4 years, no other experience [is] shown. His certificate is returned herewith.[68]

This time Reclamation Commissioner John Page wrote to Banks directly and explained, "Mr. Flowers should be interviewed, and if selected should complete application for forwarding with your report on this certificate."[69] Banks replied that he rejected Mr. Flowers and had explained why in his letter of February 21. He sent a copy of that letter and Flowers's application back to Denver.[70] A few days later Banks recommended hiring Walter F. Hill. Raymond Walter informed Banks that the Bureau held up the appointment hoping that Banks would reconsider the Flowers application.[71]

At this point Banks received a two-page letter from Raymond Walter that brought things to a head. Walter reviewed the job applications of both Hill and Flowers and pointed out that each had a high school diploma, was about twenty years old, and had no office work experience. Flowers had some training at the University of Washington but Hill had no college background. And then Walter got to the point:

> [The] Photograph attached to the application of Flowers shows that he is a Negro. While nothing is mentioned in any of your letters concerning this, it may be that it has some bearing in the matter. For your information, the following sentence is quoted from a memorandum dated October 31, 1934, addressed

to the Commissioner of Reclamation, by Mr. E. K. Burlew, Administrative Assistant: "Secretary Ickes desires to employ colored boys in messenger positions wherever possible." Since that time, it has been necessary to employ colored youths in all messenger positions in the Denver office and this same procedure is being followed in all project offices making employments under this designation.[72]

Unfortunately, Robert Flowers's file ended there and no existing record tells if he eventually got the job. But by the end of 1936 around fifty blacks worked on the project along with a few Indians from the nearby Colville Reservation.[73] Whether Frank Banks's actions reflected his personal feelings or his wish to avoid potential problems, real or imagined, is debatable. Banks insisted that no blacks live in Mason City or in the Engineers' Camp[74] and he appeared defensive about this in a letter he wrote the following summer:

Negroes have been employed on this project in increasing numbers, until at present time 45 are employed in various wage classifications running from fifty cents per hour to seventy-five cents per hour, out of a total of 4,750 laborers and mechanics employed by the contractor on this work. The percentage of negro [sic] employees is thus fixed at .97 of one per cent. The World Almanac for 1936 gives the male white population of the State of Washington as 800,924 and the male negro [sic] population as 3,797, or .47 of one per cent. This indicates very clearly that the ratio of negro workmen to white workmen on this project is double the ratio of male negro residents to male white residents in the State of Washington, and there certainly can be no complaint from the standpoint of numbers employed.

Banks concluded that 60 percent of the white workers lived away from the project, so blacks had no reason to complain that all of them had to live elsewhere.[75] Ironically, while some white workers questioned why they had to live in the contractor's camp, blacks complained that they could not.

White workers had other criticisms. Section 7a of the National Industrial Recovery Act, passed in the spring of 1933, guaranteed workers the right to organize and bargain collectively. When the Supreme Court declared that legislation unconstitutional, the Wagner Act of 1935 replaced it. That year some labor leaders, such as John L. Lewis, broke away from the more traditional American Federation of Labor (AFL), which favored craft unions, and formed what later became the Congress of Industrial Organizations (CIO). The CIO organized unskilled labor into so-called industrial unions. The two groups competed for members. Employers saw both unions as a threat and some tried to counter them with company organizations they could control. Others simply disdained collective bargaining and fought unionization altogether. Against this background events unfolded at Grand Coulee.

Early in 1936 the Building Trades Councils in Spokane complained loudly about employment practices at the dam. Its representatives investigated and in the fall of 1936 declared their first goal, to remove C. C. Beery, who still headed the National Reemployment Service (NRS). They presented evidence that Beery opposed union members and openly discriminated against them. The Building Trades Councils complained to the Department of Labor, and on March 4, 1936 that department sent a representative who, shortly after arriving at Grand Coulee, fired Beery and his assistant for mismanaging the NRS budget.[76]

Edward M. Miller, transferred from Fort Peck, arrived at Grand Coulee as the new NRS administrator. He immediately aggravated the situation when he announced that he would not allow any person owning agricultural land to work at the dam—an effort to keep farmers from taking jobs away from the "truly unemployed." Eastern Washington residents hardly saw the logic in the plan that disallowed all but 175 of the 2,200 in the area who had work applications pending. Angered by the move, Okanogan County cut its financial support to the NRS and joined with Grant and Douglas counties to protest NRS policies. Banding together in what they labeled a Tri-County League, they demanded an open shop at the dam.[77] Senator Lewis Schwellenbach and Representatives Knute Hill and Sam Hill also protested Miller's actions.

All parties concerned held a meeting with Frank Banks on June 9. The NRS people argued that local men would receive jobs if they returned questionnaire cards and pledged cooperation with local county commissioners. But within a month Okanogan County officials wired Secretary Ickes protesting the extensive hiring of out-of-state men rather than locals.[78] Discontent with the NRS then diminished but it continued through the construction period. Washington state residents were justified in their contention that the contractor and the government thwarted the rules and hired men who had worked on other projects, especially at Boulder Dam. But this disagreement also faded into the background. Both were eclipsed as the Building Trades Councils pursued its second goal.

In the spring of 1936 the Spokane labor organizations had announced their intent to unionize the Grand Coulee Project.[79] MWAK declared that under its government contract, it had to remain neutral toward the union movement. Nothing much noteworthy happened until the fall of 1936 when conflict erupted, grabbing the headlines and holding them for almost a year. It began with a demand from the International Order of Operating Engineers that MWAK recognize them as an AFL affiliate. The engineers also threatened a strike if some of their members did not receive higher wages.[80] MWAK, despite its stated neutrality, refused to recognize the union or raise anybody's pay. It did, however, agree to continue talks with union representatives, although nothing came of them.

For months, various pro-union groups had claimed that MWAK cheated workers by incorrectly classifying them or not crediting all hours worked. Obviously, the argument ran, they needed union protection. In January 1937, Reclamation Commissioner Page reported to the Public Works Administration that a special investigation showed clear evidence that the MWAK Company had indeed underpaid wages in the amount of $87,320.91.[81] The Bureau and MWAK quietly settled the matter but the union movement gained momentum. Nationally the CIO, more radical than the older AFL, moved to organize unskilled laborers at projects throughout the country. Starting early in May 1937, CIO representatives appeared around Grand Coulee distributing pamphlets attacking the AFL. As the labor force at the dam increased, CIO people lured them to meetings and tried to enroll them. The AFL fought back and leaders of its various affiliates argued that alignment with the CIO would be a mistake.[82]

Things heated in June when CIO members and their sympathizers attempted, without success, to arrange a sit-down strike. In response, the AFL held its first mass meeting. Only about 100 men attended although the union claimed it represented around 1,000 employees. The CIO then called two meetings and boasted a membership of 2,100. At the same time, a small group put together the Federation of Construction Employees, an independent union that said it wanted to prevent strikes called by outside interests without a vote of all employees. The union demanded a fifty cent hourly wage increase and told the press that it represented 1,000 men. Frank Banks commented, "Just what the outcome will be of the three-cornered struggle for membership and control is difficult to predict, but there is nothing to indicate just now that there will be any trouble in the near future."[83]

The MWAK Company, however, decided not to gamble. Fearing possible CIO success, on July 29, without warning and with fanfare, it announced,

> In view of the best interests of this project and the security of all our employees, we have today signed a firm contract with unions affiliated with the American Federation of Labor.... We ask your cooperation in affiliating with these unions at the earliest possible date. It is mutually agreed that there shall be no strikes, lockouts, or other cessation of work by either party during the life of this agreement.[84]

The CIO immediately protested to the National Labor Relations Board that the agreement was arbitrary and required a vote of the company's employees.[85] Calling the arrangement with the AFL a "scab agreement," the CIO set up pickets the next day.[86] More than anything else, CIO organizers protested the details of the AFL-MWAK agreement that all but made the project a closed shop and required employees to join an AFL affiliate.[87] MWAK officials forced the pickets outside the work area perimeter and barred them from Mason City. When some of the pickets defied those rules, MWAK and the Bureau of Reclamation charged them with trespass, local authorities arrested them, and they spent a few hours in one of the local jails until they produced $15 bail.[88]

The CIO attempted to petition President Roosevelt and Labor Secretary Francis Perkins directly, demanding that the Grand Coulee project be shut down to avoid serious trouble and to allow time for a full investigation. The National Labor Relations Board (NLRB) did agree to conduct a study but meanwhile, afraid that it would not receive satisfaction, the CIO, through its affiliate—the International Union of Dam Construction Workers—called a strike of all project workers for 8:00 a. m. on Saturday morning, August 7. It further demanded that MWAK rehire nine men, fired for pro-CIO agitation. William Hillyer, the district spokesman for the CIO, declared to the press that he represented between 2,400 and 2,500 of the 6,000 men employed at the dam.[89]

Adding to the heightened drama, the *Wenatchee Daily World* and the *Spokane Chronicle* announced that Harry Bridges, controversial radical leader of the International Longshoremen's Association, then arriving in Seattle, would travel to Grand Coulee and lead the CIO strike.[90] Having gained a reputation in the 1934 San Francisco general strike, Bridges succeeded in unionizing west coast dock workers and he often moved into confrontational situations. His detractors went so far as to label him a communist, a charge that Bridges denied throughout his eventful career.[91]

At Grand Coulee the CIO postponed its strike and Hillyer announced that Bridges would be at the dam site for a mass meeting Monday evening. Although Hillyer painted a picture of formidable support for the CIO, the *Spokesman-Review* noted that only 250 men attended the meeting, voted not to strike, and seemed reluctant to leave their jobs.[92] The *Spokesman-Review* called the proposed strike a "fizzle." In Seattle, Harry Bridges flatly denied any involvement in the labor difficulties at Grand Coulee and admitted, "We don't know what the score is [at Grand Coulee]."[93] In the midst of the fuss, the *Wenatchee Daily World* pointed out that despite the labor uproar, work at the dam had not slowed and that, in fact, MWAK poured a record 14,065 cubic yards of concrete on Saturday, the day set for the strike.[94]

Charles Hope, representing the NLRB, arrived at Grand Coulee on August 11 and things quieted down. The CIO continued demanding that workers vote to decide which organization they wanted to represent them, and Frank Banks cautioned his superiors in Denver that, while the confrontation had momentarily abated, it could flare up again.[95] When the NLRB agreed to assume jurisdiction over the dispute, the CIO claimed victory. At the same time, MWAK continued urging its employees to join one of the AFL unions and implied that it would fire those who did not.[96]

The Bureau of Reclamation remained aloof from the labor dispute and Commissioner Page declared that neither the Department of the Interior nor any of its agencies would interfere in disagreements between contractors and unions, or between unions.[97] A few days later Frank Banks wrote, "The labor situation here has been very quiet. . . . The contractor is beginning to cut his forces down somewhat."[98] Nothing more happened and the conflict withered away peacefully.

And so it ended. The threatened strike never happened. The completion of the MWAK contract undermined the CIO as available work disappeared and more men lost jobs. MWAK avoided forcing its employees to join the AFL and the workplace assumed only the appearance of a closed shop. Nobody really lost in the confrontation, including the CIO which never represented the large number of men it claimed. The labor problems during this phase of Grand Coulee construction created considerably more publicity than product.

As for the job, the second Columbia River diversion initially also appeared uneventful. Both the Bureau and the contractor were optimistic about the success of the difficult task. Then on March 2, 1937 men working in the area between the two arms of wood cribs that stretched out from the east bank noticed that one of the cells on the downstream side leaked. By March 18 that inflow increased to over 15,000 gallons of water a minute.[99] Suddenly, the inside facing of the weak cell burst.[100] Summoned from his bed, Harvey Slocum arrived on the scene still in his red pajamas. He and the others grabbed anything loose and, using dump trucks and other equipment, threw it into the river, hoping to plug the leak. They used over twenty mattresses, sagebrush, straw, various building materials, pieces of canvas, earth, and rock.[101] This temporarily slowed, but did not stop the water.

Although the contractor never abandoned work on the foundation, the engineers realized they faced a significant problem. They had not reinforced the downstream cofferdam with steel driven all the way to bedrock. Hence it lacked the strength of the first cofferdam on the west side of the river. Fortunately, when the leak began, the Columbia carried only about 27,000 cfs, as the spring flood had not yet started.[102] A leak during the flood would have created a major tragedy.

Through the rest of March and into April, MWAK people struggled to plug the leak completely. They turned to the Black Hills of South Dakota for a supply of bentonite. As the *MWAK Columbian* reported,

Among its properties is its amazing swelling propensity, its ability to form gelatinous paste with water and to form gels which remain in suspension in thin dilutions. When put into water Wyoming bentonite expands from 10 to 30 times its original volume, to form gelatinous masses.[103]

Well into April the men worked, seeing modest successes followed by a renewed gush of water. Late in May the leak opened again, water pouring into the construction area, taxing the pumps.[104] Although the river rose appreciably, the weather remained cool and the potential for a big flood diminished. Eventually the contractor drove sheet pilings deep into the river bed to fortify the cofferdams. This additional expense, just seven months before MWAK completed its contract, solved the problem and kept the company more than a year ahead of schedule.

No one expected the contractor to subdue the river and pour the concrete so quickly. But by June 1937 the two ends of the trestles met over the center of the river and the foundation held half its concrete. The river was already back in its ancient bed, but it was hardly the same. Now a growing, gray, 3,250,000-cubic-yard slab of concrete blocked its path, channeling water through a few slots. The Columbia passed over what that month became the world's largest dam.[105] Still rushing their job, MWAK officials bragged that they poured more concrete in one twenty-four-hour period than anyone in construction history. To prove the point, on August 28, workers laid 15,844 cubic yards, bringing the total for that month to 377,133 cubic yards, also a record.[106] In all, the base of the dam now covered thirty acres of bedrock. By the end of 1937, the east and west sections of the dam met.

An unexpected casualty in the midst of this success was general superintendent Harvey Slocum. On August 30, 1937 the *Wenatchee Daily World* announced that Slocum had resigned his job. Some spread the rumor that his health had deteriorated, forcing the move. And, as doctors have come to understand the problem, that claim had merit. At the time, however, people recognized only that Harvey Slocum drank too much. He loved liquor and he loved women. The combination, plus his inability to control either craving, caused him ongoing conflict with his employer. Slocum would disappear for days at a time, spending his unofficial sabbaticals in the town of Grand Coulee drinking and patronizing the "ladies" who worked there.[107] Once he sent a crew up and they installed a sprinkler system, on company time and at company expense, on the roof of one of the "houses" where some of his "girlfriends" lived. In a way the action had historical significance as the first air conditioning provided by Grand Coulee Dam.

Eventually MWAK lost patience with Slocum and fired him. Petitions supporting the popular foreman, signed by large numbers of the men, failed to win his reinstatement.[108] Then the workers chipped in and bought Slocum a new car as a going-away gift. He reappeared briefly later, but for the most part, when he drove away he ended his career at Grand Coulee Dam.

By the middle of 1937 the foundation neared completion. The contractor and the government made plans to finish the job and move to the next phase. The Bureau of Reclamation agreed to buy Mason City, the concrete mixing plants, aggregate plant, and conveyors.[109] As some workmen brought the different blocks to their full height, finishing off the foundation, others removed the rest of the cofferdams.[110] Much of the scrap steel went into shipments sold to Japan. The *Wenatchee Daily World* observed that the Japanese were then invading China and making threatening noises towards areas beyond. In one article written about the scrap metal sales, the author quoted a concerned citizen who said, "wouldn't it be an ironical thing if some steel that helped build that dam later came back to destroy it?"[111]

The *MWAK Columbian* carried a message to all employees in its November 25 edition:

> Our contract work on this project is rapidly closing down and we now believe that the main concrete will be substantially completed between December 5th and 10th. At about this time the gravel plant will be shut down, shop work generally curtailed, and the force drastically reduced for the few remaining pours which will extend after this date. Except for a small force, therefore, all men on the project should count on being laid off about this time. New job opportunities under the new contractor will remain limited until the fall of 1938.[112]

With that, 5,000 men began losing their jobs, with no prospect of more work for eight or nine months.

On a sour note, MWAK foremen noticed equipment disappearing. Tools, large and small, walked away from the project and despite attempts to stop it, the men carted off just about anything they could. Officials found one laborer with a jackhammer that he attempted to spirit away when he thought nobody was watching.[113] Exasperated, MWAK managers tried stopping and searching cars, and found both employees and ex-employees with company property. Eventually they managed to slow the thefts but never stopped them entirely.[114] In January 1938, MWAK officials received an unsigned letter from Three Hill, Alberta, Canada, containing $15, which one anonymous ex-employee sent to pay for tools he took.[115] It was the only compensation the company obtained for its losses.

The contractor shut down the east-side concrete plant on November 21 and finished up the last pours from the west side. On November 24 the Columbia River filled its old channel except for a waterfall, some sixty feet high, that spilled over the central part of the foundation. Workers completed the last major pour on December 27, and MWAK people announced they would finish on February 1, 1938, exactly fourteen months early.[116]

In April 1937, *Time* pointed out that an unusually large number of men had died on the project. Although the magazine exaggerated the seriousness of the problem, the hospital in Mason City saw a steady stream of injuries—many serious. Men and equipment plunged into the Columbia, which flowed swiftly past the site— and the odds of survival did not favor those unfortunates. If they drowned, they usually turned up downstream at Bridgeport or Brewster. The accidents continued right up to the end. On November 20, 1937, forty-seven-year-old Arthur L. Leifson and twenty-one-year-old Louis J. Rogers both died when their truck fell 200 feet down a hillside while they traveled across the highway leading to the dam. Ten days later forty-five-year-old Dan Reese fell into the river when a cable broke and struck him, knocking him off block No. 40. Reese managed to struggle toward shore, and men downstream rescued him a few minutes later. Reese considered himself extremely lucky and nobody disagreed.[117]

One final death marked the end of MWAK's effort. Roland H. Tegtmeier, a nationally known marathon swimmer from Tacoma, drowned in the river just below the dam. MWAK had hired Tegtmeier to patrol the river and watch for workmen who might fall in while removing the last part of the downstream cofferdam. One day the outboard motor on Tegtmeier's small boat stopped and the frail craft floated into the swirl caused by water rapidly flowing over the toe of the dam. While workers above watched helplessly, the force of the water tore the boat apart and the suction of the swirl pulled the champion swimmer to his death.[118]

Contractors usually expected about one death for each one million dollars spent. By that measure, during the MWAK contract period, when sixty men died—forty-five in direct employ of the company—the statistics were slightly below average.[119] Giving some perspective, the *Wenatchee Daily World* noted that as many people died in road accidents around the dam as lost their lives on the job. MWAK carried state industrial accident coverage for its employees, and saw to it that widows and children left fatherless received compensation.[120]

Construction deaths should not overshadow accomplishments. During construction, engineers, designers, and advisors developed new techniques and innovations that changed and advanced methods used on large-scale projects. The long conveyor belts stood out as the most notable example. In two years some of those belts had carried 10,000,000 cubic yards of raw stock from the gravel pit to the aggregate plant.[121]

Largely, however, work at Grand Coulee Dam picked up where the construction at the Boulder Project left off. Many of the engineers and specialists who served at Boulder found their way to eastern Washington where they continued the techniques pioneered earlier.[122] As C. Douglas Riddle, Chief Engineer for the MWAK Company wrote in 1936, "In working out construction problems presented on such a vast scale at Grand Coulee Dam, no startling novelties have been attempted. On the contrary, the selection of tools and methods was based on proven experience."[123]

Nevertheless, the diversion of the Columbia River was spectacular. Completing the foundation early, MWAK set what then seemed remarkable, but what would prove to be short-lived, records in the amount of material displaced and hauled, and the amount of concrete poured. In all, nearly 11,000 men had worked for MWAK.[124] Collectively they invested over twenty-seven million man-hours clearing overburden and placing 4,525,209 cubic yards of concrete.[125] Although no revolutionary technological breakthroughs marked the construction, thousands of small innovations and novel ideas, solutions to unique problems, and strategies never before used, marked the effort. The completion of the dam foundation was a formidable achievement worthy of the praise and the pride felt by the men and the contractors.

The change from the low dam to the foundation for the high dam forced MWAK to spend six million dollars over its original bid.[126] Further adjustments raised the total to $39,430,000. In addition, the company, in a document amounting to several hundred typewritten pages, claimed $5,267,253.55 more.[127] Much of the claim rested on MWAK's allegation that the government misrepresented the configuration of the river bottom, requiring expensive adjustments.

On January 8, 1940 the Federal Court of Claims sustained a government demurrer in the case and dismissed the MWAK claim. In its holding the court stated, "The contract is a harsh one, but its language is perfectly plain and we cannot reform it."[128] Through 1940 and over the next few years MWAK pressed its claim before the United States Supreme Court. Meanwhile, the Court of Claims entertained an appeal, held in San Francisco, at which Frank Banks testified. Again MWAK lost, and a few months later, on October 14, 1946, the Supreme Court declined the case. The company then returned to the lower courts which quickly dismissed the action. The last judgment fell on January 8, 1947, ending litigation that had drug on for nearly ten years. By then MWAK, which became one part of the entity that built the rest of Grand Coulee Dam, had finished the project and long since moved on to other jobs.

Rufus Woods. *Courtesy Bureau of Reclamation.*

# Chapter Ten

# Paying for the Biggest Thing on Earth

They are always wanting the government to spend the taxpayers' money to build something. Every congressman wants to get an appropriation to dam up his client's little stream with federal funds. If the politicians have their way, there won't be a foot of water in this country that's not standing above a dam.

*Will Rogers*[1]

Western history thrives on superlatives. Nothing, it seems, can be merely large; it must be massive.

*William L. Lang*[2]

It is man's greatest structure, and the superlatives of the English language have been nearly worn out describing it.

*Fred O. Jones*[3]

While MWAK labored to build the foundation of Grand Coulee Dam, Congress, sometimes grudgingly, found money to pay the bills. The 1935 legislation provided legal support for the project, but then the Roosevelt administration shifted the burden of financing to Congress, allowing limited New Deal relief dollars to go elsewhere.[4] This meant that the dam now appeared annually in the federal budget and provided an issue New Deal opponents used to criticize the administration. How much these people, usually Republicans in the East, really wanted specifically to stop the dam, and how much they just wanted to play politics and bedevil Roosevelt and his goals, is questionable. At times their threats appeared formidable and yet, in the end, money for the dam always came easily in the Senate and it usually showed a modestly comfortable margin of approval in the House.

The possibility, however real or imagined, that Congress might stop providing money traumatized Grand Coulee backers in Washington state. Throughout the latter 1930s they continued to scrutinize every congressional move concerning the dam, marshaling their forces as each threat appeared, ready to lobby whenever the dam seemed in jeopardy and eager to lash out at opponents. Their effect on the outcome was minimal, but certainly they made their presence felt and their feelings known.

In 1936 the Bureau of Reclamation had nineteen dams under construction around the country.[5] To continue that work, the Interior Department requested $64 million for the following fiscal year with $22 million earmarked for Grand Coulee.[6] Senate committees approved the Interior Department bill, set aside $20 million for the dam, and passed the measure on March 2, 1936.[7]

The House of Representatives, which had balked at legalizing the dam, now hesitated at providing any money. Francis Culkin of New York, Grand Coulee's perennial nemesis, joined forces with John Taber (Republican from New York) and Richard Bowditch Wigglesworth (Republican of Massachusetts).[8] Together they vehemently opposed funding the dam and moved to halt its construction. Culkin quoted the president of the American Society of Civil Engineers who said that Grand Coulee Dam was "a grandiose project of no more usefulness than the pyramids of Egypt." And Culkin himself added, "Who will buy the power, Jack Robinson Rabbit?"[9]

Washington's Sam Hill argued that the "power interests" inspired the debates against reclamation in general and Grand Coulee specifically.[10] Those debates drug on into April when a collection of Eastern representatives managed to shove through an amendment capping Grand Coulee expenditures at $63 million, a slap at the Roosevelt administration from congressmen who disapproved of the President initiating construction without their approval.[11] Shocked, and afraid that the government might stop work on the dam, James O'Sullivan, still in his capacity with Washington state's Columbia Basin Commission, consulted with Representative Samuel Hill and Senator Lewis Schwellenbach, who assured him that the amendment would have little significance in the long run.

It simply meant that Congress must agree to any further work, and the Washington congressional representatives felt certain it would—sooner or later. Unwilling to take the risk, O'Sullivan lobbied through April for amendments to the funding bill guaranteeing existing contracts. Despite urgings from the Washington delegation that he go slowly, O'Sullivan pushed. Commenting on his performance, Rufus Woods's business manager wrote to his boss in Wenatchee, "As far as James O'Sullivan is concerned, I am afraid he is making a fool of himself and endangering the Coulee Dam project by attempting to make its construction an open issue between the Democratic and Republican forces."[12]

The House cut the entire $61 million earmarked for irrigation with one representative calling reclamation "a huge pork barrel thrust out of western skies." Secretary Harold Ickes announced that if the House did not restore the Interior Department funding, work at Grand Coulee would stop on October 1.[13] On June 15, after considerable maneuvering, the House reconsidered, returning $31 million for reclamation and approving the $20 million for Grand Coulee, but taking it from the general fund. The bill then returned to the Senate which passed it on June 18, 1936.[14]

In an editorial the trade journal *Pacific Builder and Engineer* declared, "The surprising and vicious but unsuccessful attack on Grand Coulee in the House of Representatives during the last weeks of the session should be an adequate warning to reclamationists in general and supporters of Grand Coulee in particular."[15]

In a personal letter to James O'Sullivan, *Pacific Builder*'s editor, Walter A. Averill, chastised Grand Coulee backers:

> To me it is self-evident that the cause of Grand Coulee Dam could be fostered most effectively if the various factions interested could enthusiastically correlate their efforts. . . . You should have help from more people and more intelligently planned help than you have had except from a handful of wonderful stalwarts.[16]

Averill identified a problem that plagued project supporters who continued to disagree, argue, and feud with each other, leaving their efforts uncoordinated and haphazard. Some of this carried over from the gravity/pumping plan debate of the 1920s and early 1930s when pumpers often flared at each other and split over tiny issues. At times Grand Coulee succeeded despite its friends.

On another front, O'Sullivan was more successful with the long-simmering Kettle Falls issue. The Washington Water Power Company still held a permit for the site and it revived its plans to build a dam there. On May 25, 1936 the Federal Power Commission held a hearing on the Washington Water Power Company application. The question of the government's intentions at Grand Coulee became the pivotal issue.[17] O'Sullivan, who accused the company of wanting to develop the location only to limit the size of Grand Coulee Dam, pointed out that the government's high dam, if built, would eventually inundate Kettle Falls, making the site unusable.[18] After a long debate and considerable testimony from Senators Homer Bone and Lewis Schwellenbach and Representative Sam Hill, Bureau of Reclamation officials, and James O'Sullivan, on June 13, 1936 the Federal Power Commission formally denied the Kettle Falls application stating that it did not fit into the comprehensive plan for the improvement and utilization of regional water resources.[19] Washington state followed on February 16, 1937 when its Supervisor of Hydraulics withdrew all Washington Water Power Company permits on the Kettle Falls site.[20] After more than a decade, Kettle Falls no longer presented an obstacle to Grand Coulee high dam construction. It was a notable victory for O'Sullivan and the Columbia Basin Commission.

After Roosevelt decided for the high dam, Assistant Reclamation Commissioner William E. Warne announced that the Bureau of Reclamation wanted to build the large structure immediately, but that depended on congressional appropriations. Especially crucial would be the money allotted in 1937 since that would enable the start of the second major contract. Warne explained that $6 million would only finish the MWAK contract covering the foundation and that the Bureau needed $15 million to initiate the next step. Acting Commissioner John Page visited Spokane in September to initiate a campaign throughout Western states to "save the federal reclamation policy and Grand Coulee Dam." Page commented, "I've been in the Bureau of Reclamation for 27 years, but not until the last session of Congress did I fully appreciate [that] reclamation is not one of the government's favorite children. Reclamation is an orphan, or at best a step-child."[21] Page worried his audiences when he reminded them that unless Congress made direct appropriations, construction at Grand Coulee must stop when expenditures reached $63 million. "We must irrigate public opinion so we can irrigate our projects," said Washington Governor Clarence Martin when he spoke to over 100 people at the twenty-fourth annual meeting of the Washington Irrigation Institute.

In the November 1936 general election, Martin won a second term as Washington's governor and Franklin Roosevelt and James Garner carried the national ticket in a landslide.[22] Although a state measure allowing the government to enter the wholesale power production and distribution business failed, the people approved fifteen new public utility districts, most of them around Grand Coulee. These promised an increased market for the power that Grand Coulee Dam would produce, showed support for the dam, and indicated increased favor for public power state-wide. Senator Lewis Schwellenbach summed up the election's significance when he declared that the results confirmed his belief that the government would now surely carry Grand Coulee through to completion.[23]

But Schwellenbach was startled when Roosevelt delivered his annual budget message to Congress on January 8, 1937. The President requested $34,665,000 to continue Western reclamation projects, with only $7,300,000 to finish the Grand Coulee foundation. Then Roosevelt emphasized the need for a balanced budget.[24] The $7.3 million exactly reached the $63 million ceiling set by Congress. The Department of the Interior had wanted $15 million and declared as ill-advised any limit on a project already under construction.[25] Schwellenbach urged Congress to override its own limit and pass an additional $7.7 million allotment. Governor Martin rallied the entire state behind the dam, insisting that its benefits would go well beyond the Northwest. He suggested that the Washington delegation appeal directly to the President for the full amount needed to continue construction. Without the money, Martin added, all work would stop at the dam about the end of December.[26]

At this point, the Columbia Basin Commission expired. Torn by squabbles and friction among its membership, which argued over retaining James O'Sullivan as executive secretary and whether the dam was a power or an irrigation project, the commission irritated many who felt that, with the dam under construction, it had outlived its usefulness.[27] The Spokane faction never reconciled itself to a commission controlled by Wenatchee/Ephrata people, or its own diminished role. Feeling that it could best direct the future of the dam, the Spokane Chamber of Commerce exerted political pressure on Governor Martin and the state legislature until they eliminated funding for the Columbia Basin Commission, effectively ending its existence. The members held their final meeting in the Olympian Hotel in Olympia on January 28, 1937.

James O'Sullivan received assurances that he would remain a state employee, but in August James Brooke Fink, Department of Conservation and Development Director and the man now dealing with the dam, told O'Sullivan that, for political reasons, he no longer had a job.[28] While this failed to quiet the activities of Rufus Woods, James O'Sullivan, and the others who backed the high dam and the irrigation project, it did eliminate their official standing and reduced their authority and clout. Further, it showed once again how strife among the backers of the dam threatened their cause, and the continued influence wielded by Spokane.

Although the loss of the Columbia Basin Commission was not tragic, it came at an inopportune time. The congressional battle for funding in 1937 was difficult and complicated. Representative Francis Culkin again led the fight against reclamation and against Grand Coulee. For his first act, Culkin introduced a bill prohibiting expenditure of any government money for any new irrigation projects.[29] This startled reclamation interests, who increasingly saw such efforts as an Eastern attempt to cut off what the West saw as essential irrigation. On February 2 the House Irrigation and Reclamation Committee killed Culkin's bill.

At this point, Roosevelt complicated the issue by insisting on an anti-speculation bill guaranteeing inexpensive land for settlers in the future Columbia Basin Project.[30] Bone and Schwellenbach introduced the legislation. Early in May the bill passed the Senate and went to the House, followed by the Interior Department supply bill that carried money for reclamation and Grand Coulee funding. After impassioned testimony from Commissioner Page and Secretary Ickes, the House Appropriations Committee approved $40,981,000 for Western reclamation, including $13,000,000 for Grand Coulee. Then the measure went to the entire House.

Representative Culkin, backed by John Taber and Robert Fleming Rich (Republican of Pennsylvania) again rose in opposition. "Do not be fooled by the red herring of power these mad reclamationists have drawn across your path," he told the House membership. Reclamation, he declared, came from the soul-less selfishness of localities and he called Grand Coulee "a colossal un-economic folly born of deceit of the distinguished occupant of the White House." "One former Congressional representative from the State of Washington put that over on the president and then was forced to retire from Congress because it was learned some 700,000 of the acres to be irrigated were held by land sharks," Culkin charged in a thinly veiled reference to C. C. Dill.[31] "No man who ever represented the great state of Washington ever did anything for Grand Coulee which the people of that state would not support," shouted Charles Leavy in response.[32]

On May 17, 1937 the House approved the entire $40,981,000 requested for reclamation, including the $13 million for Grand Coulee, along with the anti-speculation measure. Rufus Woods, who had spent two months in the capital working for passage of both bills, proclaimed victory in the "biggest battle."[33] Because of congressional preoccupation with Roosevelt's Supreme Court reform proposal, the House allowed less time than usual for debating reclamation cuts. For the same reason, there was little discussion on the anti-speculation measure. On May 27, 1937 the anti-speculation bill went to Roosevelt, who immediately signed it.[34]

The funding bill, however, went back to the Senate where Schwellenbach led an unsuccessful fight to restore one million dollars to Grand Coulee. The senator struggled to animate support for reclamation, federal flood control, and water conservation, but he met formidable opposition.[35] Again, other concerns distracted and held the attention of the senators. In the end, the Senate left the bill unchanged, retaining the $13 million figure for Grand Coulee. On August 9, 1937, Roosevelt signed the bill that also sanctioned high dam construction and removed the $63 million Grand Coulee spending limitation.[36]

While Congress debated Grand Coulee funding, those supporting the dam, both in government and out, waged a propaganda campaign of their own. To counter adverse sentiment by Culkin and other detractors they worked the background to create a picture of a huge, wonderful dam— the "biggest thing on Earth." Over time they built an image, like that of Paul Bunyan or Davy Crockett, based loosely on reality but inflated so that it became a caricature overshadowing truth. The fables of Grand Coulee Dam and the

Columbia Basin Project, spun out into feats of rhetorical hyperbole, became the myths of the public power and reclamation domains.[37]

In the midst of the Great Depression of the 1930s, New Dealers promoted Grand Coulee Dam as something big and wonderful, the largest structure on the planet, a reassuring reality built by the government despite the economic calamity. Federal agencies, particularly the Bureau of Reclamation, repeatedly pointed to this as proof that the nation, and certainly its government, was healthy and getting healthier. In magazines and newspapers, especially in then popular rotogravure sections on Sunday, frequent stories about the enormous Grand Coulee Dam became a focus for optimism about the country's future.[38]

The result was an unorganized and unplanned conspiracy of publicity in which Grand Coulee Dam and the Columbia Basin Project became beggars of superlatives. In 1942, when free-lance writer and future Oregon senator Richard Neuberger said, "Every one in America has heard of Grand Coulee," he may well have been correct. In the 1930s, with the attention given it in the mass media, including radio, it was difficult for anyone to miss mention of Grand Coulee. After all, as Neuberger himself wrote many times, it was the "biggest thing on Earth."

Historian Stewart Holbrook wrote, "Possibly nothing since the Erie Canal received comparable notice in the press which was kept more than well informed by a small army of Federal publicists."[39] Newspapers throughout the Northwest and elsewhere gave construction at Grand Coulee nearly daily coverage, often on the front page.[40] By World War II, the image of Grand Coulee Dam was significantly larger than the structure it represented. During the congressional funding battles of 1935-39, that image, combining the notion of a big government success story and aid to the jobless, was aimed at convincing Congress to spend the money. Publicists repeatedly badgered congressional representatives with the notion that, if nothing else, they were building "The Biggest Thing on Earth."[41]

*Newsweek*, in its first Grand Coulee story, called it "The greatest power, reclamation, and flood control project ever conceived." An article that labeled the entire undertaking a "crazy idea" went on to say that it will make "Muscle Shoals and Boulder Dam look like a set of child's building blocks." The authors included a map that showed Grand Coulee at about the location of Wenatchee, indicating some unfamiliarity with Western geography.[42]

In 1935 the *Seattle Daily Journal of Commerce* called Grand Coulee the "Largest Masonry Structure in the World" and added composite drawings showing the dam together with the diminutive looking Washington Monument and the national capital.[43] *Fortune* labeled it "The World's Most Monstrous Dam."[44] The *Seattle Times* announced that "The Great Pyramid of Cheops is Dwarfed by Coulee Dam." The pyramid, the article explained, held only 3,335,000 cubic yards of material whereas the low dam, then under construction, would hold twice that amount, and the high dam, when built, would double that again.[45] Other publications often imposed three or four renderings of the Great Pyramid in front of or near the finished dam. The same special issue of the *Times*, aimed at tourists traveling to the construction site, declared that "a tremendous cataract will be created when [the] river hurls itself over a precipice 550 feet high and 1,650 feet wide." Grand Coulee, it announced in large headlines, would unquestionably be "The Eighth Wonder of the World," and "The Greatest Structure Built By Man."[46] Picking up the theme, the *Saturday Evening Post*, in an article titled "The Eighth World Wonder," declared, "Superlatives founded in fact come readily to the tongue when speaking of Grand Coulee." That article also used the Great Pyramids as a comparison and added that, "unlike the pyramid, there is a useful purpose for Grand Coulee Dam."[47] "The Eighth Wonder of the World" became one of the dam's most familiar sobriquets.[48]

The Roosevelt administration struggled to convince everyone that although the dam was in a remote corner of Washington state, its construction, funded by federal dollars, benefited the entire nation. Government publications, such as *Reclamation Era*, regularly released data showing the construction contributions made by almost every state. In 1937, for example, of the forty states providing materials for the dam, figures ranged between a high of $10,501,131.22 from Washington state to a low of $5.40 from Louisiana.[49] Whether this made it a "national" effort was a matter of definition.

Of course Rufus Woods and his newspaper continued to exalt the dam. And when *Collier's* ran two very negative articles, one in 1934 and the second fifteen months later, Woods reacted.[50] He used the *World*'s front page, and most of two inside pages, to rebut the 1934 piece. "The man ought to be sent to jail save only that he can prove himself ignorant instead of malicious," penned Woods.[51] And the Wenatchee editor offered the author $100 if he could find an article in any reputable publication within the previous ten years that contained more untruths than his. The next year, when the second article in *Collier's* declared that the Columbia Basin held 1,250,000 acres of dead land bitter with alkali, Woods asked,

> The question is how does Collier's get away with it. The way some of its writers handle the truth is amazing. Such a policy wouldn't be tolerated by the management of even the smaller newspapers of the country for fifteen minutes.[52]

But Woods himself toyed with truth when he wrote that the Columbia Basin would shortly be a "Garden of Eden—a New Empire—the Most Modern Community in the World."[53] As they had during the struggle for the dam in the 1920s, both sides exaggerated just a little every now and then.

By 1935 the Bureau of Reclamation fully realized the good will potential presented by facilitating and entertaining tourists. To accommodate the 3,000 cars expected daily, government officials built parking lots and vista points, one on each side of the Columbia. The Northern Pacific Railroad included in its timetable the comment that those traveling in the Pacific Northwest should consider a side trip to the dam. There will be "no sight like it for the next 100 years," penned the author. By the start of the 1936 tour season, the Bureau of Reclamation had built grandstands at vista points on each side of the river. It provided places where some of the 40,000 monthly visitors could learn from guides who interpreted the swirl of activity below and delivered lectures on the art of dam building.[54]

The public provided a seemingly insatiable appetite for statistics about "the eighth wonder of the world." People loved to hear how many pancakes the 3,000 to 6,000 workers ate each morning at breakfast or how many miles of tubes ran through the dam. They devoured pictures of the great structure, as high as a forty-six-story building, just five feet shorter than the Washington Monument, and they saw drawings of the 12.5 million barrels of concrete or envisioned them together in a train 500 miles long. Most popular were comparisons with the Great Pyramid of Egypt, or two, or three, or even four of them. The number of pyramids depended on which article the public read.

When Franklin Roosevelt returned to Grand Coulee on October 2, 1937, he reflected on the statistics, saying,

> The easiest way to describe those figures is to say that this is the largest structure, so far as anybody knows, that has ever been undertaken by man in one place. Superlatives do not count for anything because it is so much bigger than anything ever tried before that there is no comparison.[55]

The President, harried by the flap his plan to expand the Supreme Court created, but buoyed by passage of the anti-speculation bill and congressional funding for reclamation, had begun his tour of the Western states in September.[56] He used the excursion to highlight New Deal projects including Bonneville Dam and Timberline Lodge, both of which he dedicated. When he arrived in Ephrata, 10,000 people greeted him. At Grand Coulee, Frank Banks showed the President a model of the dam and explained how it would look and function when complete. Bureau officials showed him around, driving in a large open touring car with Banks sitting next to the chief executive. Roosevelt took special interest in the residential communities around the dam, especially one named Delano.[57]

Roosevelt told the crowd,

> I think in the State of Washington there is a splendid understanding of one of the objectives in the development of these acres that are going to be irrigated. There are thousands of families in this country in the Middle West, in the Plains area, who are not making good because they are trying to farm on poor land. I look forward to the day when this valley, this basin, is opened up, giving the first opportunity to these American families who need some good farm land in place of their present farms. They are a splendid crowd of people and it is up to us, as a Nation, to help them to live better than they are living now.

Of course Roosevelt's trip, and the publicity it generated, were part of the propaganda effort and the President used the opportunity to emphasize the dam's national significance,

> It is a great project—something that appeals to the imagination of the whole country. There is just one other word that is worth saying from the national point of view. We think of this

as something that is benefiting this part of the country primarily, giving employment to a great many people in this neigh-
borhood. But we must also remember that one half of the total
cost of this dam is paid to the factories east of the Mississippi
River. In other words, it is putting to work in the steel centers
and other great manufacturing centers of the east thousands of
people in making the materials that go into the dam. So, in a
very correct sense, it is a national undertaking and doing a
national good.[58]

Roosevelt spoke from a special platform in the center of Mead
Park in Mason City, built so his car could drive directly onto it.
From there he gushed about the progress made in the three years
since his first visit. Bureau of Reclamation employees told Roosevelt
everything they could about the dam; small wonder that later he
remarked, "my head is full of figures."[59]

Harold Ickes came next, declaring that Grand Coulee "will virtually open up a new empire in the West."[60] Reclamation Commissioner John C. Page visited in December and recalled Roosevelt's
second inaugural when he declared: "President Roosevelt has said
that one-third of the Nation is ill-nourished, ill-clad, and ill housed.
Let us add also one-third is ill-watered."[61] Clearly the President and
his administration went out of their way to contribute to the publicity that helped build Grand Coulee's image.

The man who most enlarged and publicized Grand Coulee
Dam, however, was not a member of the Roosevelt team or a Reclamation Bureau employee. Already excited by New Deal goals, free-
lance writer Richard Neuberger produced dozens of articles
praising, explaining, and commenting on the successes of various
government programs in the Pacific Northwest.[62] Even before the
presidential trip, Neuberger had embraced Grand Coulee in a number of pieces. He hit his stride with an article that appeared in
*Harper's Magazine* in February 1937. Titled "The Biggest Thing
on Earth," it was one of two articles by him published that year
under that title.[63] He exaggerated a bit when he wrote later,

> So much electricity will be turned out that the plant will have to
> be run by remote control, so that the operators won't be elec-
> trocuted. Enough power will be generated at Grand Coulee to
> light two-thirds of the farmhouses of the nation.[64]

That year *Life Magazine* published a feature story by Neuberger, and his work often appeared in the *New York Times Magazine*.[65] In *Survey Graphic* he wrote,

> No country except the United States would have undertaken Grand Coulee Dam: a gigantic concrete barrier to make the upland desert bloom and provide farm homes for 50,000 families now on the move from submarginal acres. . . . Grand Coulee will irrigate land for as many farms as there are in the entire state of New Jersey. All this will be controlled by the government with speculation and profit subordinated to the general welfare.[66]

Other publications picked up Neuberger's enthusiasm and helped build the image of Grand Coulee Dam. *Popular Mechanics* called it, "The Biggest Thing on Earth."[67] In a later issue they called it "The Largest Thing ever Built."[68] *Scientific American* labeled it the "Monster Dam."[69] Carried away with his own rhetoric, another writer concluded, "It is so big you cannot see it. Can the ant see the elephant?"[70]

Most amazing, however, was an article that *China Weekly Review* ran in May 1941. It reprinted a story that appeared in Shanghai, although from an American source, which said, "From the standpoint of general interest, the Grand Coulee Dam is the largest monument ever made by man on this earth."[71] More than anyone else, those editors should have realized the truth in what Murray Morgan wrote when he said that Grand Coulee Dam is "the largest concrete structure in the world, one of the biggest things built since the Great Wall of China."[72]

To dramatize the dam's size, the Bureau of Reclamation made sensational comparisons. For example, the dam held enough concrete to duplicate the Empire State Building two-and-one-half times. If made into a sixteen-foot-wide highway, Grand Coulee would stretch from New York to Seattle, then to Los Angeles, and then back to New York. All of that would fill 50,000 box cars forming a train over 500 miles long.[73]

From the press coverage it appeared that Grand Coulee was the only significant dam being built by the government, but there were others. While one set of workers struggled in Washington state, another group worked equally hard on Fort Peck Dam in

Montana. Although an earth-fill rather than a concrete dam and less controversial than Grand Coulee, it was many times larger than its Washington sibling.[74] Fort Peck, by comparison, received almost no publicity. The copy writers dwelled on Grand Coulee, drawing it out until it was larger than life. Still today the exaggeration persists, part of the wonderful myth and image created in the 1930s. Even now historians and others label Grand Coulee the biggest thing on earth.[75]

Accomplishing the goal of those generating it, the barrage of publicity swamped the country and its legislators, possibly influencing the latter when they legalized and funded Grand Coulee Dam. By 1938 all that remained was for Congress to continue providing money. Representative Charles H. Leavy sounded a warning in October 1937: "The Grand Coulee Dam fight is not over." Selling cheap power would be the only thing that would convince the East that the project had merit. "FDR loves Washington and Grand Coulee Dam," he added, "but money comes from Congress."[76] Six weeks later James O'Sullivan, supported by the Quincy Chamber of Commerce and a few other backers, left for Washington, D. C., and another appropriation fight.

Criticism of Grand Coulee had hardly abated. *Public Utilities Fortnightly* published an article openly opposing "ill-considered Federal public works projects." In considering the economic justification of Grand Coulee the author wrote,

> It becomes very evident on a study of available data that the great cost of these structures cannot possibly be justified or repaid by irrigation alone and that irrigation must be subsidized by the sale of immense quantities of electrical energy at a profit.[77]

In January 1938, President Roosevelt, in his annual budget message to Congress, seemed to back away from the dam and reclamation. He asked for just over $14 million for Oregon and Washington irrigation ventures and announced that finishing reclamation projects currently underway would cost $600 million. The government should complete them, he declared, before any new starts. Still, Roosevelt called for greater flood control activity and less reclamation.[78] Supporters of the dam immediately switched and stressed Grand Coulee's flood control potential.[79]

Reclamation Commissioner Page requested $22 million for Grand Coulee over the 1938-39 fiscal year, but Roosevelt's budget director advised him to be satisfied with $13 million.[80] Toward the end of April, Congress passed the budget with the $13 million for Grand Coulee Dam. Project supporters had grasped at flood control, planning, resettlement, and just about anything else they could to justify the dam. But Grand Coulee benefited more from the economic downturn that started in 1937. Pressed by increasing unemployment, Congress approved the funding with less debate than in the past. Secretary Ickes then asked for an additional $15 million for the dam to "prime the pump" and Roosevelt requested a $3 billion relief bill.[81] Eastern Republicans, including the usual group led by Representative Culkin, unsuccessfully proposed amendments to exclude public power or government-sponsored electrification efforts from relief spending. Rufus Woods announced that work at Grand Coulee would stop on June 30 unless the dam received more money, and he urged people in Washington to write congressmen asking that they pass the pending relief bill.[82] Finally, on June 17, 1938, Congress approved the measures along with a $375 million flood control program. A few days later Secretary Ickes announced that $10 million would keep work going at Grand Coulee and later the administration increased the amount to $13,005,000, making a total of over $26 million available for 1938-39.[83]

In November 1938, Harold Ickes told the press that he wanted a large appropriation for Grand Coulee from the next Congress. Ickes hoped to secure as much as $28 million to carry construction through 1939-40. Three considerations motivated the Interior Secretary. First, conditions in Europe and the possibility of war presented a previously unforeseen need for the power.[84] Second, completing the dam more quickly would eliminate the potential for continued criticism by New Deal adversaries. Third, the sooner the dam produced power, the sooner it could start repaying its cost, which would also soften criticism. The 1938 elections brought more Republicans to Congress, giving added impetus to this strategy.[85] When the administration released its budget, Roosevelt requested over $49 million for reclamation with $23 million going to Grand Coulee. Rufus Woods urged the government to start serious work on the irrigation aspect of the project and speculated that the Works Progress Administration could put as many as 10,000 men to work digging ditches as soon as Grand Coulee started producing electricity.[86] But Woods quickly realized that the prospect of war, now used to promote the dam, meant postponing serious reclamation work. Although nobody knew it then, the war set back irrigation almost a decade.

As late as 1939, Grand Coulee opponents refused to admit defeat. Republican Culkin asked the Appropriations Committee not just to reduce funding, but to abandon Grand Coulee altogether. "We might build a fence around it and charge admission," he declared. "The pyramids of Egypt have been quite a popular tourist attraction for a good many years."[87] Nevertheless, in March 1939 the committee approved the $23 million request.[88] Culkin then took on Grand Coulee before the entire House. In a debate over the Interior Department appropriations bill he introduced an amendment to strike the $23 million for Grand Coulee. Charles Leavy pointed out that so far the government had spent nearly $100 million on the dam and that an incomplete project would be useless. He asked Culkin what he intended to do about that. "Well," shot back the New Yorker, expanding on his new theme, "the pyramids have been useless for a long time too." Culkin added that if completed, the 1,200,000 irrigated acres would result in a "merciless and killing competition" for Eastern and Midwestern farmers. And he attacked the power that the finished dam would generate. "There will be no one but jack rabbits and coyotes to sell the electricity to," he shouted.[89]

Representative John Elliott Rankin, Democrat from Mississippi, accused Culkin of receiving support from the power trust. On March 17, 1939, after a fight that lasted almost the full day, the House of Representatives rejected Culkin's amendment and preserved the $23 million for Grand Coulee. Eight Republicans from the West lined up with the Democrats in the voting. The *Seattle Post-Intelligencer* declared that in winning the vote, Grand Coulee had "surmounted its greatest obstacle."[90]

The House approved the funding bill in its entirety and sent it on to the Senate. On May 11, 1939, President Roosevelt signed the measure that seemingly guaranteed Grand Coulee funding through June 1940. By December, however, accelerated work at the dam rapidly exhausted the $23 million and the Bureau requested an additional $8 million to finance construction through the end of the fiscal year. The contractors declared that they would not carry the cost with the hope that Congress would provide the money later. When the money ran out, work would end.[91]

Representatives Leavy and recently elected Washington state Democrat John Main Coffee met with Roosevelt and pointed out that activity at the dam would stop on or about March 15, 1940 unless Congress provided more money. Further, they argued that if a flood should occur on the Columbia River before workers finished pouring concrete, it could damage the unfinished structure. Five days later Roosevelt asked Congress for a special immediate appropriation of $35 million for various projects with $8 million earmarked for Grand Coulee Dam.[92] The House Appropriations Committee cut the request by one million dollars and the full House, followed by the Senate, easily passed the measure.[93] The money sustained work until the next congressional appropriation took effect July first,[94] when Congress provided an additional $12 million. Michael W. Straus, then director of information for the Interior Department, predicted that Grand Coulee Dam, which he now included as a "defense project," would not have any difficulty with future appropriations.[95] In Washington state, Grand Coulee supporters echoed the sentiment. A group of Ephrata businessmen formed the Northwest National Defense Council and pressured the government to prepare for war by bringing vital industries into the "secure" Columbia Basin—an area obviously safe from enemy attack.

As Franklin Roosevelt won a third term, and Arthur Langlie became Washington's new governor, work continued at Grand Coulee Dam. In January 1941 Roosevelt said that the government must give preference to projects that could generate power for defense. He asked $8 million for Grand Coulee over the 1941-42 fiscal year.[96] In March, Reclamation Commissioner Page approached the House Appropriations Committee and asked for not only the $8 million, but also an additional $7.5 million for generators. Every day lost in installing the generators cost the country 165,000 pounds of aluminum for defense, Page declared.[97]

In April the Reynolds Metals Company announced it would start work on a large aluminum facility near Longview, Washington, the second major aluminum firm to build on the Columbia, lured by the inexpensive hydropower promised by the dams at

Bonneville and Grand Coulee.[98] Incredibly, that week the *Grand Rapids Herald* in Michigan carried an editorial stating that no market existed for any electricity generated at Grand Coulee Dam. The *Seattle Post-Intelligencer* rebutted the article saying that the Bonneville Power Administration already used all available power and needed more quickly.[99] The House Appropriations Committee then recommended $8 million for Grand Coulee, but it limited the amount available for the project's reclamation aspect.[100] The Senate increased the Grand Coulee appropriation to $11 million and included the first three generators that Ickes declared critical for defense. Late that spring Congress passed the bill. That appropriation, in the spring of 1941, was the last before the start of World War II. It completed the dam. Workers finished "the biggest thing on Earth" just in time for the war.

So the government built the dam—first using relief funds to create a small power project that provided work for the unemployed. Next Congress paid the bills. New Deal visionaries hoped eventually to turn the dam and its reclamation project into a model controlled economy that would demonstrate the benefits of planning. Finally, Grand Coulee became a defense project as the nation entered World War II. Whether or not the dam was ever really in danger of losing funding is doubtful. Certainly the dam had its enemies. But the Bureau of Reclamation frequently requested more than it needed, compensating for anticipated cuts. While political opponents loudly criticized the dam, their noise exceeded their clout. It is also true that had the Congress, from the first, been shown the high dam and its full cost, it might have balked at the project. Perhaps sensing this, Roosevelt deliberately began with the low structure and then slowly moved to the big dam that he wanted all along. Roosevelt would have dealt harshly with any who scuttled his pet project. Although at times, for political reasons, he appeared to distance himself from the dam, as the President must have realized, once started it would be almost impossible to stop the "Biggest Thing on Earth."

The towns of Grand Coulee and Coulee Center. *Courtesy Bureau of Reclamation.*

# Chapter Eleven
# Building a Community

Have you ever seen the American Dream walking? Well, I have. I saw it walking up the side of the Columbia River canyon, scribbling its puny etchings of squalor and cupidity against an austere backdrop of leaning cliffs and sudden chasms, and crooning the old American theme songs of Get Rich Quick and Something for Nothing. The dream is a town. It calls itself Grand Coulee; it is built of faith, hope, barn siding, and paper board; when I was there it was inhabited by about 1,500 people. It had twenty eating places, as many saloons, at least a half-dozen wide-open brothels, five grocery stores, two jewelry stores, a furniture store, two drugstores, two ladies' wear shops, three beauty shops, a proportionate quota of painless dentists and radio-repair shops, and six real estate agents.

*James Rorty*[1]

On November 23, 1936 the first issue of *Life* magazine appeared on the nation's newsstands. Its cover featured a picture of Fort Peck Dam in Montana while the story described not the dam but the construction community that surrounded it. "Franklin Roosevelt has a Wild West" and "The Cow Towns that Get Their Milk From Kegs," declared headlines over copy that depicted the six "frontier towns around Fort Peck" as being "wide open and as rickety as git-up-and-git or Hell's Delight."[2] The article's pictures looked exactly like a place just over 600 miles away in Washington state where similar activity encircled Grand Coulee Dam. They also resembled the government compound near Boulder City in Nevada in the late 1920s and early 1930s, or even Page, Arizona, near the Glen Canyon Project, in the late 1950s and early 1960s.[3] During their halcyon years all of these places embodied and epitomized the mood and ambiance of the "Old West." Similar to thousands of frontier mining camps, those communities grew quickly, thrived on money, or the hope of money, and withered when the money disappeared. Their citizens built shanty houses, drank and played in main street saloons, patronized stores that flourished during the good days, and raised their families as best they could. Most of them were good hard-working people often stained by the excesses of the few among them.

Not all frontier towns grew entirely overnight, prospered briefly, and then died forever. Many, like Grand Coulee, experienced cycles of expansion and contraction over the decades. Each new construction effort brought workers and periods of high living that briefly belied the depression. At the close of each job or during the winter, layoffs decimated the region's economy. This pattern marked Grand Coulee from 1933 through completion of the third powerhouse in the late 1970s.

The residents of existing communities that antedated Grand Coulee Dam saw its construction as an opportunity to improve their own economic conditions and sought ways to draw some of the money in their direction. This occasionally led to anything from friendly, to outright nasty, competition. The first towns affected by Grand Coulee Dam construction were the established communities of Almira, Wilbur, and Hartline, twenty to thirty miles away. They pictured themselves as access points on the way to the dam, and the Wilbur Commercial Club spent $200 on a huge sign that declared it the "Gateway to Grand Coulee Dam." Somebody who disagreed sawed the sign in half one night. Almira merchants declared a boycott on goods manufactured in Wilbur. The rivalry between those towns continued through the decade.

Coulee City residents, farther southwest, argued that freight routes to the dam must pass through their town.[4] When the government began building its railroad there, the town spruced up its historic buildings and attempted to make the place more attractive. Over 100 Wenatchee merchants took exception to Coulee City's gains and protested that freight routes to the dam should pass their way, as they offered more services.[5] In Ephrata, the locals began an annual celebration called "New Deal Day" that lasted for a few years. Frank T. Bell, Commissioner of Fisheries in Washington, D.C., who came from Ephrata, built the Bell Hotel there in February 1938. It boasted fifty rooms with air conditioners, and was full almost every night during the boom period. From 1933 until 1938 the town's population doubled. Down the road, Quincy paved its streets. In the other direction, Soap Lake experienced a building boom that stretched from 1933 to 1940.[6]

Greater changes occurred around the dam itself. The first influx of job seekers arrived well before work started. "We have received thousands of applications for employment and are getting them classified," wrote Frank Banks in August 1933.[7] "Drawn by the possibility of employment, hundreds of laborers have trekked to the site by automobile, train, horse-drawn conveyances and on foot, and most of them have been disappointed," commented the *Wenatchee Daily World*.[8] Ida Fleischman, who owned land in the coulee above the dam, found one miserable man, his wife, and baby, huddled together in a small cave. Others lived in tents, cardboard boxes, or tarpaper shacks.[9] The Columbia Basin Commission ordered unwanted vagrants to leave, but most stayed close, hoping to snatch any available job or to establish residence, giving them a hiring priority.[10]

Finding a place to live and buying necessities became a challenge in such an isolated area. During the first months of construction the preliminary contractors built bunkhouses, a cookhouse, and other buildings. Entrepreneurs added a few shops, a commissary, and even two service stations.[11] Mike Carr ran the Damsite Shoe Shop in an old cabin near the river just above the spot where Sam Seaton operated his ferry.[12] Clay Frank had a "cabin on wheels" from which he sold box lunches to the workers. Fred Warner's barber shop was so small that any customer not sitting in the chair had to wait outside.[13] Those early ventures disappeared when MWAK arrived and cleared the construction site.

The Bureau of Reclamation decided to regulate land use only in the town housing its own employees, which would eventually be called Engineers' Town. There the government imported truck loads of top soil, so that its village would become a "garden spot."[14] During the construction period professionals and their families populated the small community of around 500. With its green grass and tree-lined streets, and a reputation for being "spotless," it gained an air of superiority.[15]

At first the government employed no women at Grand Coulee. Banks argued that the community lacked suitable quarters for "female employees," but eventually he hired women to work in the project's administrative center as stenographic help and secretaries.[16] A notch below the engineers, but still above the people fending for themselves, were the residents of Mason City on the east, or Okanogan County side, of the Columbia. After the experience at Boulder Dam, where regulations forced all workers to live in the government compound, rumors spread that everyone at Grand Coulee must eventually reside in Mason City, where authorities could keep an eye on them. Frank Banks quickly announced that such would not be the case, although, as things turned out, the contractors did pressure some men to occupy the bunkhouses.[17]

The MWAK Company built Mason City. The northern half of the town held houses with one and two bedrooms. The southern half consisted of bunkhouses, a hospital, and cookhouses. In 1934 the contractor charged twenty-five cents daily for a bed in one of its forty-five bunkhouses, which held twenty-four men each. Nine smaller units, holding eight each, provided space for foremen. The housing assessment came on top of $1.15 for meals served in the cookhouse. Dues of forty-five cents weekly covered medical services, including a stay in the hospital for any illness or accident not connected to the job. State industrial accident funding took care of the rest.[18] In time the facilities grew and by mid-1935 over 100 dormitories lined the streets while the mess halls swelled to accommodate 900 men eating family-style at each sitting.[19] At first MWAK put up 220 "modern electric homes" for its employees with families. By the end of the second year that number increased to 360 houses, mostly prefabricated units. With power supplied by the Washington Water Power Company, Mason City became, in the language of the Bureau of Reclamation, a million dollar all-electric city, the first in the nation, and they built it in only three months.[20] The *Walla Walla Sunday Union Bulletin* reported, however, that just a short distance from the edge of town one easily found jackrabbits and rattlesnakes.[21]

When Consolidated Builders, Inc., took over the city during the second contract period, from 1938 until 1941, the managers made some changes. They added a department store, home delivery service from the grocery store, a laundry, a coffee shop, a new high school with a larger gymnasium, a hotel, and a service station. They also added "No Trespassing" signs along with regulations limiting access. In a letter to residents, the company urged people to "Keep the area secure." While they did not require employees to shop in the area, any stranger entering the town needed a permit, especially if he wanted to sell something.[22] They also put up a sign near the bunkhouses that read "Women not allowed."[23] Only the select few lived in Mason City. A point system based on seniority, size of family, rank, years of service, and occasionally influence with the company, determined who lived in what house.

Mason City changed the politics of traditionally Republican Okanogan County. In those early construction days, knowing well the hiring preference shown party loyalists, few admitted membership in the Grand Old Party and the Republicans had trouble finding someone to represent them on the local election board.[24] The Bureau of Reclamation interceded in one case, forcing the Democrats to allow both political parties to solicit campaign contributions from Mason City residents.[25] In time Democratic domination weakened and in 1940 the *Wenatchee Daily World* reported, "In the September 10 primary election Republican candidates showed surprising strength in Mason City which has traditionally been considered one of the strongest of the nation's Democrat strongholds."[26]

The towns in the valley below the dam were staid and organized. Above them, in the Grand Coulee, communities grew free from government planning or regulation. As the number of workers employed at the dam rose from 2,000 to around 8,000, Mason City held only a fraction of the total and the rest fended for themselves. While the New Jerusalem took shape in the valley, Sodom and Gomorrah flourished on the cliffs above, where life took on a very different flavor.

Most famous was the town of Grand Coulee. Its merchants officially dedicated the still-unincorporated community on September 15, 1933—the first of the "instant" cities recognized, although not yet authorized, by the Grant County commissioners.[27] A few days later work started on a new hotel and general store. Bootlegging flourished from the start, and when national prohibition ended late in 1933, a number of taverns quickly appeared. In April 1934, Grand Coulee opened its new jail and the authorities announced that they discouraged drunks in what seemed to them a "pioneer atmosphere." Still, the *Wenatchee Daily World* reported, "It is estimated that more liquor is consumed in Grand Coulee townsite per square foot than in any other part of the state." Rather than condemning this, however, the article concluded that it showed the need there for a large liquor store.[28]

Washington governor Clarence Martin announced that the authorities needed to curb gambling and vice. He added that many residents had children needing schools and protection from exposure to unsavory influences. A few days later, during the opening of the $25,000 Columbia Cafe, the state police "swept" the town, arrested four gamblers, and hauled them off to the Almira jail.[29] The *Spokesman-Review* reported,

> Echoes of the raids here Saturday night were heard at a meeting of the Grand Coulee Business Men's association last night when the visit of state patrolmen and the subsequent arrests were denounced in a resolution protesting the action. A second resolution criticized the "unfavorable and unwarranted publicity given Grand Coulee as a wide open town."[30]

A special Sunday edition of that newspaper carried an article with the headline, "Night life Festive":

> The streets, of course, are ungraded and minus sidewalks. The buildings are of frame with false fronts and paint is a refinement yet to come. . . . A few squatters are living in tents. . . . Law and order have not yet descended upon these dam site towns to any appreciable degree. They are literally "wide open." That may change one day soon, but at present one's conscience is one's guide and there is a moratorium on consciences. . . . Cafes abound. Beer parlors are everywhere. At night most of them blossom forth as cabarets, night clubs and dance halls. Each has an orchestra of sorts, perhaps only a piano and a banjo. A piano and a bass drum in the front window of one establishment might be regarded as the symbol of the unrestrained night life. The management employs hostesses as dance partners and dancing is free. . . . Picturesque names are bobbing up—such amusement resorts as the Mint, Frisco, Amuso, Purple Sage, Big Three and others of the ilk. A little grocery, much smaller than a moving van, carries the sign on the outside wall: "The biggest little grocery by a dam site."[31]

A few months later the *Seattle Times* reported, "Dancing girls—young, eyes alive to carefree life—await with their alluring evening gowns. Some have been here since December, not impatient, but knowing that sooner or later men must come by the hundreds."[32]

As the population jumped to over 3,000, the *Walla Walla Sunday Union Bulletin* commented that Grand Coulee was a place filled with "Tar paper dens of iniquity ready to aid boisterous, brawling recreations typical of workers in the 'wild west.'" It added, "proxided young women are standing around waiting for heavy construction work to begin."[33] The *Wenatchee Daily World* reported, "Colored entertainers seek night spot," and added that "The latest thing in the offing is a night club with a darky orchestra."[34]

With the MWAK contract, large-scale construction began late in 1934. Suddenly 2,000 men arrived, and they earned good wages; government inspectors took home $1,600 to $2,300 yearly and engineers collected up to $4,600.[35] Laborers pocketed from fifty cents to just over a dollar an hour.[36] As merchants in the gold camps mined the miners, entrepreneurs in Grand Coulee did the same—making profitable livings by providing whatever the men needed or wanted.

Grant County and the Columbia Basin Commission reluctantly paid for a small school. Students immediately overwhelmed the facility and the teachers held classes in staggered shifts to accommodate the overcrowding.[37] In contrast, on Main Street entrepreneurs put up a lavish ninety-room hotel that boasted twenty-two baths. Governor Martin showed up in 1936 and officiated at the opening of the competing Continental Hotel which operator Fred M. Weil billed as the "Newest small hotel in the state." Up the street, the less elegant "Dreamland" flophouse took in men for fifty cents nightly. Unconcerned with the drinking, gambling, and vice, which lubricated the town's economy, or with the shortage of schools, Grand Coulee residents paid more attention to the dust that rose from the unpaved streets in the summer, and turned to mud whenever it rained.[38] Merchants agreed to raise $1,000 for gravel, but it was hardly enough.

By the middle of 1935, the still-rising population of the entire area topped 12,000. Missing sewers and minimal sanitation concerned state and county authorities, while the residents still battled the mud. A local joke had a man, mired in the goo of B Street, complaining to a nearby friend, equally trapped, that he was stuck up to his waist and had no idea when he might pull himself free. The friend replied that the first man's situation hardly seemed so bad and pointed out that he himself was sitting on a horse! They called the mud "gumbo," and one enterprising resident made a comfortable living just pulling cars out of it. In the summer the dust was so bad the dance halls had to pause every hour or so and sweep the stuff out to keep it from choking the patrons. In 1937 the state budgeted $10,000 for more gravel and local officials went after additional funds. Two councilmen, carrying pillow slips, visited the famous B Street brothels wringing donations from merchants, girls, and customers.[39] The town council wisely sent the men in pairs, fearing that a lone agent might abscond with some or all the money.

As the dam's payroll increased, Grand Coulee's reputation worsened. The *Wenatchee Daily World* wrote early in 1935,

> The dam has attracted bootleggers, cheap crooks and dope fiends. There are shady women, certainly in the minority, and men who work off of them. Card sharks and men who play billiards better than anyone else here. Loud talkers and soft mix with one another. Grand Coulee is the melting pot of the West.[40]

Health officials found the "social disease problem at Grand Coulee 'deplorable.'" Dr. George Sparling, head of the local office of the state health department, announced that "everyone found with disease will be sent to the county jail in Ephrata and kept there until cured."[41]

Occasional waves of moral outrage swept the town. The State of Washington installed a prophylactic station and sent trucks with loudspeakers through the street asking people to fight social disease and to be more quiet in the bars.[42] Raids continued on illegal drinking and gambling spots and those arrested often found themselves in Ephrata, or even Wenatchee if the local jails were full. Grand Coulee's commercial club again protested that the situation was hardly as bad as the press indicated and argued that sixty-nine cases of venereal disease was hardly unreasonable in a population of over 10,000.[43]

In time the pattern of life in Grand Coulee took on a routine. Things were dull on Thursday because nobody had any money. But Friday was payday. Friday and Saturday nights saw the greatest activity, with things tapering off from Sunday through the next Thursday. When around-the-clock construction began, the town's "night life" never ended and one of the entertainment spots installed showers so that the men could come directly from work without having to stop and clean up first.

The most famous character in Grand Coulee was undoubtedly Tillie Pozar. Everyone called her Mom and she called the place she ran Mom's. Tillie Pozar was born in 1895 in Austro-Hungary. She immigrated to the United States in 1912 and married John Pozar in Nevada a few years later. They arrived in Grand Coulee early in 1934 and built a tavern that, at first, was quite primitive. Over time things improved and Tillie Pozar became one of the town's stock characters, complete with a bevy of fifteen to sixteen girls, called taxi dancers, entertaining her customers at the going rate of a dime a dance. Four of Mom's employees, whom she always claimed were only "good" girls, actually met their future husbands and married them in her establishment. Long after the boom period ended Tillie Pozar lived and worked in Grand Coulee. John Pozar died in 1955. Tillie continued, through hard times when her tavern did little business and when only a few old-timers sought her out to reminisce. She died on January 7, 1971 while on a business trip.[44]

During the summer of 1937 the morality issue reappeared in the news. Merchants along S Street, in what the *Wenatchee Daily World* called a rival with the older and more successful B Street, complained to Governor Martin about the situation in town. B Street, they claimed, was nothing more than a "vice belt."[45] Grand Coulee merchants argued, as always, that things were hardly as bad as the critics claimed and Mayor Frank Tierney declared the charges grossly exaggerated. But Governor Martin, feeling the pressure, appointed a commission to investigate. Chief William Cole of the Washington State Patrol visited Grand Coulee one Friday night but found little of concern.

More serious was free-lance journalist Richard Neuberger's visit. He told those he talked to that news about B Street and its "wide open" status had reached the east coast. One paper in New York, Neuberger revealed, referred to Grand Coulee as "The Cesspool of the New Deal."[46] Neuberger hardly helped defuse the situation. His first article about Grand Coulee in the Portland *Oregonian* called the place the "Toughest Town in North America," and he claimed that B Street brothels operated wide open:[47]

> Salvation and sin stroll on opposite boardwalks on the same street. . . . Along B Street fantastically named brothels operate

wide open. . . . On one side of the rutted road an evangelist cries out that he is saving souls; on the other a wizened little fellow with a whiskey breath whispers that he knows of a shack where for the modest price of one dollar, you can watch a young lady dancing in the nude. . . . Social diseases are a problem in Grand Coulee. Professional brothel owners complain that cut rate "taxi dancers" have made prostitution a sub-marginal occupation in Grand Coulee. "The damned dames who 'chippy' for 35 cents have wrecked it."[48]

When Neuberger reprinted the material in his book, *Our Promised Land*, published in 1938, residents who fumed at him in 1937 lost their tempers all over again.[49]

Some of the people who lived there called Grand Coulee the "Hidden City." It sat on the south side of the coulee itself, up a bit from the Speedball Highway that led to the southwest, toward Coulee City. Tourists seldom saw the place, as going there required a detour of a few blocks. Despite its location near the main road, Grand Coulee remained isolated and a place largely for the locals.

The town also retained a rustic flavor because some of the buildings never received much paint, and shacks, housing for those people looking for jobs or the newly arrived, always flourished on its edges. "For two to three years after work began at the dam, you could hear someone with a hammer building day or night," remembered Dave Rawe who ran a grocery store just off B Street.[50] Rawe had been in the hardware business in North Dakota when he heard that work in Washington paid as much as fifty cents an hour. He and his brother-in-law drove two truck-loads of merchandise cross country and started selling it the minute they arrived in Grand Coulee. They built their first store on C Street. Later Rawe went into the grocery business and in time he had three outlets going at once, operating under the motto, "If we don't have it—you don't need it."[51]

One part of Grand Coulee held its black population. "It wasn't planned," commented Rawe, "but they just lived together in that place."[52] White people came to call that area "Nigger-Town" and it also included a few of Hispanic background. One or two of the brothels provided black women for their patrons. The male residents worked at the dam and maintained a low profile, allowing them to coexist with their neighbors.

Like most boom towns, Grand Coulee had spectacular fires now and then. The Frontier Tavern burned down on a hot August day in 1936.[53] One week later a big fire took a cheap rooming house called Norman Rooms. The following July a Saturday night fire wiped out the *Grand Coulee News*, the public market, and nine other businesses. Hot weather and an acute water shortage exacerbated the problem, inhibiting firefighters and allowing the flames to spread. Exactly one year later, on July 28, 1938, one complete block along B Street burned, destroying every business there. Owners complained that high rates prevented them from buying fire insurance and the Reconstruction Finance Corporation refused money to help rebuild. After the height of construction, two more fires destroyed much of old Grand Coulee. In August 1946 two full blocks burned, destroying the west end of B Street, and on December 2, 1948 fire took four more buildings there. As the historic district disappeared, the *Wenatchee Daily World* lamented that "all of the old 'Hidden City' of the construction days is endangered."[54]

In 1938 the Works Progress Administration appropriated $108,000 to pave the streets. Seventy-five men started work early in March and by the middle of May, 280 poured asphalt and concrete. When the town could not provide matching funds the work stopped. In the end, the government spent only $34,000, leaving only some streets with a hard surface and the city in debt.[55]

Everyone who knew anything about Grand Coulee acknowledged B Street's unsavory reputation. But there was validity in the claim of the merchants that the stories exaggerated the truth. Typically, boom towns had districts where saloons and brothels operated and where the town's reputation rested on gambling and hard drinking. But in the larger picture, those isolated areas hardly represented the entire town and all of its residents. Behind the brash false fronts of the taverns and dance halls lived the bulk of the people, and most of them wanted only to earn a living and live peacefully.

Grand Coulee contrasted with Engineers' Town/Mason City, and the two communities displayed considerable enmity over the years. Those up in the coulee were jealous of the free electric power, water subsidies, and pampered living conditions provided for some, but not others, by "everyone's" federal government. At times this spent itself in athletic contests between teams from the various communities. The Grand Coulee Engineers sent their basketball team to play neighboring communities in friendly rivalries that masked considerable jealousy. The *hoi polloi*, relegated to the contractor's town on the east side of the river, and the unwashed masses who populated the boom towns up in the Grand Coulee, saw the engineers as aloof and snobbish. The groups divided into a caste system of sorts, with little social mixing. Clearly those who lived on their own in the coulee sat at the bottom of the hierarchy, while those who worked for the Bureau of Reclamation and commanded quality houses in Engineers' Town resided at the top. As Murray Morgan wrote in 1954,

> Relations between the two towns are somewhat strained. The government town has always been decorous to the point of dullness; the free town once was gay to the point of scandal. ... It achieved a certain celebrity as the site of the longest line of crib-houses known to the West since the gold rush.[56]

In addition to Grand Coulee and Engineers' Town there were two adjacent communities: Coulee Center and Coulee Heights.[57] These were located south of the highway while Grand Coulee, the "Hidden City," nestled on the northern or opposite slope. The other two settlements, which maintained a fierce independence for a few years, guarded their reputations, and saw themselves in competition with their less savory neighbor, despite vital economic links. Because of this political division, the smaller and slightly isolated Grand Coulee seemed even more the sin center catering exclusively to the prurient interests of wild, drunken, and lust-filled men. That reputation had minimal validity when applied to the larger community. Grand Coulee, always by far the smallest of the three towns, became the soiled step-sister living next to a pair of goodie-two-shoes with even more "holier-than-thou" neighbors down in the valley. The merchants in Grand Coulee justifiably felt threatened by their neighbors and for two years fought off consolidating the three towns for fear that, once in control, the more staid majority would shut down, or at least severely curtail, their livelihood. Most realized, to one degree or another, that it was the night-life, as much as anything else, that kept all of them going. And they used the age-old argument that if the men did not find what they wanted in Grand Coulee, they would find it somewhere else.

Inevitably, residents began to argue that the three communities should combine, form one government, and save money. Rival petitions circulated in the spring of 1935. One called for incorporating Grand Coulee as a separate city, while the other would unite all three towns.[58] The second finally gained enough signatures and Grant County commissioners scheduled an incorporation election for July 16, 1935.[59] Fearful that such a union might curtail business, Grand Coulee merchants went to Ephrata, the county seat, to block the election. They succeeded when, on July 10, the county court enjoined the commissioners on the grounds that the petitions contained technical errors.[60] In August, Grand Coulee, Coulee Heights, and Coulee Center officials formed a committee and attempted to reconcile differences, but the deliberations ended with each faction calling the other "obnoxious."[61]

At about that time Coulee Center suffered a bad fire that almost destroyed the Roosevelt Theater. None of the three entities could support an adequate fire department, and more than anything this helped draw them together despite their differences. A movement to install sewers also encouraged the sides to harmonize, although Grand Coulee resisted.

When officials raided gambling joints along B Street and seized eleven illegal slot machines, the harassed tavern owners suspected that those backing unification deliberately reported malfeasance to the authorities. When Grand Coulee businessmen heard that the harassment would continue indefinitely, they capitulated and dropped their opposition to incorporation. The state supreme court approved an election.[62] Emotions showed during the campaign, but the October 30, 1935 vote was amazingly light with 268 supporting incorporation and 223 opposed. In Grand Coulee 167 voted no with 48 favoring the move. In Coulee Heights and Coulee Center, 220 voted yes with 56 opposed.[63]

The election hardly settled the matter. The two factions fought again when they attempted to form a unified chamber of commerce. The move failed and two chambers emerged, one representing only Grand Coulee while the other claimed to speak for all, calling itself the Greater Grand Coulee Chamber of Commerce. A new contretemps centered on a move by Grand Coulee business people to have the city open up additional roads from the Speedball Highway directly to B Street. Up to this point, only one road conveniently left the highway and the merchants hoped that improved access would increase business. For exactly the same reason, business owners in Coulee Heights and Coulee Center opposed the move. Now that the towns had combined, they would have to pay for the very roads causing that loss. In the end, Grand Coulee won and a second entrance near Delano linked their town to the highway.

Surprisingly, city officials in the new town decided not to curtail gambling along B Street, but rather to legalize slot machines that they then hoped to tax. Even though Governor Martin and the State of Washington announced that such a move would be contrary to state law, the new Grand Coulee city council passed the slot machine ordinance anyway.[64] The state prevailed when patrolmen confiscated all the evident slot machines, thus nullifying the city ordinance in fact, if not on Grand Coulee's books. Grand Coulee tried again to legalize pinball machines in 1939, but that effort also failed.

One incident dramatizes the problems residents around Grand Coulee Dam faced as they attempted to form stable communities. On Memorial Day 1938 veterans groups decided to erect a monument dedicated to the workmen who died building the dam. They installed a tall granite shaft on a knoll just behind the community of Delano. A brass plaque listed the fatalities through May 1938. Grand Coulee city officials made an initial payment of $200 to the Colville Monument Works and planned a three-day celebration in Basin City to raise the remaining $1,500. Unfortunately, the fund-raiser lost money. For two years the Colville people waited, somewhat impatiently, for their money and then in November 1939 they threatened to repossess the monument.

In the spring of 1940 a business and professional women's organization announced it would save the monument and began a drive with considerable spirit, but few volunteers. As work on the dam neared completion and World War II started, workers left for more lucrative jobs elsewhere. In the end, the women raised none of the money. On Monday, June 3, 1940, the monument maker dismantled the column and reclaimed his stone.[65]

The trivial incident points up the problem that plagued Grand Coulee. Workmen came and went throughout the construction periods. Few considered it their permanent home and consequently most never developed any loyalty to the place. Of course others saw things differently and took considerable pride in the towns. Some remain even now, living near the dam they built, fiercely proud of their accomplishment and their community. But they are the small minority. The rest came and went, leaving when it suited their purpose. They wanted to earn money, perhaps sample the pleasures of B Street, and then go elsewhere. Under those conditions, the community never really gelled and it dissolved when the flow of construction money ended.

During the two extraordinary construction periods, roughly from 1934 through late 1937, and again from mid-1938 through mid-1941, the towns around Grand Coulee did flourish. The *Wenatchee Daily World* reported that 20 percent of the men came from out of state, but many claiming Washington residency had only recently established it.[66] The average worker was thirty years old with one-third falling between twenty and twenty-four. Fifty-seven percent were single. Of those who started at the dam in 1934, 689, or about 10 percent of the total then working, remained in 1937. Only 24 percent of those hired during 1936 still lived around the dam two years later.[67]

Many left during the hiatus after the MWAK contract ended late in 1937. With renewed work in 1938, nearly 2,000 had jobs again and the Grand Coulee commercial club staged a jubilee to celebrate the renewed prosperity.[68] Housing shortages returned and a building boom restored everyone's confidence.[69] It did not last. By 1942 the schools around Grand Coulee Dam reported fewer students.[70] Various groups petitioned the government to establish industries near the dam during World War II, and one company contemplated a magnesium plant three miles away, but it never materialized.[71] Transmission lines carried away the power generated at the dam, so no industry came to the immediate vicinity and few to the region. As the remaining residents planted victory gardens, the army took away many of the bunkhouses for use elsewhere. To simplify administration, Mason City and the Engineers' Town consolidated and together finally became Coulee Dam.[72]

Employees living in Coulee Dam complained about government control and bureaucratic restrictions. Murray Morgan wrote,

> In Grand Coulee a man can own his own house, but the housing is not as good as in Coulee Dam. So most of the bureau

people continue to live in the government town, a fact which led one Grand Coulee merchant to say, "They won't shoot Santa Claus down there, but they can sure bitch about some of his presents."[73]

In other words, the people in the valley wanted all the benefits of living in the town subsidized by the government without, at the same time, having to abide by any government-imposed rules and regulations. Up in the coulee, they wanted the government to evacuate Coulee Dam. In 1956, Coulee Dam won as Congress allowed the Bureau of Reclamation to sell the homes to their residents. On February 16, 1959, Coulee Dam voted to incorporate as an independent city. But it is not now nor has it ever been truly independent. The two dissimilar communities, Coulee Dam and Grand Coulee, always complemented and relied on each other despite the jealousy and contention. Neither could have existed without the other.[74]

When construction ended, the struggle to survive at all overwhelmed both communities. Up above, B Street shut down, more or less. Fires destroyed many of the old buildings, an irrigation canal took more of the town, and the main business district shifted to what had been Coulee Center and along the highway. On September 11, 1964 the Green Hut Cafe, a landmark that had stood just below Grand Coulee overlooking the dam, burned, symbolically marking the end of the construction era. The area around the dam underwent one more revival while contractors put up the third powerhouse, but that ended in 1980. Today only employees of the government and a few retired old-timers live in the communities of Coulee Dam, Grand Coulee, and the residential, modern, and prosperous newer Electric City, west of Grand Coulee and south of the highway. On B Street, the building that housed Mom's remained for years. Two dusty mannequins sat in the window, dressed like workers lounging next to a broken-down table having a drink and endlessly recalling the old days.

Right and left sides of the dam nearing completion, with the spillway under construction, 1939. *Courtesy Eastern Washington State Historical Society, no. L87-1.16775-39.*

# Chapter Twelve
# Henry Kaiser Builds the Dam

A visiting Egyptian engineer once asked a Bureau of Reclamation man how long a record of flow of the Columbia River the Bureau had. Thirty-eight years. The Egyptian showed consternation. How long did they have on the Nile at Aswan? Six hundred years! "Well," the reclamation man said, only briefly taken aback, "we did not feel that we could wait that long to build the dam in the Columbia River Basin."

*William E. Warne*[1]

Spanning the Columbia with the underpinnings for Grand Coulee Dam upset the predictions of those who argued that the engineers would never contain the river. And MWAK surprised itself and everyone else with how quickly and successfully it diverted the water and poured the foundation. All of that, however, paled in contrast to the second contract. From 1938 until 1941 Consolidated Builders Incorporated built Grand Coulee Dam, pouring well over six million cubic yards of concrete. In doing so, the contractor completed the most massive block of concrete ever laid. When finished, the job left the Pacific Northwest with the biggest hydroelectric development on the face of the planet and a facility that would eventually irrigate the largest single reclamation project ever attempted.[2]

Through the spring and summer of 1937 government officials worked on specifications for the second stage. They outlined completion of the dam and the left, or west, powerhouse; a foundation for the right, or east, powerhouse; and the foundation for the irrigation pumping plant. In September, Frank Banks announced that the Bureau of Reclamation would soon put the new job out for bid.[3] Then the American Federation of Labor (AFL) and the Congress of Industrial Organizations (CIO) demanded that the government negotiate a union contract and set wages before awarding the job.[4] The Interior and Labor departments worked to establish scales satisfactory to the unions without, at the same time, saddling any prospective bidders with unreasonably high wages. Unlike the first contract, under which MWAK supplied most of the materials used in the dam, the Bureau of Reclamation decided to be the supplier on the second job. The bid, then, covered labor almost exclusively, highlighting the need to settle the wage issue. To some observers it began to look as if nobody would tackle the difficult task.

Things broke on December 9 when Henry John Kaiser met with MWAK head Guy Atkinson in Atkinson's Spokane hotel room. They realized that no company alone could manage the bid within the limits set by the government and still pay the wages demanded by the union. Kaiser proposed a spectacular merger of companies that would yield a 20 percent cost reduction and meet the Bureau's estimate. He and Atkinson quickly formed what they initially called the Interior Construction Company, and that day they struck an agreement with the AFL.[5] Under that contract unskilled laborers would receive seventy-five cents an hour, or fifteen cents above the government minimum. Skilled laborers would receive up to $1.65 an hour, again fifteen cents above the government minimum.[6] This satisfied AFL requirements.

The arrangement between Kaiser and Atkinson, which trade journal *Pacific Builder and Engineer* called a "stunning last minute amalgamation," included MWAK (Silas Mason Company of New York, Walsh Construction Company of Iowa, and Atkinson-Kier Company of California); the Kaiser Construction Company of Seattle, Washington; the Morrison Knudsen Company of Boise, Idaho; the Utah Construction Company of Ogden, Utah; the J. F. Shea Company, Inc., of San Francisco, California; the Pacific Bridge Company of San Francisco; the McDonald and Kahn Company of San Francisco; and the General Construction Company of Seattle.[7] With Henry Kaiser as its president, the so-called Six Companies (really seven) now united with the very firms that had snatched away the first contract four years earlier. Their bid of $34,442,240 won the contract to finish Grand Coulee Dam. As it turned out, a second company also submitted a bid. The Pacific Construction Company, also a conglomerate, presented an offer of $42,185,802.50.[8]

As before, the government held the bid opening in the Spokane Civic Building's auditorium. The *Spokesman-Review* reported:

> The crowd which packed the auditorium of the Civic building was electrified when Mr. Banks announced that the first bid opened was by the Interior Construction, Henry J. Kaiser, President, Guy F. Atkinson, vice president, Charles A. Shea, vice president, and Tom Walsh, chairman of the board. . . . Final details of the merger were not completed until a few hours before the bids were opened.[9]

The Six Companies, which had built Hoover Dam, the Golden Gate Bridge, Bonneville Dam, and which would eventually build the Alaska military highway during World War II, finally became a part of the Grand Coulee effort. At last Henry Kaiser captured the prize he narrowly missed a few years before.

Kaiser's son, Edgar Fosburgh Kaiser, became the general manager and Clayton P. Bedford assumed the role of general superintendent. Horace A. "Happy" Parker continued as construction engineer for the Bureau of Reclamation, working directly under Frank Banks who remained as the Bureau's supervisor. These were the men who oversaw the actual day-to-day construction of the dam.

To complete legal formalities, officials of the Interior Construction Company, formed under Nevada law, held a meeting in Reno early in February. There Kaiser and his fellows complied with a request from Harold Ickes, who feared that people might confuse the corporate name, Interior Construction, with the Department of the Interior. Ickes asked Kaiser to rename the corporation and after a brief discussion they selected, and for the next four years used, Consolidated Builders Incorporated, or CBI.[10]

The Department of the Interior and the Bureau of Reclamation accepted the foundation from MWAK on March 12, 1938,[11] but the government did not officially order CBI to proceed until March 18, 1938.[12] The AFL assured CBI that it had over 3,000 men standing by, but the company announced it would hire only 500-700 for the first eight months, disappointing the unemployed laborers who suffered during the hiatus between contracts. State authorities worried about living conditions at Grand Coulee if there was no work until the fall.

A few laborers were lucky. Small crews undertook preliminary tasks such as dismantling the suspension bridge that had carried aggregate across the Columbia River. CBI took apart and, on the east side of the river, joined the two concrete mixing plants, creating a new facility almost twice the size of the originals. In this modified plant, that trade journals called a "Siamese twin," they would produce all the remaining concrete.[13] A new railroad bridge provided easier access to the east side. CBI then removed the original construction trestle and built a new one at a higher level.

In Mason City, CBI added, among other things, a high school building. It refurbished some of the dormitories and added fireplaces and chimneys to fifty-nine houses, ending the community's distinction as an all-electric city. CBI explained that the Washington Water Power Company, supplier of power during MWAK construction, had raised its rates, making electric heating less desirable. CBI managers encouraged residents to reduce costs by finding wood for household heating.[14]

Despite gloomy early predictions, CBI had 1,665 men employed by mid-May 1938. But the accelerated activity hardly accommodated everyone who wanted a job. Bitterly disappointed, many blamed the union for their troubles, and a few expressed their feelings in letters to various government agencies or to President Roosevelt directly. This one to Secretary of Labor Francis Perkins is typical:

> Dear Madam, I am writing you in regards to the deplorable conditions on the Grand Coulee Dam at present. I went to work as a jackhammer man here October 3, 1935 and worked about 8 months as a jackhammer man and then went to pipefitting. Finally working up to Foreman of the pipefitters. I was laid off some time in Oct. 1937. In the meantime I built a home here. For seven months I have been out of work. We have lived up what little we had saved and have exhausted our credit for groceries. At this moment we are on the verge of starvation as we haven't enough groceries in the house for one meal. Now here is where the trouble comes in. The Company has signed a closed shop agreement with the American Federation of Labor and all men are hired through the union. The A. F. of L. wants $25.00 for me to join the union. Can you tell me Madam how a man that has a family on the point of starvation is going to raise $25.00 to buy his self a job? I understand the Dam was started

to give men unemployed employment. If this is the case why is the A. F. O. L.. allowed to make us buy our jobs from them?[15]

Most of the complaints went to the Reclamation Bureau's office in Washington, D. C., and each writer received a reply. Reclamation Commissioner Page advised Frank Banks to prepare a "brief summary of the facts" that the government could send to each correspondent.[16] Others received letters from an assistant commissioner expressing regret and explaining the Bureau's inability to help as "the matter of employment is the sole responsibility of the contractor and the government cannot interfere as long as the contractor complies with the provisions of the contract."[17] Despite the often-repeated promise that the dam would provide work for tens of thousands, the project never achieved such levels of unemployment relief. When challenged, the government sidestepped the issue. Although it dramatically aided the economy of the immediate area around the dam, albeit temporarily and sporadically, Grand Coulee's wider impact on the depression, despite propaganda to the contrary, was a major disappointment.

Clearly the contractors favored some men over others, especially rejecting those active in the CIO. There is evidence that, as before in the MWAK contract, trusted employees leaving jobs in other areas, such as Bonneville Dam or the Boulder Project, found it easy to secure positions at Grand Coulee while people living near the dam remained jobless. Despite how things appeared to men living near Grand Coulee, however, the problem lay not entirely with the union or the contractor but with the fact that, at least during 1938, there were fewer jobs available. CBI foresaw this and repeatedly advised the unemployed to look for work elsewhere.[18] For those who had jobs, working conditions were excellent. For those who did not have jobs, life in remote eastern Washington was miserable.

Nevertheless, men and their families stayed near the dam. After 1937 the national depression worsened, adding to the number of jobless both regionally and across the country. Rufus Woods and others then urged the government to direct the Works Progress Administration to create jobs around the dam to alleviate unemployment. Woods, who had often predicted many jobs while promoting the dam, now estimated that the government could create anywhere from 1,000 to 10,000 positions digging ditches and building roads and trails. But Frank Banks rejected the plan, stating that he already "had entirely too much trouble" with WPA workers.[19] Despite the frequently repeated myth that building the dam would create jobs, unemployed men lingered around Grand Coulee throughout its construction and until World War II ended the depression.

Building Grand Coulee Dam remained essentially a task for men. An unwritten but often-stated rule denied women access to the construction area for any reason; men then considered their presence around heavy work a jinx. For the most part, women living in the area passively accepted what, in later years, they would call discrimination. But in the late 1930s Americans did not frequently challenge the lines that separated the realm of men from that of women.

This tacit acceptance of separate domains led to a minor but amusing squabble in the late summer of 1938. The huge construction area required an extensive telephone system linking its various parts. A total of 145 lines led to the central switchboard in the field office. Women operated these switchboards and eventually some seventy-five served either as operators or in other office jobs. When CBI needed a secondary switchboard inside the work area, the company sent a woman to operate it. The men resented the incursion into their space, but when the company suggested replacing the woman with a man, the other women employees argued that the position was a woman's job and, despite the location, a woman must fill it. The contractor agreed so, despite male grumbling, the one female, complete with hard hat, continued to operate the switchboard. Others joined her over the years. That one exception is the only recorded example of women working within the heavy construction zone on Grand Coulee Dam.

As had MWAK, CBI hired a number of blacks despite resistance from foremen and other workers. The National Association for the Advancement of Colored People (NAACP) monitored employment at Grand Coulee on into the CBI contract period. It occasionally filed complaints with the Interior Department's Adviser on Negro Affairs in Washington, D.C. Generally the comments alleged that some of the foremen at Grand Coulee strictly segregated the crews and fired black workers first during slack periods. When work resumed, white men often replaced blacks.[20]

Presented with the charges, the AFL denied any discrimination:

> The colored men employed on the Dam are members of and have been cleared through the Hodcarriers [*sic.*] Union Local #1151. The Business Agent of this local, W. B. Cook, has told me several times that he considers the welfare of these men just as important as any of his other members, and according to the constitution of their International Organization.[21]

The Business Representative of the Hod Carrier's Union added,

> I do not care to leave the impression that I think the colored folks are wrong in their contentions entirely but I do wish you to know that I have formed the opinion by personal contacts with the colored people that they are overly sensitive in regard to the complaint voiced by their Association.[22]

Frank Banks arranged for meetings between representatives of the NAACP and the contractor, who agreed to give black workers equal employment opportunities.[23] One year later about twenty blacks worked at the dam among the 3,000 to 4,000 whites.[24] While that hardly represented anything approaching the percentage of blacks in Washington state, it did, as Frank Banks pointed out a few years before, seem equitable to CBI, considering the small black population in the eastern part of the state. It complied, at least, with the contractor's definition of "equal opportunity." At Grand Coulee the government improved on its record at Hoover Dam, where blacks had been excluded. In Washington state, the Bureau of Reclamation and the contractors made an effort to hire both blacks and Indians. But the subtle resistance, the friction between the races, and the rationalizations showed that much more would be needed before minorities received fair treatment.

Despite the discontent among the jobless and the racial friction, CBI proceeded with the task of building the dam. It erected a new high trestle that allowed cranes access to all parts of the structure below. When complete, the single span stretched 3,600 feet across the Columbia river and stood over 200 feet above the completed dam foundation, with its top deck at 1,180 feet above sea level. It required 5,700 tons of steel besides parts salvaged from the old upstream structure. The trestle deck held four standard-gage railroad tracks eventually carrying seven large cantilever cranes of the hammerhead type that operated on gantries supplemented by three smaller whirley derricks. Bethlehem Steel completed the trestle on September 20, just over twelve weeks after starting what by itself amounted to a million dollar project.[25] The *Wenatchee Daily World* dubbed it the costliest bridge in Washington state and pointed out that concrete would eventually bury much of it, as it had its predecessors.[26]

With some fanfare, CBI prepared to place its first load of concrete. Company officials scheduled the pour for Thursday, July 21, 1938, but technical problems with the rebuilt mixing facility caused a two-day delay.[27] Limited amounts of concrete emerged over the next week, and then CBI closed down the operation and made changes to improve delivery system efficiency from the plant to the cranes to the waiting blocks below. They resumed pouring late in August and moved from two to three shifts daily.[28] Company officials predicted that they would have 17 percent of the dam complete by the end of the year, and 55 percent by the end of 1939. With a new plan that called for rapid pouring, the company estimated finishing the dam sometime in 1941, over a year ahead of schedule.[29] Not only the size, but also the speed of the operation would be unprecedented.

By the end of July, twenty-one subcontractors worked at different tasks around the dam. One of the largest, Western Pipe and Steel Company of San Francisco, began installing penstocks and pump-inlet pipes around which CBI poured concrete. The huge steel penstocks would eventually carry water from behind the dam to the turbines in the left and right powerhouses. On the west side, behind a wing dam, inlet pipes would take water to the pumps that would someday push it up the hill above the dam and into the equalizing reservoir, the first step on its way to the waiting Columbia Basin Project. Besides supplying the material, Western Pipe's $1.5 million contract also required over nine miles of heavy welding needed to hold the rings of steel together and withstand the pressure of the surging water.

Consolidated Builders Incorporated concentrated on mixing and pouring concrete on the blocks, following roughly the same routine as its predecessor. Different was the fact that during the

MWAK contract period, cofferdams diverted the Columbia River, allowing access to large sections of the foundation. On the second stage, that process changed. MWAK had left diversion slots in the surface of the foundation and CBI built eight large portable gates, each fifty-two feet wide and thirty-five feet high and weighing seventy-five tons. These gates, rather than cofferdams, now controlled the flow of the Columbia River. They cut off the water and allowed the contractor to pour concrete into the depressions, raising the individual blocks there in standard five-foot lifts until they were level with or higher than existing blocks. This created new depressions through which the river then flowed.[30] Nine sets of steel guides, also portable, fixed to the front of a given slot, channeled the gates when an A-frame steam hoist, on a barge behind the dam, dropped them into position, cutting off the flow of water through a particular slot. Engineers, moving up from level to level, carefully raised and lowered the gates to avoid sudden changes in the river below the dam that would have inconvenienced ferries, power plants, and intakes for domestic and agricultural water and irrigation systems. Water from behind the dam also moved through outlet tubes—tunnels that penetrated the structure from the upstream to the downstream side. These, and the water tumbling through the slots, provided an ongoing display for thousands of tourists who watched from the vista houses above.

Thirty-six hours before any scheduled pour in a given block the workmen stripped away the forms from the previous pour, cleaned the concrete surface, and placed bolts to hold the next tier. Carpenters fastened the remaining bolts, around which they fixed the new forms. First they cleaned the wood surfaces and applied a special oil that prevented a bond between the wood and the concrete, allowing easy removal later. Next workmen laid the cooling, grouting, and drainage pipe. Then they poured the next five-foot-thick layer of concrete. At forty-five degrees, cold Columbia River water ran through the pipes and lowered the temperature of the setting mass in three to six weeks rather than the many years required without it. In all, the entire structure liberated heat equivalent to burning 30,000 pounds of coal.[31]

Two government inspectors oversaw each stage of the work and filled out "clearance cards," with one checking thirty specific items guaranteeing quantity and location while the other did the same for quality and workmanship. A government inspection clearance was good for three hours, and if no new pour occurred in that time, the inspectors repeated the procedure.[32]

Ironically, the cooling process began about the same time that the weather turned cold. It was, by then, a familiar routine, with winter ending the ability to lay concrete. By Christmas the contractor had laid off 1,600 men and on January 10, 1939 all work stopped until the return of warmer temperatures. Only a few crews, employed directly by the Bureau of Reclamation, continued placing drainage tunnels on the west side, near the vista house, where slides still threatened the dam below. Ice covered parts of the new lake, now sixty feet deep and extending twenty miles upstream.

After the winter, CBI resumed pouring concrete on February 20, 1939. By April the contractor employed over 5,500 men and the pace of work increased dramatically. Electric shovels began lifting seven to eight cubic-yard-loads of aggregate onto the waiting conveyor belts that carried the material to the concrete mixing plant. At peak production, the equipment handled over 3,000 tons hourly. Compressed air shuttled up to 1,000 barrels of cement an hour at over 100 miles per hour through 6,000 feet of tubes to the same location where the ingredients came together and, after the addition of water and sand, became the finished concrete.[33] Each mix initially took five minutes and produced four cubic yards. The contractor's engineers reduced the mixing time, despite concern among concrete industry experts about the quality of the finished product. Through modifications in the equipment, such as changes in the shape of the mixer blades, they eliminated fifteen seconds here, ten seconds there, and eventually brought the time down to an amazing two minutes.[34]

The contractor and the Bureau of Reclamation selected May 25, 1939 as a special day. By reducing the volume of concrete laid during the previous week, most of the dam's surface was ready for a new layer. Laborers began at midnight and worked without stopping for twenty-four hours. Every 4.18 seconds one new cubic yard of concrete rested somewhere on the dam which, by the end of the day, had accumulated a total of 20,684.5 cubic yards. The record achievement made headlines in most of the trade journals and local newspapers.[35] Of course, it was an unusual event set up under extraordinary conditions and CBI never normally approached that volume. But the contractor and the men considered the trouble and effort worth it to add one more "greatest" accomplishment to Grand Coulee's record.

Not long after the "big pour," some of the laborers had misgivings about the feted event. The government already had trouble securing enough cement and the unusually large pour exhausted the ready supply. Within a week CBI laid off 200 workers with more being discussed over the following days. By June 12, work stopped almost entirely.[36] The Bureau of Reclamation began searching for cement outside of Washington state, bringing protests from local cement producers who argued that the shortage was temporary and that the government should keep its promise to buy exclusively from them. The AFL countered that the government should go wherever necessary to maintain the work and sustain jobs.[37]

By the middle of July shipments of cement arrived from California, augmenting those coming from western Washington state. In the meantime, however, the Columbia River rose and inundated the dam's spillway section making work there impossible. The engineers opened all the diversion slots, providing tourists with fifteen waterfalls, each fifty feet wide, tumbling over the face of the center section. CBI reduced pouring to five days a week through early August, and then increased to six days.[38] On August 23 the company resumed the original seven-day schedule. By the end of August CBI was swamped with cement and had trouble finding places to put it.

The most dramatic single year at Grand Coulee was 1939. The dam steadily grew in size as the blocks rose from the foundation to their ultimate height. Crews worked around-the-clock moving the loads of concrete and completing the other tasks required in building the great structure. The sight of over 2,000 men (on each shift) coming and going, doing different jobs, each purposeful, appeared uncoordinated to the casual observer. Although it is cliché, the din of machinery and the movement of workers and equipment throughout the project area, creating an incessant pitch of activity, resembled a giant pulsating ant hill. At night the scene took on an ethereal and psychedelic quality. Thousands of floodlights illuminated the dam and its surroundings, producing new and unfamiliar images and shadows, an almost unreal fantasy land. The east and west sides of the dam rose while the spillway remained low until winter when the Columbia carried less water. During September 1939, CBI added over 400,000 cubic yards of concrete, breaking MWAK's record for a thirty-day period. In 1939 CBI placed 3,650,000 cubic yards of concrete, or roughly four times the amount poured during all of 1938.[39] In winter the diminished volume of water, diverted through the lowest level of outlet tubes, allowed unrestricted access to the spillway section.[40] Viewed from the vista houses above, the panorama, with the oversized dam in its center, never failed to impress visitors.

While the bulk of laborers continued laying new concrete, the grout crews worked behind them, filling in the cracks and welding the independent columns into one solid mass.[41] They pumped grout into expansion joints, spaces between blocks, and any openings between the five-foot lifts. They also placed drainage pipes to carry away seepage, for designers had accurately predicted that the dam would leak, despite the grouting.

Whether speed contributed to the accident and death rate is debatable. Every fatality, felt by the men and their families, added to the human dimension that accompanied the technical success and became part of the incalculable cost of the dam. In a 1938 article the *Spokesman-Review* captured the personal drama and tragedy that haunted every large construction project:

> Virgil L. Smith, 33, Mason City, CBI concrete walking foreman, died from the result of his injuries and R. A. Dadka, 31, Bureau of Reclamation junior engineer, Electric City, and Ed Nanpooya, 27, Nespelem, CBI Indian concrete laborer, lie seriously injured at the Mason City hospital following an accident which occurred at block 35 on the dam at 3:20 a.m. today. A huge concrete bucket had been emptied and was being hoisted up to the deck of the trestle when it struck a catwalk bridge upon which the men were standing, turning it over and hurling them to the concrete 25 feet below. The three men were rushed to the hospital where Smith died at 6:40 from a fractured skull and internal injuries. Smith left a widow and two children aged 2 and $3^1/_2$.[42]

The construction company held safety meetings and stressed the need for care on the job. Nevertheless, men fell from catwalks and cement ledges, or were struck or crushed by falling objects.[43] By far the most dramatic single accident happened on the afternoon of November 1, 1939. Albert A. Priest had just finished the day shift operating one of the "whirley" cranes. Below him his son, twenty-eight-year-old Alton E. Priest, along with forty-year-old Henry Malloy, also finished and stood together on a catwalk near the center of the dam. Two other men joined them and talked quietly as they prepared to go home. The younger Priest and his father worked together as a team with the older man operating the crane while the younger man served as signalman indicating when to raise, lower, or maneuver the bucket from side to side, a job usually indicated by the term "whistle-punk."

Albert Priest stood and relinquished his seat at the crane controls to his replacement. But Priest had forgotten that the bucket, then hanging from the crane, was full of concrete and ready to drop into position below. An empty bucket weighed three tons, while a full one exceeded eleven tons. As the new operator released the controls, assuming that an empty bucket would rise, the horrified pair watched the heavy mass plunge downward toward the top of the dam. In agonizing helplessness, Albert Priest stood next to the crane window knowing exactly what would happen. The bucket crashed into the catwalk on which his son and the three other men stood. The force broke the wooden members and catapulted the four men thirty feet onto one of the adjacent blocks. Alton Priest, a bachelor, died instantly as did Henry Malloy, who left a widow and two children. The other two men survived, but with severe, debilitating injuries.[44]

Eighteen men died during 1938 and 1939. Of the 499 accidents reported from the start of the CBI contract through January 1, 1940, 140 resulted from falling people and 71 from falling objects. By the end of 1939, about sixty-five people had died building Grand Coulee Dam.[45] The old stories of men buried in the concrete resurfaced with added vigor despite Bureau of Reclamation guides who tried to deny them. A new rumor claimed that white survey markers, actually targets used to insure that the finished dam would be straight, visible to onlookers on the hills around the construction site, were memorials to the many who lay inside the massive structure. Others said they were Indian graves.[46]

The contractor took over the hospital in Mason City. There a plan to guarantee medical care for employees, initiated by CBI, formed the basis for and start of the Kaiser Permanente Medical Care Program (KPMCP).[47] Seldom did CBI fill all sixty beds in its hospital in Mason City. Nevertheless, a spate of accidents now and then taxed the facilities. Despite the carnage, the Bureau of Reclamation argued that its safety record at Grand Coulee was better than the average of all the other work it oversaw. As at Boulder Dam, the huge size of the project, involving so many people, made the accident statistics appear so dreadful. Taken on the basis of deaths per million dollars spent, or the average per worker on the job, the Bureau was correct in stating that at Grand Coulee it achieved a notable safety record.[48]

Consolidated Builders constructed Grand Coulee Dam so quickly that the Bureau of Reclamation revised its schedule. Over a year earlier than anticipated it placed orders for the west powerhouse's first turbines and generators. Newport News Shipbuilding and Dry Dock Company in Virginia received the $1.5 million contract for three record-sized turbines. For $2,627,970 each, Westinghouse put together the accompanying 108,000-kilowatt generators of equally unprecedented capacity. Looking small fifty years later, the machines at the time dwarfed anything ever built.[49] In addition, Westinghouse also constructed two smaller 12,500-kilowatt generators, known as house units, which would supply the power needed to operate equipment at the dam itself. By December 1939 the empty shell of the west powerhouse stood essentially complete and ready for the machines.

The winter of 1939-40 was unusually cold and, once again, canvas canopies covered the freshly poured concrete. A barge, dubbed the "Iron Duke," with nine fuel oil-fed boilers, floated on the lake behind the dam, pumping steam into the tents. CBI began laying off workers, dropping the force by one-third early in November.[50] On December 16 the company further reduced concrete pouring and a few days later suspended it entirely until the following March. Total employment dropped to 500 but when the weather unexpectedly warmed it climbed back to 2,000. Some of the workers fought new slides on both the west and east sides of the river. Through 1940 and into the winter of 1940-41, crews removed another million cubic yards of the temperamental viscous earth. More slides required relocation of the west vista house in September 1940.[51] Despite the ongoing effort, the engineers never fully subdued the slides that had plagued construction from the beginning.

That winter a few men built a small final trestle, higher than the main unit, needed to top-off the dam.[52] They resumed concrete work on February 20, 1940, beginning first with the west powerhouse and the pump plant. Throughout 1940 the men concentrated on the spillway crest, the bridge piers that would carry a road across the top of the dam, and placing the drum gates that would regulate the spillway.[53] Through June and into early July high water poured over the center sections, forming a waterfall now higher than Niagara and, in the opinion of some, equally spectacular.[54] Workers poured the ten millionth cubic yard of concrete on June 21, 1940, and later CBI announced that only about 500,000 cubic yards remained.

As water rose in the reservoir, it exerted pressure on the dam's upstream face causing a deflection or bend of five-eighths of an inch, an amount that increased to just over one inch when the water reached its highest level. To allow for the shift, or so-called "twist effect," the designers required the builders to leave five narrow slots that ranged from a foot to just over a few inches wide running through the width of the dam. As the structure settled into its final position, work crews removed sand that temporarily filled those slots and replaced it with grout.[55]

That June the government began expediting delivery of the electrical equipment as a national defense measure.[56] Reclamation Commissioner John Page announced that despite rumors in the press, there was no special significance in the increased work, but few believed him. By August 2, 1940 thirty-two blocks on the east side of the dam and two on the west side had reached their ultimate height, an elevation of 1,311.08 feet above sea level. By the end of the summer, CBI had poured most of the concrete, leaving only small jobs such as finishing the powerhouse, installing the four elevators inside the dam, and placing the aluminum railings along the top. In September 1940 the contractor reduced the work schedule to five days a week. Many workers left for Friant Dam in central California, just as they had come from Boulder and Bonneville dams. Others went to west coast shipyards and similar defense-related positions. The contractor closed the mess hall in Mason City. With the coming of the cold weather, concrete placement stopped altogether on November 10 with the dam lacking only 35,000 cubic yards.

Through the late fall and winter of 1940-41, 1,000 men installed the eleven drum gates, each 115 feet long and 28 feet high, across the top of the spillway section. On January 1, 1941, CBI had completed 93 percent of its contract in 70 percent of its allowed time. Invigorated by the seeming inevitability of war, and the prospect of lucrative defense contracts, CBI stepped up the pace to complete the job.[57]

The smaller station service generators, or house units, produced the first power. On January 21, 1941, Frank Banks unceremoniously started the first machine on its maiden run. The second generator followed on Tuesday, February 15.[58] "Coulee Dam Power for Defense to Start Saturday, March 22," declared a headline in the *Grant County Journal* when it announced the formal dedication and official start of the two generators.[59] "The Biggest Structure Ever Built By Man Will Begin Functioning," added *Business Week*.[60] With around 8,000 watching, a band from Grand Coulee High School led a long parade down the road and into the new left powerhouse. Microphones from various national radio networks stood waiting. Officials made their speeches followed by a congratulatory message from Interior Secretary Ickes who wrote, "the dam alone comprises the greatest single structure man has built."[61]

Then, at a little after one in the afternoon, Chief Jim James of the Nez Perce Indian Nation, in full regalia and accompanied by other members of his tribe, pushed a button that officially started the first house unit rotating. James followed with a speech in his native language which the radio affiliates carried. Reporters asked the Indians to pose for pictures by climbing onto the big generator that stood unfinished nearby. James and his fellows reluctantly complied. They must have had some misgivings about the event celebrating completion of the dam that flooded Native American ancestral hunting grounds and from which their tribe would hardly benefit. But for better or worse, two years ahead of schedule, the first electric power flowed from a generator at Grand Coulee Dam.

Initially the electricity from the house units provided only limited power for needs inside the powerhouse and dam.

On May 22, 1941 the town of Coulee Dam and sections of Mason City received electricity from the small generators.[62] Installation continued on a 6,900-volt line fifteen miles to the Nespelem Rural Electrification Association. On September 12 the Nespelem

REA became the first direct customer to buy power generated at Grand Coulee Dam.

But those were only the two small units. The turbines that caused the big generators to spin began their journey from the east coast. Their size required special packing cases and the trains that hauled them had to follow a peculiar routing that avoided passing under low viaducts and over narrow bridges. In some places the weight of the turbines required reinforcing the roadbeds.[63] On January 15, 1941, the first generator began its 2,500 mile trip to Grand Coulee from Pittsburgh. The journey required eleven days and the parts, later assembled at the dam, filled thirty-eight flat cars.[64] On February 3 crews from the East began putting together the thousands of pieces. Labeled unit "L-3," indicating that it occupied the third position in the left powerhouse, it turned for the first time in a test run on September 28, 1941.

On September 28 the first large generating unit ran successfully under a nearly full head (that is, the highest water pressure it would experience) and electricians began testing links between it and the Bonneville Power Administration transmission system. On October 4, 1941 the Columbia Broadcasting System's announcers returned to Grand Coulee to witness and document the official run of the world's first 108,000-kilowatt generator.[65] Two days later the generator successfully synchronized with the BPA system and began supplying power.

The ceremonial start of the first generator symbolically, if not actually, marked the completion of the dam. CBI announced that it would finish the remaining tasks in about a year, with three months more for clean-up. On December 12, 1941, CBI announced that it had placed all the concrete required and that Grand Coulee Dam then held almost exactly eleven million cubic yards.[66] The dam stood 550 feet high from foundation to top and stretched three-fourths of a mile across the Columbia River. Materials for the dam had filled over 60,000 railroad freight cars through the nearly ten-year construction period. On January 1, 1943 the Bureau of Reclamation assumed possession and took over operation of the dam. Laborers dismantled what remained of the trestles and CBI sent the scrap metal to west coast shipyards where it became part of Victory Ships.

On June 1, 1942 a crowd of around 15,000 gathered on the hill above the dam, behind barriers erected by the government. As predicted, the rising reservoir reached the crest of the dam, 1,290 feet above sea level, at 2:30 p. m., sending first a trickle and then a full flow down the spillway face.[67] The *Spokane Spokesman-Review* called it "The birth of one of the world's greatest waterfalls—a mammoth cataract, the greatest falling torrent ever created by man."[68] Behind the dam the reservoir, now full, stretched 151 miles to the Canadian border.

On January 31, 1943 the last few CBI men completed all work and the company received the final installment of the $41,366,900 owed it by the government. Due to changes in the design and additional work, this was almost seven million dollars over CBI's $34,442,240 bid. Again, considering cost overruns in later years on almost any large project, especially those managed by the government, the increase was negligible. In all, the bookkeepers claimed that through June 30, 1943, Grand Coulee Dam cost $162,610,943. As the government had allotted $179,477,675, the balance on hand, which went back to the treasury, was over $16 million.[69] In post-World War II years, such an occurrence would be astounding.

The *Wenatchee Daily World* estimated that 152 million man hours went into Grand Coulee, with well over 56 million of them at the dam site and the rest in various mines, factories, and incidental work around the country.[70] Either directly or through the various contractors, the Department of the Interior estimated that it had paid more than $54 million in wages from 1933 through the end of 1941.

Well over 12,000 men and women worked at the dam at one time or another from 1933 through 1943. At one point an estimated 7,400 held jobs at the same time. On average, 57 percent of the labor force was single. Over 60 percent were somewhere between the ages of twenty to twenty-nine. They represented no particular nationality or religious groups. Residents from all forty-eight states found jobs there; 40 percent of them worked as unskilled laborers. King County, Washington, supplied more men than any other county, statewide or nationally.[71]

*Time* dubbed Grand Coulee Dam "the biggest thing man had ever made on earth," and called it a monument to the seventy-two men killed during construction.[72] The exact number of men who died will never be known. Hubert Blonk, the on-site and then novice reporter, was the only one who kept records.[73] He decided which deaths to include and which did not apply. Blonk's figure of seventy-two, which the Bureau of Reclamation has accepted and which *Time* reported, is as good a guess as any.

In 1940 the *Wenatchee Daily World* reflected on the effort to build Grand Coulee Dam, then still in progress. The editor wrote:

> Never in the history of this ol' world have the amazing construction feats performed on the shores of the Columbia river been equaled. Engineers say that never again, well anyway, not for 100 years, will the equal be seen.[74]

Earlier, the trade journal *Pacific Builder and Engineer* noted, "Construction of the great concrete structure has involved to date a steady succession of construction feats without precedent in the annals of the industry."[75] In 1946 the government itself wrote, "all previous construction records were broken and unique construction methods were introduced and perfected."[76]

What had been true of the MWAK contract, however, was also true of the CBI effort. Each dam is a custom job requiring innovation and accommodation depending on existing conditions. Without a doubt the engineers and workers at Grand Coulee deserve considerable credit for their achievements. But in the overall analysis, the construction at Grand Coulee was remarkable only in its magnitude and not for any dramatic originality. The plans followed accepted dam-building practices. Simply put, it was big, and that was and remains its singular claim to fame.

Such a fact should not detract from the accomplishment. That the government financed and erected it during the Great Depression of the 1930s was, in itself, notable. That so many men and women participated in the effort and successfully stemmed the flow of the Columbia River, despite what a later generation of environmentalists might think, was also worth noting. In the end, what Woody Guthrie wrote in 1941 summed up Grand Coulee better than had just about anyone else. For, once completed, it truly was "Grand Coulee Dam, The Mightiest thing ever built by a man."[77]

Kettle Falls, 1939. *Courtesy Bureau of Reclamation, Columbia Basin Project Office.*

# Chapter Thirteen
# Land, Roads, Graves, and Salmon

It is not to be denied that water resource development on the
Columbia River system has had an adverse effect on the fishing industry. Similarly indisputable is the fact that fishing practices of the industry itself have also contributed significantly to
the reduction in size of Columbia River salmon runs.[1]

Dropping eleven million cubic yards of concrete across the Columbia River and creating a lake stretching over 150 miles involved
major dislocations both behind and below Grand Coulee Dam. The
initial preparations took place upstream from the construction site
when teams of workers cleared the soon-to-be flooded basin of trees
and undergrowth. Others relocated roads and railroads. Isolated
farms and whole towns moved to higher ground as residents left
long-established family homes. The Bureau of Reclamation, the
United States Department of Fisheries, and state agencies attempted to preserve, or compensate for, damage to the annual runs
of salmon. Finally, the government dealt with resident Native Americans whose individual and tribal economies often centered on the
river. The job of clearing, removing, relocating, and adjusting, which
went on largely behind the scenes, in its way equaled the task of
building the dam.

To prepare for the new lake the government did a number of
things at the same time. It secured title to the land, surveyed and
placed elevation markers, and began clearing trees and underbrush
up to the high water mark. Anything left would later float to the
surface and become a hazard both for people navigating on the
lake and at the water intakes behind the dam. Through late 1933
and into 1934, survey teams charted the first seventy-three miles
upstream. There were dangers. At Hellgate, not far from where
the Spokane River enters the Columbia, surveyors came on what

they dubbed a "rattlesnake paradise."[2] Despite the snakes and the remote terrain, by December 1935 they reached the international boundary.[3] Through 1936 the surveyors finalized markers and indicated the high water line.[4] They ran a second line twenty feet higher at 1,310 feet above sea level and that became the "taking line," indicating the land the government would acquire, even though clearing would only be up to the 1,290 foot level. The General Land Office encountered problems with Indian reservation lands within the reservoir boundaries.[5] The property came under the jurisdiction of long-established treaty rights and negotiating a settlement took time and patience since it involved a number of government agencies including the Bureau of Indian Affairs.

Even without that agreement, in December 1937 the government announced a land-clearing plan combining the efforts of the Works Progress Administration (WPA), the Civilian Conservation Corps (CCC), and the Bureau of Reclamation. The first phase of the program involved a $9 million effort on 28,448 acres, including removing or relocating 1,200 buildings, seventy-seven miles of primary state highway, fifty-eight miles of county roads, twenty-six miles of Great Northern Railroad tracks, twenty-five miles of telegraph lines, and three cemeteries.[6] Later phases greatly increased those numbers. The first CCC crew started work on December 21, 1937.[7]

The agreement with the WPA temporarily fell through when government officials argued that it gave too much money to Washington state, which already had received more than its share. Eventually, however, considering the employment it created, federal and state agencies agreed that the project merited funding. In July 1938 the *Wenatchee Daily World* announced that President Franklin Roosevelt gave his blessing to the project, that clearing of the reservoir would be the nation's largest WPA project, that it would employ well over 1,000 men, and that it would cost between $4 and $5 million.[8] The WPA submitted its final program for the clearing on August 11.[9] After that the Bureau of Reclamation rapidly called for bids. It projected a number of camps, barges, and floating facilities to house the laborers. By October the first camp operated at Lincoln near the confluence of the Spokane and Columbia Rivers.[10] Once started, the clearing continued apace, spurred by the rapidly rising water behind the dam. By the end of 1939, 2,000 men worked out of seven major camps and a few smaller locations.[11] The second encampment was at Keller, but it moved in June 1939 to a new location near Kettle Falls. The team at the Lincoln Camp followed suit and moved to Gerome a month later.[12] The Keller and Kettle Falls camps were the larger settlements, eventually holding 450 men or more. After its stay at Kettle Falls, that facility moved to Gifford. These "portable" camps had a dozen or more buildings housing forty men each. Camps Spokane, Detillion, and Little Falls were small, with no more than 200 workers using two-man tents.[13]

During 1938 and 1939 each WPA crew worked about seven hours a day for a two-week period, then a new group replaced them. This rotation system maximized the total number employed.[14] For many, two weeks was about as long as they wanted in the isolated upper reaches of the Columbia River. The floating camp, dubbed Camp Ferry, consisted of two-story superstructures built on three large barges.[15] Smaller than the land-based camps, it initially housed seventy-five men and served the more inaccessible spots. By the fall of 1939, 125 men shared the floating bunkhouses.[16] Unlike their counterparts in the land-based camps, the float camp crews did not come and go as often, sometimes being stuck at a remote spot for up to three weeks.[17]

In February 1941 the Bureau of Reclamation urged the WPA to increase each laborer from seven to eight hours daily since Grand Coulee Dam was then complete and the lake would soon make upstream access more difficult. And every high water period brought into the reservoir bothersome brush, branches, and debris that floated downstream and accumulated behind the dam. "By completing such clearing, a considerable saving will be made in both time and money in the final completion of the project," assistant Reclamation commissioner Harry Bashore argued in his February 1941 letter to Malcolm Miller of the WPA.[18] Two months later the government declared the basin-clearing effort a national defense project, further intensifying the push to complete the job.[19]

Without question the clearing started too late. The contractors built the dam much faster than the Bureau anticipated. In August 1939, with the second phase of construction well under way, the reservoir was only 20 percent clear. This increased to over 80 percent by early 1941. In the process, the Bureau of Reclamation sold, through a bidding process, 32,337,284 board feet of lumber.[20]

In all, the government cleared something just short of 54,000 acres, with over 11,000 of that freed of stumps.[21] This involved about two-thirds of the total reservoir area.[22]

On July 15, 1941, in a symbolic ceremony, Frank Banks and Washington state WPA director Carl M. Smith, felled what they identified as the "last tree" in the Columbia River Reservoir.[23] Although the crews would later cut more than eighty additional acres, and even though the work continued until November, the act by Smith and Banks officially marked the conclusion of the clearing.[24] After that, others labored for a few years placing riprap—a coating of rock and gravel designed to prevent erosion—along the shore of the lake.

There were thousands of cases involving private land that would be flooded. The government easily resolved some land-acquisition problems, while others required complicated litigation. For years landowners living around Grand Coulee hoped that a dam would eventually enhance the value of their land. When the federal government offered to buy the land it needed, most owners refused. Then the government exercised its declaration of taking, leaving the courts to determine compensation. Sam Seaton, who ran a ferry at the dam site, was typical of others. He asked considerably more for his land than the government wanted to pay.[25] Seaton combined his case with some of his neighbors and together they appeared in court on July 15, 1935. Altogether, their disputed property totaled 1,101.63 acres that the government appraised at $15,443, while the owners demanded $5 million.[26] The trial ran until August 2, 1935, when the jury awarded the land owners $17,338.92.[27] The *Spokesman-Review* estimated that their lawyers charged them over $10,000.[28] Infuriated, the landowners took their case to the United States Circuit Court of Appeals, which upheld the jury.[29] Not letting the matter drop, the landowners and their lawyers petitioned the United States Supreme Court, but it refused the case.[30] Seaton moved to Elmertown and for the rest of his life willingly told anyone who would listen about the rotten deal he received at the hands of his government.[31]

Upstream, none argued that their property was worth millions of dollars, and this reduced the task to one of settling claims and moving on. Nevertheless, this disrupted the lives of people living in small towns and farms, with most feeling that the government did not pay enough for their homes and property. One T. S. Moore, an Indian who owned a farm and who spoke for many, told the *Wenatchee Daily World*, "What we ought to do is all close up and go on relief and make them pay us."[32]

By the end of 1939 the government owned over 90 percent of the land held by non-Native American residents. Water eventually covered Peach, followed by Keller, Lincoln, Gerome, Gifford, Inchelium, Daisy, Kettle Falls, Marcus, and Boyds. The people of Kettle Falls moved to Meyers Falls, a smaller incorporated community. With their larger numbers, the displaced Kettle Falls residents then voted to annex Meyers Falls to Kettle Falls and changed the name of the new entity to Kettle Falls. Meyers Falls people never forgave them for that.[33] The residents of Lincoln decided to relocate on the site of old Fort Spokane but discovered that it was Indian land, part of the Spokane Indian Reservation. They tried petitioning Senator Lewis Baxter Schwellenbach for permission to move there anyway. The attempt failed and, in the end, Lincoln, Peach, and a few other places ceased to exist.

Some of the dislocated used their government settlement to buy back their houses at auctions and then move them to new locations.[34] The *Wenatchee Daily World* reported that people relocated over 5,000 structures when 400 to 500 farms vanished under the water.[35] The Bureau of Reclamation received adverse publicity when a nationally distributed newsreel pictured an elderly couple leaving their home, which was then shown burning in the background as the wife wept. In its own defense the Bureau pointed out that it paid the couple well and that they had one full year to leave, but had waited until the last minute.[36]

The concerns of Native American residents received considerably less media attention. At first the Colville Indians, whose reservation bordered the Columbia River at Grand Coulee and for a distance above the dam, saw the construction as an economic windfall. They sold food and other materials to workers and they anticipated, as did their non-Indian neighbors, a rapid increase in the value of their land. But it quickly became apparent that the Colville people would not become rich.[37] Neither would the Spokane Indians who owned land farther upriver. Because treaties with the government controlled use of the land, the authorities agreed that all involved agencies had to negotiate a settlement which Congress then needed to approve. One issue complicated the discussions. For centuries Indians buried their dead near the Columbia River. Late in 1935 representatives of the various tribes met at Nespelem to discuss removing all threatened graves. Divided on what to do, the Indian leaders did agree, over the next two years, that the government had already violated their treaty rights and that they needed to organize with other Northwest Indians to fight for justice and fair compensation.[38]

The talks stretched through the fall of 1938, when all parties agreed to move all located graves. That decision became part of the bill being readied for Congress that also gave the government permission to flood parts of the Colville and Spokane reservations.[39] Although the final terms of the bill remained vague, workers began identifying and relocating the graves when water began covering the land. Newspapers around the Northwest and nationally carried a few stories about the grave relocation effort, and museums and private citizens contacted the Bureau of Reclamation asking how to help or acquire artifacts for their collections.[40] Despite the offers, the Bureau of Reclamation had trouble finding someone suitable to provide what Reclamation Commissioner John Page called "adequate archaeological supervision" over the grave relocation project.[41] The Bureau of Reclamation approached the Smithsonian Institution asking if it had an archaeologist available, but it declined due to limited funds and a feeling that "no discoveries of primary importance would be made."[42]

In cooperation with the Eastern Washington State Historical Society, a group of interested citizens formed the Inland Empire Indian Relic Society. It offered assistance to the Bureau of Reclamation and offered the Historical Society museum as a place for "Indian artifacts and skeletons and other material . . . preserved."[43] The Bureau had no idea if the Indians would object to the removal of such objects. Commissioner Page wrote to the Office of Indian Affairs and asked for an opinion.[44] Sadly, no record of the reply survives.

The Bureau finally allowed Dr. Alexander D. Krieger from the Department of Anthropology at the University of Oregon and Dr. Phillip Drucker, an anthropologist from the University of California, to supervise the grave removal, with Drucker in charge.[45] Krieger used between ten and twenty people for about one year locating and moving graves. The Bureau of Reclamation approached the WPA for additional manpower, but it refused.[46]

In 1939 the Bureau of Reclamation opened bids on the job of relocating between 600 and 800 graves.[47] In addition the National Youth Administration assigned a group of young men to the project and the Bureau provided them with a truck and some equipment.[48] Then the Colville Indians objected to anyone removing artifacts and the Bureau halted work while consulting with tribal representatives.[49] It is impossible to determine how much material people carried away either before or after the Indians complained. Clear, however, is that tribal representatives objected to what they saw as the looting of their ancestral lands. Clear also is the fact that this is exactly what many of those who offered to help had in mind.

By the end of 1939 the government had relocated 915 graves at a cost of $19,642.60.[50] By the fall of 1940 the number rose to over 1,200. Indian leaders then indicated discovery of over 2,000 additional sites, with more turning up daily. The Bureau, seeing no end to the effort, stopped it entirely.[51] Despite protests from the Colville Nation and the agreement the Bureau had made with the Indians, it relocated no more graves, and water shortly rose over those both discovered and undiscovered.

As for the land, like its neighbors the Colville Nation expected more money than the government wanted to pay. By March 1939 some Indian land was already under water, and tribal representatives realized that this weakened their bargaining position. Although the final agreement required an act of Congress, once the rising water covered their property, they would find holding out for a higher price increasingly difficult. Furthermore, many individual owners settled independently. By 1939 the government had paid almost forty of the fifty-four families involved.[52]

At last an agreement, worked out late in 1939, gave the Indian people $174,000 for their land. The Bureau of Reclamation pointed out that it had already paid $54,000 of that amount to private owners, and that it would use the rest to settle outstanding claims, with any remaining funds going to the tribal council.[53] This ended the discussions and Congress passed the necessary bill legalizing the transaction that fall.[54]

In 1941 the Bureau of Reclamation and representatives of the Colville Tribal Council opened discussions concerning the use of

Columbia Lake—the reservoir that became Franklin D. Roosevelt Lake. The Indian representatives wanted to control development along the lake shore bordering their lands, a demand they still make. In addition, the Colville Confederated Tribes and the Spokane Tribes never reconciled themselves to the loss of their property or fishing rights in the wake of Grand Coulee construction. In 1951 they filed a claim against the government which included damages and complaints about Grand Coulee Dam.[55] In 1975 the tribes presented the federal government with claims of $100 million and 185 megawatts of power annually to adjust for the damages done them during construction at Grand Coulee, and later at Chief Joseph Dam. In 1978 to support those claims, the tribes produced a twenty-eight minute color film called *The Price We Paid*. In it they argued that during the 1930s the government told them to accept the offer and the government would take care of them later. In return for lost salmon fishing at Kettle Falls, the flooding of many home sites, root digging areas, traditional burial sites, and the destruction of their way of life, the film pointed out that the government never justly compensated the Indian people, despite promises made decades ago.[56]

For five years a federal task force looked into the claims of the tribal councils. The Indians acknowledged their position's legal weakness, but insisted that "historical inequities" needed adjustment. In the summer of 1980 the United States government, following the recommendation of the task force, rejected the claims. Little happened until 1991 when tribal representatives again approached the government. Talks continued until March 1994 when a surprise announcement in the *Spokesman-Review* revealed that federal officials had agreed to pay the Colville Confederated Tribes the lump sum of $53 million and a minimum of $15.25 million annually thereafter. The yearly payments would represent the tribes' share of the power revenues generated at Grand Coulee Dam. On April 16, 1994 tribal members voted overwhelmingly to accept the offer.[57]

Through the 1980s and 1990s members of the Colville Nation and other residents around Grand Coulee often disagreed over recreational development on Roosevelt Lake. Each group sought the advantage in exploiting the reservoir. Since there has been no appreciable construction at the dam since the early 1980s, the local population has dropped and accommodating tourism appears the best hope to revitalize the sagging economy. Whether the Indian entrepreneurs will be part, all, or none of this venture is unclear. What promises the government once made or implied can never be determined. It is certain that if white residents felt the government cheated them, their complaints wither in comparison to those of their Indian neighbors.

Along with securing Indian and non-Indian land, the Bureau of Reclamation relocated roads and railroads. In 1935 the Bureau anticipated rebuilding seventy-two miles of primary roads, fifty-four miles of secondary roads, and twenty-five miles of track.[58] That work extended through 1940 with the bridges and track alone costing nearly two million dollars.[59] The government finished relocating 180 miles of highway in 1943.[60]

It took longer and proved more difficult to deal with the annual salmon runs. Rock Island Dam, built by the Puget Sound Power and Light Company, was the first hydroelectric project on the main stem of the Columbia River. Started in 1930 and finished in 1933, it preceded by only a few years work at both Grand Coulee and Bonneville. These became what salmon expert and author Anthony Netboy called the first of "The Killer Dams."[61]

Netboy alleged that the Chief of the Army Corps of Engineers, when asked why the early plans for Bonneville Dam showed no fishways, commented, "We do not intend to play nursemaid to the fish."[62] That incredible and probably apocryphal comment, if actually made, was ill advised because the Federal Power Act of 1920 required the licensee of a hydroelectric project on a public waterway to protect migratory fish or compensate by building hatcheries. At Bonneville the Corps of Engineers built fish ladders that allowed at least some of the fish to find their way upstream. At Grand Coulee the Bureau of Reclamation chose other alternatives.

Although the Bureau took steps to preserve the salmon, had its measures failed, it was ready to see the salmon disappear, deeming the economic contribution of the dam as far more important than the fish. This attitude showed from the start. The Bureau of Reclamation took control of Grand Coulee late in 1933, but squabbled for over three years with the U.S. Department of Fisheries and the State of Washington over the preservation of fish runs before taking any action. The delay was almost too long, and overcoming the difficulties it created required experimental, expensive measures.

At the outset the Bureau of Reclamation recommended a flume and a mechanical elevator to transport fish around Grand Coulee Dam.[63] Later it added plans for a fish hatchery.[64] United States Commissioner of Fisheries Frank Bell wrote Reclamation Commissioner Elwood Mead that a ladder would not be satisfactory at Grand Coulee due to the height of the dam. He added,

> Any system of raising fish over the dam in elevators or by other mechanical means would probably fail because of [the] great mortality of young fish. The only alternative is to trap the runs and artificially propagate the eggs, replanting the young below the dam.[65]

Frank Banks began exploring alternatives. Personally convinced that the salmon deserved preservation, despite the expense, Banks considered capturing salmon below Grand Coulee and transferring the eggs to the San Poil River, above the dam, or to a nearby hatchery.[66] But the plan required transporting both adult and young fish and it looked very complicated.

The Bureau of Reclamation and the United States Fish Commission studied the problem through 1935. At this time nobody knew exactly what affect Bonneville Dam would have on the salmon. Thomas A. E. Lally, chairman of the Washington State Game Commission, predicted that Bonneville would eliminate 90 percent of the salmon and that Rock Island would do more damage. James O'Sullivan, then Executive Secretary of the Washington State Columbia Basin Commission, suggested that if the other dams did that much damage, it would be questionable logic to spend excessive time and resources salvaging a lost cause at Grand Coulee.[67]

At about the same time, Canadian officials discussed the impact of the Columbia River dams. Looking at reports by the United States Army Corps of Engineers (especially the 308 report of 1932), members of the Canadian Legation in Washington, D. C., realized that Grand Coulee Dam would block passage of salmon from their extensive spawning grounds north of the international border. In October 1934 they notified the Minister of Fisheries in Ottawa. Eleven days later the reply concluded, "The assumption that there is no commercial salmon fishery on the Columbia River in Canada is correct, and hence Canadian interests in that respect will not be affected if the dam at Grand Coulee is not equipped with fishway facilities."[68] In other words, Canadian officials wrote off salmon in their stretch of the Columbia River because they did not see any financial impact at that time and they did not consider significant the maintenance of the species in those waters.

On the American side, however, the government did continue exploring ways to sustain at least some of the salmon run. Frank Banks announced that the Bureau would build a single $500,000 hatchery near the small construction town of Mason City. At this point considerable jurisdictional disagreement and argument among the various government bodies hindered any progress. Through 1936 the agencies debated and Frank Bell called for a conference to settle the matter. Meanwhile Bureau of Reclamation engineer Sinclair O. Harper wrote to acting commissioner John C. Page,

> The Bureau of Reclamation has no one in its employ familiar with the solution of problems of that nature, and considering the importance of the matter, it is desired to secure the advice and assistance of at least two engineers with knowledge and experience in the work of fish conservation, control, and protection.[69]

In other words, after nearly three years on the job, the Bureau still had not hired anyone who knew how to handle the salmon. Bell's conference only highlighted this lack of knowledge as it deteriorated into wrangling and indecision.

Meanwhile workers at Grand Coulee continued pouring concrete and finished placing the dam's foundation. With the Columbia blocked and no facilities for salmon in place the Bureau announced plans for temporary fish ladders in the twenty-five-foot-wide slots left in the dam through which the river flowed. These ladders would allow fish to swim through the now-formidable barrier by raising them forty feet into the reservoir. Near the end of December 1937 workers finished the first ladder, made from a series of seven log obstructions, each creating a pond at a higher level.[70]

On January 2, 1938 the Washington State Department of Fisheries completed its detailed report with recommendations for migratory fish control. All the government agencies adopted it and called it the North Central Washington Upper Columbia River Salmon Conservation Project. The Bureau of Reclamation estimated its cost at $2.6 million and said that the money would come from the Grand Coulee Dam building fund.[71] Under the plan, fish traps at Rock Island Dam would snare ripe salmon, which trucks would transport to the new hatchery at Leavenworth and later to other hatcheries. Biologists would plant young hatchery fish in streams below Grand Coulee with the hope that they would identify these as their new homes and return to them. The idea was speculative and experimental.[72]

In January 1939 two things happened that affected the project. First, Secretary of the Interior Harold L. Ickes, who had expressed some interest in the salmon, appointed Dr. W. F. Durand and Dr. Willis H. Rich of Palo Alto, California, and Robert D. Calkins of Berkeley, California, as consultants to investigate and report on the migratory fish problems on the upper Columbia River.[73] Second, the Bureau of Fisheries and the Bureau of Reclamation realized that they could not complete work on the Leavenworth facility in time to handle the upcoming fish run, and they put together a temporary plan. Commissioner Bell announced that biologists would capture salmon at Rock Island and transport them to their traditional breeding sites above Grand Coulee.

In January the Secretary of the Interior ordered eight custom-built tank trucks to carry chinooks and bluebacks from Rock Island Dam to various spawning creeks, some above Grand Coulee, but most below the new dam.[74] Now working together more successfully after over two years of contention and controversy, the Bureau of Reclamation, the State of Washington, and the U. S. Department of Fisheries agreed on the emergency plan. The Bureau coordinated the experimental salmon transplant effort during 1939 and agreed to complete the Leavenworth hatchery and three other facilities, at a cost of $2.5 million.[75]

On May 1, 1939 the fish traps at Rock Island began operating and on May 3 the first truck load of salmon began its trip around Grand Coulee Dam. The operation continued through the summer and by the middle of July workers moved over 6,000 fish, some as far away as the Canadian border, others into the Wenatchee, Methow, Okanogan, and Entiat rivers where Ickes's experts reasoned they might survive and even adopt as permanent spawning grounds.[76] During 1939 the operation moved an estimated 36,000 salmon at a cost of about $2.70 each.[77]

On March 27, 1940 the trucks began hauling fish for the second year. By May 1 the ponds at Leavenworth were ready. The Bureau of Reclamation then transferred operation and maintenance of the facility to the Bureau of Fisheries.[78] The Washington State Game Commission announced that in 1941 the government would build a hatchery at Chamokane Creek, a tributary of the Spokane River twenty-five miles northwest of Spokane, to stock the Grand Coulee reservoir. It also built another hatchery near Ford, Washington, for the same purpose.

With Grand Coulee Dam nearing completion, people upstream watched the salmon preservation effort with considerable misgiving. During May 1939 the people of Keller, forty miles above the dam, celebrated their last annual Salmon Day Program.[79] On June 14-16, 1940, in a far more emotional atmosphere, representatives of the Native American population from throughout the Pacific Northwest gathered for a three-day "Ceremony of Tears" at Kettle Falls. Legends said that at one time men could walk across the river at Kettle Falls on the backs of the fish.[80] With the dam, not only did the salmon disappear, but also not long after the event the rising water submerged the falls too.[81] For the Indians, the end of the salmon presented a devastating economic and cultural loss.[82]

Everything now depended on the hatcheries and on the experiment to rear the fish artificially and transplant them. Along the way Department of Fisheries people met with unforeseen problems. For example, early in 1942 an ice jam formed in the Entiat River cutting the flow of water and causing the loss of several thousand young fish. Transporting the fish by truck proved less successful than hoped as many of the fish died.[83] While eggs hatched satisfactorily, an inadequate water supply plagued the Leavenworth facility and some breeding ponds remained empty.[84] Five to 30 percent of the young fish died on their downstream migration when they passed through the turbines or over the spillways of each downstream dam. In 1965 scientists found that water pouring over spillways becomes super-saturated with nitrogen, which kills fish.[85]

Considering these problems resulting from the limited knowledge of salmon biology, and the increasing threat posed by the dams, the wonder is that the salmon survived as well as they did.

On August 6, 1941 the first salmon, 50,000 Chinooks reared at the Leavenworth hatchery, entered the Entiat River and began the experiment.[86] From 1942 through 1945 officials watched the salmon counts to determine whether any of the hatchery-reared fish survived, how many returned, and if any returned to the new streams. In 1943 the Bureau of Reclamation declared the salmon transportation and relocation process a success. Adult fish entered the four rivers below Grand Coulee Dam, where as hatchlings they began a few years before.[87] The Bureau then moved to relieve itself of further responsibility. Frank Banks wrote to Commissioner Page,

> Bureau representatives have tacitly understood that as soon as the operation of the hatchery facilities succeeded in transferring the salmon run formerly passing Grand Coulee Dam to the four tributaries, the Bureau's responsibility would be ended and if either the State or the Wildlife Service desired to continue hatchery operations in future years, it would be at the sole expense of the organization continuing to operate. In as much as the cyclical span of a salmon's life is ordinarily 4 or 5 years, this period of the Bureau's financial operation has been indefinitely assumed to be from 4 to 6 years.[88]

Banks pointed out that December 1944 marked six years since the Bureau began the hatchery and he recommended turning all remaining responsibilities over to the Fish and Wildlife Service in July of that year.[89] In its "Annual Project History" for Grand Coulee Dam, the Bureau reported,

> Operation of the hatcheries and related facilities constructed in connection with the Columbia Basin Project by the Bureau, were continued by the Fish and Wildlife Service and the Washington State Department of Game. While the results of the actual operation during the year were somewhat disappointing, there were many factors pointing towards obtaining improved results in the future.[90]

In fact, the young fish in the hatcheries suffered from diseases and significant loss of life until after 1960 when scientists better understood their nutritional needs.[91] In 1944 the government altered the fish ladders at Rock Island allowing salmon into the river above the dam and eliminating the expensive trucking. By 1949 the Bureau had turned over full ownership of all the hatcheries to the Fish and Wildlife Service.[92]

On January 26 and 27, 1944 a meeting of the Army Corps of Engineers, the Bureau of Reclamation, the Oregon and Washington game departments, and representatives of commercial fishermen, held in Portland, Oregon, discussed the status of migratory fish in the Columbia River.[93] Along with other data, attendees considered the counts of fish taken over the years at Rock Island Dam. In 1933, 51,879 salmon moved through the ladders there. That number dropped to 9,759 in 1934, and then stayed at between 20,000 to 35,000 every year until 1943 except 1942, when it dropped to 7,086.[94] Based on the figures, Frank Banks pronounced the salmon relocation experiment a success and noted that no salmon appeared at the base of Grand Coulee Dam in 1943.[95]

In 1946 the Bureau of Reclamation announced that the salmon transplant experiment showed gratifying results. The Bureau predicted that the artificially enhanced runs might actually increase the volume of fish.[96] In 1948 it announced the largest fish count at Rock Island since 1933, stating that the hatchery and fish transplant program appeared an unquestionable success.[97]

How much the salmon transplant experiment really succeeded in the long run is unknown. Between 1953 and 1968 new Columbia River dams downstream at McNary, The Dalles, Priest Rapids, John Day, and Wanapum, along with Ice Harbor, Lower Monumental, Little Goose, and Lower Granite on the Snake River, further disrupted the annual fish runs. The addition of extensive irrigation works on the Columbia Basin Project in the 1950s and 1960s, not to mention other projects elsewhere, removed large volumes of water from the river at exactly the time of year when migrating fish most needed it. In addition, the irrigation facilities returned to the river salts, fertilizers, pesticides, and other pollutants.[98] In 1988 an article in the *Wenatchee Daily World* pointed out that since 1972 the government had spent $127 million dollars, or about $8 million a year, preserving the salmon for an industry that yielded $17 million annually. The article asked if the power ratepayers in the Pacific Northwest thought the result justified the expense.[99] With increased concern in the 1990s for the salmon as an endangered species, the government and Northwest power users will spend considerably more if they preserve what is left of the runs.

The dams on the Columbia were not the first to damage the salmon. Over-fishing, waste, and pollution threatened the salmon by 1900 and irrigation projects took an additional toll after the turn of the century.[100] The dams, however, accelerated the process. Grand Coulee eliminated over 1,000 miles of rivers and streams, nearly one-half of the watershed, that the salmon had traveled.[101] Certainly it has been the combination of forces, and not any one factor, that has brought the salmon to the edge of extinction and led to concern about their fate. In the 1930s, when the dams were the highest priority, the salmon survived more through good luck than government efforts.

Providing for the salmon, the roads and railroads, clearing the reservoir, and acquiring land was a big job. Grand Coulee Dam caused appreciable ecological changes to the Columbia River and its basin both up- and downstream. Under laws of the 1990s, with requirements for complicated and detailed environmental impact studies, concern about endangered species, and greater interest in litigation, the rights of minorities, farmers, job protection, and the demands of other special interest groups, it is doubtful that Grand Coulee Dam, or anything like it, could be built. In vivid contrast, looking back at the 1930s, when the promise of abundant cheap electric power held hope for a better life and jobs for thousands during a terrible depression, it is surprising that the government and the general population showed as much concern for conservation and preservation as they did. In terms of society's values over half a century later, the environmental damage done by Grand Coulee Dam may appear deplorable. But considering what might have happened, the effort appears more benign. There were trade-offs and there are strong differences of opinion about the outcome as each generation weighs the benefits and the losses the dam has brought.

The first three generators in the west powerhouse, 1942. *Courtesy Wenatchee World.*

# Chapter Fourteen
# Selling Grand Coulee's Power

The Northwest is certain to become the greatest power empire in the nation.

*Harold L. Ickes*[1]

To build such an empire is the stated purpose of the Senate's most handsome bachelor, Warren Magnuson of Washington, and to effect it he introduced on April 19, 1949, Senate Bill 1645: to create a Columbia Valley Administration.

*Albert Williams*[2]

Once the government began building Grand Coulee and Bonneville dams, proponents and opponents alike fought over control of their electricity. The winner would gain formidable authority over future regional development and potentially huge profits. Because hydroelectric energy became integral to the Pacific Northwest's economy, its politics were, and are, an enigmatic, convoluted local and national battlefield.[3] Private power companies, various government agencies, and regional interest groups all vied with each other for the authority, influence, and especially the money connected with having a finger on the electric switch. Grand Coulee Dam complicated the issue because of its enormous power potential and its link with reclamation.

The New Deal of the 1930s brought with it an emphasis on economic planning, the Tennessee Valley Authority (TVA) being the most successful example of New Deal wide-spectrum resource management. President Franklin Roosevelt and his minions projected a similar undertaking for five or six additional areas, including the Columbia River watershed. The plan would have combined, under one agency, all facets of management including reclamation, land use, power generation, fisheries, industrial development, urban growth, and agriculture. It would have overseen or even controlled the production, distribution, and sale of power throughout the region and might have eliminated the political and economic confrontations that followed.

Still, whether such a scheme would have benefited the Pacific Northwest is debatable. Jealousies among federal bureaus, notably the Army Corps of Engineers and the Bureau of Reclamation, hostility from private power interests, and reluctant businessmen and politicians who feared seeing their influence diminish all conspired to defeat at least five attempts, starting in the early 1930s and lasting into the early 1950s, to create a TVA-like agency in the far West. What emerged was the Bonneville Power Administration as it co-exists today along with the often-confusing mishmash of other state and federal agencies that share overlapping national and regional responsibilities.

Controversies first arose with the start of construction at Grand Coulee and Bonneville. No one had decided who would control the electricity generated at the federal dams, how much it would cost, how to transmit it, or who could buy it. The Reclamation Act of 1906 and the Water Power Act of 1920 gave preference for electricity generated at federal facilities to publicly owned distribution systems.[4] Grand Coulee supporters realized this and, in January 1932, Washington Representative Sam Hill and Senators Wesley L. Jones and Clarence C. Dill introduced a bill requiring the government to sell power at the switchyard first to municipalities, then political subdivisions, and finally to private corporations, in that order. Bitterly fought by private power interests apoplectic over the prospect of such competition, the bill died. Largely private interests, including the remains of the Spokane-dominated Columbia Basin Irrigation League that worked so hard in the 1920s and early 1930s against Grand Coulee Dam, then formed the National Reclamation Association (NRA).[5] Throughout its life the NRA opposed regional authorities and worked to maximize the cost of federal power to dilute its affect on private power rates.[6]

On January 3, 1935, Representative Knute Hill of Prosser, Washington, and eleven days later Senator James P. Pope of Idaho, introduced the first of the Columbia Valley Authority bills.[7] For the most part they projected a regional planning unit closely patterned after the TVA. The Columbia River dams would become the property of the new CVA, which could then control their power. Further, the CVA, administered by a three-person board, could condemn existing private power distribution systems and absorb them into its new network. And, unlike the TVA, the CVA would govern regional irrigation. The Bureau of Reclamation, the Army Corps of Engineers, public power people in North Central Washington, and irrigation interests all opposed the bill.[8] They feared that Oregon and Idaho would dominate a CVA and that the agency would divert to other projects money intended for Grand Coulee Dam and the Columbia Basin Irrigation Project.[9]

Senator Charles Linza McNary of Oregon also opposed the CVA idea, instead advocating what he called a Bonneville Power Authority which would only sell power, exactly the plan that ultimately succeeded. Better yet, McNary spoke of allowing the Bureau and the Corps to divide the region with the Bureau selling its power to Seattle and Spokane interests and the Army dealing exclusively in the Portland area.[10] This seemed to be what the two agencies wanted and both appeared opposed to any suggestion that they link the power from the two dams in some sort of pool.[11] The long-standing contention became part of a larger ongoing feud between the two agencies during the second half of Roosevelt's first term which further muddied the already complicated issues.[12]

The Senate held hearings on the CVA bills, which later died in a House committee. Meanwhile President Roosevelt's appointees on the National Resources Board set up a water planning committee that re-examined the CVA idea along with other methods of comprehensive regional planning. They recommended forming a transmission and marketing system for Grand Coulee and Bonneville energy on a central grid or single super power network linking Spokane, Puget Sound, and Portland, then selling the electricity at a uniform rate, something urged by J. D. Ross. To govern this "power pool," the National Resources Board suggested a new independent commission with corporate powers as a marketing agency. The War Department, speaking for the Army Corps of Engineers, strongly opposed this.

On July 29, 1935, Senators McNary and Frederick Steiwer introduced a compromise CVA-type bill that left the Bureau and the Corps in control of their respective dams and their power but included aspects of regional planning. Increased bickering, turf guarding, and mutual suspicions prevented progress on any of the bills. At that point, the White House temporarily lost interest in the CVA issue.[13] For a while after mid-1936, Roosevelt looked seriously at the limited marketing agency plan proposed by the National Resources Board and seconded by the Pacific Northwest Regional Planning Council.[14] Meanwhile, Congress adjourned.

On January 18, 1937, safely past the 1936 election and with a clear public mandate, Roosevelt appointed a special committee to suggest legislation to handle Bonneville Dam's power, leaving Grand Coulee for later.[15] Roosevelt gave the group two weeks to report. Drawing from the previous ideas of the National Resources Board's committee, it urged putting power development, generation, marketing, and transmission under a civilian administrator who would direct construction of transmission lines and guarantee uniform power rates regionally.[16] Roosevelt, apparently tired of the bickering and aware that power from the nearly finished Bonneville Dam would soon be available, accepted the compromise, although he saw it as a temporary administrative organization serving until a permanent plan, involving regional planning and some kind of TVA-like agency, replaced it.[17]

On June 24, 1937, Representative Joseph Jefferson Mansfield of Texas introduced the bill creating an independent administration, under the Interior Department, controlling the sale and distribution of power generated at Bonneville Dam. This became the Bonneville Power Act. It was a temporary solution and the language of the bill clearly explained that point.[18] By August 12, 1937 the bill had passed Congress and on August 20 Roosevelt signed it.[19] More ecstatic than the circumstances warranted, Homer Bone wrote to Washington Governor Clarence Martin, "It is by long odds the best piece of power legislation ever enacted and it goes far beyond the TVA act. Public power friends here think it is the foundation for legislation to follow all over the country."[20] In fact, when compared with the ambitious plan for an all-encompassing CVA, it was a feeble stopgap, although the best solution possible at the moment. The fight for a CVA continued.[21]

The Bonneville Act created the Bonneville Project, the name the new organization held until 1940 when it became the Bonneville Power Administration (BPA). The word Bonneville appeared in the name because the agency was to deal only with power from Bonneville Dam. Like many government agencies, it quickly gained a life and goals of its own, not always in harmony with other federal agencies. In time, as it achieved permanence, the BPA entered and added a new dimension to the feud between the Army Corps of Engineers and the Bureau of Reclamation. The Bureau, which wanted to control Grand Coulee power, watched creation of the BPA with great disappointment and it has been cool toward the agency ever since.[22] As for the name, as things turned out, it has become misleading and should be the Pacific Northwest Public Power Marketing Agency or something of that sort.

The BPA's first controversy was over its single administrator, to be selected by the Interior Secretary Harold Ickes. The word quickly circulated that J. D. Ross wanted the job. Livid, *Wenatchee Daily World* editor Rufus Woods, who remembered Ross as an enemy of Grand Coulee, wrote every newspaper editor in eastern Washington and told them they must join in the fight to keep Ickes from selecting Ross.[23] When Ross publicly urged a Grand Coulee/ Bonneville interconnection, Oregon opposition mounted. Washington people also opposed Ross's so-called "postage stamp rate"—another way of saying that the cost of electricity would be uniform throughout the region. They feared this would allow transmission lines to take energy elsewhere, away from eastern Washington, killing the dream of an agricultural/industrial empire. Despite the deep feelings, on October 10, 1937, Secretary Ickes announced that Ross would be the first BPA administrator. Woods called it a travesty.

Ross stated that he saw his job and the agency as temporary expedients and recommended a national power bill to replace the Bonneville Act. Then he called for an immediate link between Bonneville and Grand Coulee, with both coming under one authority.[24] Ross set about assessing regional power needs and promoting Public Utility Districts (PUD; People's Utility Districts in Oregon).[25] He implemented the postage stamp rate, the kilowatt year through which industry could buy large blocks of power for a guaranteed length of time, and the idea of interruptible power.[26] Then he died suddenly on March 15, 1939.[27] On May 4, President Roosevelt temporarily appointed Frank Banks to replace Ross. But Banks showed no zeal for either bureaucratic administration or the expansion of public power. A Republican with conservative political inclinations, he hardly equaled the crusading Ross, who had waged a fierce struggle against private power interests throughout his career. Further, Banks represented the Bureau of Reclamation, which had a low standing with New Dealers and public power advocates due to its ties with private business.[28] On September 15, Roosevelt appointed Dr. Paul Jerome Raver as permanent BPA administrator and Banks returned to Grand Coulee.[29]

Through the short-lived efforts of J. D. Ross, the worst fears of North Central Washington residents became reality. The postage stamp rate and the Bonneville/Grand Coulee connection allowed their electricity to slip away. The first transmission line from Bonneville Dam to Portland went into service on December 1, 1939.[30] On August 26, 1940, President Roosevelt made the Bonneville Power Administration the marketing agency for Grand Coulee.[31] The BPA and its transmission lines ended any hope of local control over the power generated at Grand Coulee Dam and upset residents of eastern Washington. As Kirby Billingsley, longtime Chelan County PUD official said some years later,

> We are fighting the plundering of our region; hauling away our power, our minerals, our produce and our men. . . . Grand Coulee Dam, which we fathered, has become a Frankenstein—its power is building industries elsewhere and they are draining our manpower to run those industries.[32]

J. D. Ross, the very man Rufus Woods feared and fought, in the end, did the most to destroy Woods's dream of the agricultural/industrial empire.

Although the public/private power debate of the 1920s and 1930s had passed its peak, the CVA struggles of the late 1940s and early 1950s provided an exciting follow-up. Through those battles the Bureau of Reclamation and the Corps of Engineers each protected themselves while the new BPA worked to carve out its territory, thus creating a three-way conflict.

Although the CVA idea persisted, after the demise of New Deal idealism and the start of World War II, its backers no longer desired a full planning-and-regulatory agency. Through the early 1940s advocates emphasized creating only a power authority. On September 30, 1940, Senator Bone, then backing Secretary Ickes, called for a CVA with a single administrator. That appointee, answering to the Secretary of the Interior, could acquire the properties of private companies on a system-wide basis. Opponents saw this as an attempt at regional power control.[33] Ickes believed that all power in the West should eventually come under the direction of the Interior Department in some way.[34] He felt that without such control, a future president could undo the damage done by Roosevelt to the power trust monopoly.[35] If there was to be no CVA, then Ickes wanted a super power agency with himself at its helm. In this way he would control the electricity and use its cheap cost as the "yardstick" that both Roosevelt and Ross hoped would tame private power.

Private power companies fought the CVA idea more fiercely than they had in 1935.[36] Their trade journal, *Electrical World*, published an article picturing Ickes dominating a giant government-owned power system covering the entire trans-Mississippi West. Surprisingly, the PUDs agreed and fought every CVA proposal for fear of losing their bit of local control. Together they rebuffed all attempts to establish a unified authority. So the effort in 1940 died.

Senator Bone and Representative Martin Fernard Smith, Democrat from Hoquiam, Washington, prepared a new TVA-type bill which they introduced on August 14, 1941.[37] Now opposing Ickes, the new bill proposed a three-person governing board answering directly to the President.[38] In November, Representative Knute Hill, with McNary's support, introduced Ickes's bill for a CVA within the Interior Department. Rufus Woods and members of the newly formed Northwest Public Power Association, which represented municipally owned distribution systems like Seattle and Tacoma, opposed this bill, fearing control by Ickes.[39] Those who backed the CVA idea split over the McNary/Ickes plan (with its one person responsible to the Interior Department) and the Norris-Bone plan (with a three-person board responsible directly to the President).[40] With CVA support so badly divided, the bills all failed.

The Seventy-seventh Congress, which began its first session in January 1941, eventually withstood both Pearl Harbor and a flood of CVA proposals. But Roosevelt seemed again to lose enthusiasm for CVA and even avoided the subject.[41] Preoccupied by the war and peeved by the squabbles among government agencies and Northwest residents, Roosevelt's support was critical. Without it, all the CVA bills of that session failed.

Over the remaining war years the government concentrated on a formidable side issue: the degree to which power sales would subsidize irrigation. This had significance for Grand Coulee Dam where irrigation facilities would, when fully developed, cost twice as much as the power plants.[42] Business interests argued that federal projects should return the principal invested in power and reclamation, plus interest. That would keep the cost of public power high, competitive with private sources. Those who sought public benefit through cheap power and cheap irrigation maintained that the government should collect interest only on the power investment, making public power inexpensive and driving down the cost of privately generated electricity through competition. At the same time the government could use the interest from power to help pay the reclamation costs, on which it charged no interest.[43] Private power concerns feared the competition—exactly what Franklin Roosevelt wanted. They searched unceasingly for ways to increase the cost of public power and diminish the competition. Since no overriding CVA existed to determine these amounts, they became critical issues to be fought out among the various agencies and interest groups, with benefits or penalties distributed depending on who won or lost.[44]

Reclamation proponents looked ahead to the day when the government would irrigate the Columbia Basin, a task postponed by the war. In allocating project costs the government would then determine the price that farmers receiving water would eventually have to repay. That would also affect the rate base for Grand Coulee power. Working toward a solution, the 1943 Columbia Basin Project Act required that the total cost of each project be divided into its various aspects, such as flood control, aid to navigation, power production, and irrigation.[45] This would at least determine the value of each part, indicate how much users needed to repay, and identify the amount absorbed by the taxpayer. It did not, however, settle the matter.

The debate became one between the Bureau of Reclamation, the Bonneville Power Administration, and private power. The Bureau wanted power to pay the costs of power facilities along with the interest, and to subsidize farmers' irrigation costs for forty years. It also argued that the 3 percent interest collected on the power component should help offset reclamation costs. The BPA said it should not pay the 3 percent interest component "to the Treasury to the extent that power subsidy equaled or exceeded the interest figure."[46] In over-simplified terms, the BPA worked to keep the cost of power as low as possible while the Bureau wanted it high enough to cover interest-free reclamation. For Grand Coulee irrigation, the long-run difference amounted to $71 million in repayment charges that the interest on power facilities would or would not eventually defray from the reclamation total.[47]

Ickes finally submitted the issue to the Interior Department's solicitor. On September 29, 1944 the solicitor sided with the BPA, but also stated that the Secretary of the Interior could, if he wished, add to the cost of power all the costs of irrigation beyond the ability of the farmers to pay.[48] The government could apply the interest component of power revenue to repayment of irrigation costs, but it did not have to.[49] As BPA's Paul Raver pointed out to Rufus Woods, with power priced at $17.50 per kilowatt year, Grand Coulee power would easily provide sufficient revenue to pay off each part of the power investment in fifty years, the interest on power at 3 percent, any difference between the water user payments and the cost of irrigation for fifty years, and have a surplus to boot.[50]

The National Reclamation Association, which sought competitive federal power rates, took issue with the opinion and argued that it would lead to extended government control in the West. It vowed to reverse the opinion, although it never succeeded.[51] The debate continued for years with the BPA always trying to lower power rates, the Bureau of Reclamation insisting on higher rates and power-subsidized irrigation, and the National Reclamation Association and private power concerns wanting all facilities to be paid off rapidly to raise rates even higher.

On October 30, 1944 the Bureau of Reclamation and the Bonneville Power Administration issued their *Joint Report on the Allocation and Repayment of the Costs of the Columbia Basin Project.*[52] It stated that irrigation costs would be interest free and the sale of power would pay 3 percent interest—going to offset irrigation—until the year 2017.[53] That settled project component cost allocations and set the conditions that the BPA used to determine public power rates in the Pacific Northwest.

After the war the quest for a CVA resumed. On January 12, 1945, Senator Hugh Burton Mitchell of Everett, at Roosevelt's urging, announced that he would offer a new CVA bill as his first piece of legislation. It called for a weak CVA controlled by a three-person board. That spring Representative Walter Franklin Horan of Wenatchee introduced a second bill to create a Columbia Valley Co-operative Authority with a board of five directors appointed by the President. It would have jurisdiction over flood control, domestic water, navigation, soil conservation, power, recreation, and wildlife conservation. Each of the five board members would be from one of the Pacific Northwest states. They would work within the region and guarantee local control.[54] With two bills specifically aimed at the Columbia Basin, and more for other river valleys, the Seventy-ninth Congress had until January 1947 to decide what to do. To urge them, Harry Truman, shortly after becoming President following the death of Roosevelt, endorsed the CVA idea and recommended fast action. But the National Reclamation Association still vehemently opposed regional authorities.[55] In December 1945, Senator Mitchell introduced a third CVA bill which increased state control over water rights and irrigation. With World War II now over, CVA backers saw this as a crucial moment in their struggle, but it slipped away when Congress took no action.

Two important developments occurred in 1946. First, various federal and state agencies active in the Northwest managed to form the Columbia Basin Interagency Committee (CBIC). The committee included the Department of the Interior, Department of Agriculture, Federal Power Commission, Army Corps of Engineers, Bonneville Power Administration, and over twenty lesser state and local agencies. With the chairmanship rotating among the five major federal bureaus, it was a start toward some type of coordinated cooperation among what had been jealously autonomous bodies. Some speculated that the specter of a CVA someday replacing all of them had considerable bearing on their agreeing to work together.[56]

Second, as anti-communist sentiment emerged following the war, talk of planning and large-scale regional government authorities became unfashionable. Typical of the emotional feeling against such was an anti-CVA editorial in the trade journal *Pacific Builder and Engineer* which read,

Personally, we prefer red in our nation's flag only when it is in close juxtaposition to white stripes, and to white stars in a blue

field. We prefer to support candidates for Congress who believe in the Constitution of the United States. We oppose candidates who would attempt to dilute our constitution with the
ideologies of the socialists or of the commies.[57]

In other parts of the country congressmen, hostile to the cheap
federal power available in the Northwest, questioned the entire BPA
rate structure and, lobbied by the National Reclamation Association, pushed for upward revisions to eliminate differences between
rates from region to region. One plan, backed by the influential
power lobby, sought to shorten the repayment period on public
projects.[58] And then the 1946 election ended congressional control
by the Democrats and brought into office what Harry Truman later
criticized as the Republican "do-nothing Eightieth Congress." It
hardly favored any type of CVA.

On June 25, 1947, Representative Horan introduced a bill to
establish a five-person Columbia Interstate Commission.[59] It repealed the Bonneville Act, took all responsibilities away from the
Corps of Engineers, and reduced the influence of other federal agencies, placing development of the region in the hands of the states.[60]
As an editorial in *Pacific Builder and Engineer* commented,

> When Congressman Walt Horan's Columbia Interstate Com-
> mission bill reaches the hearings stage, watch the fur fly. If
> adopted it would put the quietus on the recurrent "authority"
> thing. It would also kill the Bonneville Power Authority [*sic*]
> and, as far as the Columbia Basin area is concerned, reduce
> the Bureau of Reclamation, Corps of Engineers, and other fed-
> eral departments and bureaus to the status of construction and
> operating agencies which would retain even that status only on
> sufferance of the Commission.[61]

Despite the predictions that "fur would fly," Congress simply
took no action and the bills died. By this time increased demand
for power, the fact that the BPA paid off Bonneville Dam faster than
anticipated, and a power shortage in southern California quieted
critics of the federal dams and their cheap power.

Through 1947 the Bureau of Reclamation fought congressional
attempts to undo the 1944 solicitor's opinion and to disconnect the
link between power and reclamation.[62] Responding to this challenge,
the Bureau developed its so-called Basin Account Plan. Simply put,
it lumped together the over 200 individual reclamation projects

throughout the greater Columbia River Basin. The plan then tied the repayment cost for all the irrigation to the sale of power from all the federal dams. The result, as a Bureau report stated, would be "one harmonious whole" making "full development of the Basin feasible."[63] Power users throughout the Northwest would pay for 92 percent of the costs of irrigation.[64] The account plan made little headway then, survived the 1950s, and finally became a reality in 1966.

The disastrous Columbia River flood of 1948 led to closer cooperation among government agencies. Even before the flood, the Bureau and the Corps had begun working, through the Columbia Basin Interagency Committee, on a revision of the famous "308 Report" of the late 1920s, striving toward an overall development plan for the region.[65] Released on November 10, 1948, the program outlined a $3 billion package including seven major dams on the main stem of the Columbia, irrigation, navigation aids, and additional power for the region. More importantly, embarrassed by the flood and haunted by CVA, the agencies proved that they could work together as they produced the most coordinated plan for the basin yet seen.[66]

Despite the successful "308" revision, the CVA idea refused to die. After his stunning and unexpected election in November 1948, Harry Truman announced that he would back a new CVA effort. On a visit to the Northwest, Truman urged the region's residents to arm for the "toughest kind of a fight" in Congress to see the development of the river basins, to stop floods, and to provide cheap power. He positioned himself against what he called "private power lobbyists," and "men of little faith," who do not trust such development. Truman pointed to the flood and said it "should reinforce our determination to build the dams and other structures needed to control the nation's river basins."[67]

On April 14, 1949, Representatives Hugh Mitchell and Henry M. Jackson and Senator Warren Grant Magnuson introduced a bill to create a Columbia Valley Administration, a name that sounded less threatening, and perhaps less socialist, than the term "authority."[68] The legislation did not eliminate the existing agencies, as had previous CVA attempts. But in Washington state the voters had elected Republican Arthur Langlie as governor. He opposed creation of a Columbia Valley Administration and, like Rufus Woods, favored local control of natural resources.[69] Under Langlie's direction, Washington state authorities investigated the state's power production and ways to acquire and operate Grand Coulee Dam.[70] This got nowhere, but it demonstrated the state's increasingly hostile attitude toward CVA.

Although CVA was not completely dead, the 1949-50 effort was the last with any real conviction behind it. Not a subsistence farming region like the Tennessee Valley, the Northwest did not see profound benefits from a TVA-like governance and it rejected the idea.[71] And locals like Rufus Woods, hammering on the note of local control, persuaded many to oppose CVA. After 1950 the few half-hearted attempts to create the agency amounted to little more than a death rattle. By fighting against it, and finally resorting to cooperation, the existing federal agencies held their ground. Despite all of the effort expended on its behalf, a TVA-like agency in the Pacific Northwest probably never really had a chance.

The CVA failure ended any hope that a comprehensive planning and regulating body would oversee development of the Columbia River and its resources. The Bonneville Power Administration, created as a temporary expedient, became a permanent agency charged with marketing the power generated at federal dams, including that produced at Grand Coulee. Once in place, BPA made itself, if not indispensable, at least vital in the overall power picture. The squabbles between the Bureau of Reclamation and the Army Corps of Engineers, and the reluctance of Northwest residents to surrender their economic future to some unknown and untested body, helped kill the CVA idea.[72] Post-war McCarthyism tainted centralized planning ideals. The Korean War shifted attention away from the CVA issue as well.[73] All of these factors conspired to end any hope for a long-range coordinated regional planning vehicle, or even a lesser agency dealing comprehensively with only power issues.

By the 1950s the government had finally decided who or what would control the power generated at Grand Coulee Dam, at least for the time being. But supervision of that power and the decisions on how, where, and when to generate it remain vital to the Pacific Northwest's economy. The CVA notion is dead, but the struggle over usage of the river and its power goes on.

The completed Grand Coulee Dam, minus the third powerhouse. *Courtesy Wenatchee World.*

# Chapter Fifteen
# World War II and the Break-in Period

The men who conceived the Grand Coulee project wanted to make the wilderness flower. Their brainchild came of age just in time to be drafted.

*Murray Morgan*[1]

Now in Washington and Oregon you can hear
    the factories hum
Makin' chrome and makin' manganese, and
    light alum-i-numm
And the roaring Flying Fortress wings her
    way for Uncle Sam
Spawned on the King Columbia by the big
    Grand Coulee Dam.

*Woody Guthrie*[2]

Like any new piece of equipment, there is a break-in period for a dam. Grand Coulee's lasted over twenty years including the interval of World War II. Then the engineers had to master the idiosyncrasies that give personality to any apparatus. By the early to mid-1960s, Grand Coulee entered its mature phase and began a long period of middle age that some calculate might last centuries. Getting to that point, however, involved contending with a national emergency and a few unexpected bumps and surprises.

Nothing enhanced the image of Grand Coulee Dam as much as World War II. As late as 1940, opponents insisted that it would become a "white elephant . . . unproductive and wasteful."[3] But as Franklin Roosevelt entered his third term and German troops marched through European countries, the nation's newspapers paid more attention to the need for military preparedness. Bureau of Reclamation publicists enhanced their press releases with wartime metaphors including, "Dams can be our forts," while they pictured irrigation canals as "a network of trenches."[4] "Columbia's Power Turning Defense Wheels Here," and "Basin will feed Big Fighting Army" appeared as headlines in the *Wenatchee Daily World*.[5] It was fortuitous that the contractors finished the dam just as the war began. The immediate success of the structure meant that no future Congress, out to embarrass politically Roosevelt or the Democrats, would investigate the dam as a waste of money or a boondoggle. And yet, the dam was not really as critical to the war effort as many believed. Only after the war did it become fully productive and indispensable.

During those war years, and certainly after them, people wrote that Grand Coulee's generators, yielding amazing amounts of aluminum and the atomic fuel created at Hanford, won the war. "Six generators are working at overload to win the war," one publication reported.[6] *Scientific American* gushed,

> The equivalent of a new army of more than six and a half million husky workmen, laboring 40 hours per week, 50 weeks per year, went into action to speed up production of our materials when the Bureau of Reclamation placed in service the third main generator at Grand Coulee Dam power plant.[7]

In an article titled "White Elephant Comes Into Its Own," the *Saturday Evening Post* declared, "Ironically, the power from Coulee and Bonneville that is indirectly winning the war for us today was so much a drug on the market three years ago."[8] *Pacific Builder and Engineer* crowed, "Air power will win the war—and aluminum builds planes. Thanks to Grand Coulee and Bonneville, the Pacific Northwest is now producing approximately one-third of the aluminum produced in this country."[9] The *Wenatchee Daily World* added, "with nearly every kilowatt of this power channeled directly into plants which produce warships, war planes and war munitions, Grand Coulee dam is probably contributing more force directly into the war effort than any other single plant of any kind in America."[10]

In his annual report to the President in 1942, Interior Secretary Ickes wrote, "It is no idle assertion to say that the great war output from the Northwest—planes, metals, chemicals, ships—

would have been virtually impossible had it not been for this progressive power system."[11] Six years later Ickes added,

> The Lord must have been with the United States on this day of our great decision [to build the dam]. Although we continued to be criticized and even ridiculed for building in that great desert area, the greatest structure so far erected by the hand of man, in order to generate great quantities of power, we of course kept steadily ahead and even pushed the work to the best of our ability. . . . If it had not been for Grand Coulee during the war, to say nothing of our other great hydroelectric enterprises, we might still be fighting that war.[12]

President Harry Truman, in Pocatello, Idaho, during the campaign in 1948, said, "Without Grand Coulee and Bonneville dams it would have been almost impossible to win this war."[13] Earl Warren, running for Vice President on the opposing Republican ticket in the same race said,

> Probably Hitler would have beaten us in atom bomb development if it had not been for the hydro-electric development of the Columbia, making possible the big Hanford project which brought forth the bomb.[14]

Germany's surrender before use of the bomb hardly mattered. Grand Coulee provided for the bomb, and the bomb ended the war. Hence, Grand Coulee won World War II. The logic was irrefutable and the myth persists. Recently journalist Marc Reisner wrote of Grand Coulee, "It probably won the Second World War."[15]

Yet it was all an exaggeration. Grand Coulee Dam's contribution augmented those of Hoover Dam, the Tennessee Valley Authority dams, and other hydro and non-hydroelectric projects nationally.[16] The nation as a whole never suffered extensive power shortages or brown-outs. Grand Coulee allowed the government to produce the aluminum and run Hanford, while not disturbing the day-to-day lives of most Americans. The government could have diverted power from domestic uses but Grand Coulee, among other projects, made this unnecessary. Except for inconveniencing the civilian population, little would have changed had Grand Coulee not existed during World War II.

Although Grand Coulee did not win the war, the reverse is true. The rapid industrial expansion and population growth that the war evoked in the Pacific Northwest made a success out of Grand Coulee Dam. Historian Gerald Nash wrote that the war did for the West in four years what might otherwise have taken forty, rapidly building its cities and its industry.[17] The power demand created by the war guaranteed a market for the electricity that some had predicted would become an embarrassing surplus. Even so, most of the original eighteen Grand Coulee generators began to turn only after the guns fell silent.[18] It took a full decade to launch all the units and the war actually slowed the process. Paradoxically, World War II, which accelerated and guaranteed the demand for Grand Coulee's power, also retarded completion of its powerhouses.

Its connection with aluminum has brought much credit to Grand Coulee. The federal government had, as early as 1939, urged expansion of aluminum plants in the West, hoping to increase annual yield from 400 million to 2.8 billion pounds. With the availability of low-cost energy provided by Bonneville and with Grand Coulee eminent, the Reynolds Metals Company began work on a plant in Longview, Washington, while the Aluminum Company of America (Alcoa), which started building in 1939, soon was active in nearby Vancouver.[19] Had the dams not been present, these companies would have built elsewhere. It was the cheap abundant power—not just the power alone—that drew aluminum to the West.

From no aluminum ingot capacity in 1940, the Pacific Northwest moved to 36 percent of the nation's total by 1946, and 40 percent by the Korean War.[20] The aluminum industry developed an almost unquenchable thirst for the power that Grand Coulee and Bonneville dams produced. Despite eventual complaints that the industry was a "power hog," the government all but pandered to its needs, linking the lavish sale of blocks of power to national defense.[21] But the abundance of aluminum came late enough that much of it never saw combat use. By the end of 1943 the War Production Board declared it had aluminum "running out of our ears." It scheduled cutbacks in production and closed a few pot lines.[22] A good guess is that the aluminum for about one-third of the planes built during World War II came from power produced at Grand Coulee Dam.[23]

The power from the two federal dams produced more than aluminum. It supplied energy to the Boeing aircraft works near Seattle, and to the shipyards in Portland, Oregon, and Vancouver, Washington. At the shipyards, Grand Coulee electricity helped transform the steel girders and parts of the trestles once used in building the dam into ships destined for the Pacific. The power also lit and heated the homes of thousands of new residents who immigrated to the Northwest during the war.[24]

Finally, there was the atomic reservation at Hanford, Washington. During the war the whole thing was a big secret. In January 1943 the Bonneville Power Administration received an urgent request for a sizable block of power for an unspecified "mystery" load in the desert between Hanford and Richland. Started in the late fall of 1942, the project eventually employed tens of thousands of workers, inaugurated its first reactor on September 26, 1944, and delivered the first plutonium on February 2, 1945. All the while the army pretended that nothing was happening at Hanford although, by the end of the war, well over 100,000 people had directly or indirectly worked at the facility.[25] It was a use for Grand Coulee's electricity that none of the visionaries who promoted the dam could have imagined. Still, while the Hanford site was conveniently isolated, the job could have been done elsewhere.

To supply these heavy consumers of electricity, the Bonneville Power Administration advanced transmission line construction by two years or more.[26] In October 1940, BPA administrator Paul Raver approved the start of work on a line to link Grand Coulee with the Puget Sound area. Pleading for more money, Raver appeared before a congressional committee in the spring of 1941 and stated that the BPA had already committed the entire anticipated output from Bonneville and Grand Coulee through July 1942.[27] Congress eventually provided money to finish the Seattle link and build a line to Spokane. After Pearl Harbor, Raver announced that a second transmission line would link Grand Coulee with Bonneville and Seattle, resulting in a loop from Grand Coulee to Seattle to Bonneville and back to Grand Coulee.[28]

The big generators that produced the power flowing through those lines were the largest in the world, rated at 108,000 kilowatts (Kw) each, and capable of churning out more if pushed. Despite their somewhat experimental nature, the Bureau of Reclamation planned to have fifteen of the original eighteen installed by the end of 1948. After passage of the Lend-Lease Act in March 1941, orders for needed material rapidly increased nationally and the Bureau agreed to meet the 1948 goal by the end of 1945. The BPA, suddenly short of power, waited for more generators at Grand Coulee and began buying from private or municipal systems.

The first three units in the left powerhouse, L-3, L-2, and L-1, came on line quickly. Unit L-3, the first, began production on October 4, 1941, L-2 on January 29, and L-1 on April 7, 1942. At the same time, the Office of Production Management approved a top rating for the next six generators, and Raver urged Congress to consider quickly installing all the huge units.[29] With three generators running, Grand Coulee became one of the largest power plants in the world.[30]

Congress approved the second set of three generators and then the third.[31] By early 1942, however, planners in Washington, D.C., realized they needed to emphasize increased production at existing power plants rather than building new ones that might take years to finish. Although the War Production Board eventually granted a high priority for the second three generators at Grand Coulee, it withdrew its support for the last three.[32]

At the same time the War Production Board authorized a rather unorthodox proposal. The Bureau of Reclamation was working on the Shasta Dam in California. Two of its five 75,000 Kw generators sat in storage awaiting completion of that facility. Two generators in hand seemed better than three on the drawing board, so the Bureau diverted the units to Grand Coulee.[33] The plan called for them to fill the spaces designed for the seventh through ninth generators in the left powerhouse. The generators in Grand Coulee's left, or west, powerhouse let water enter their turbines from the left, rotating them clockwise. The Shasta generators allowed water to enter their turbines from the right, spinning them counter-clockwise. The engineers put the first Shasta unit in the place for generator L-7, but it received water from the neighboring penstock intended to feed the generator in pit L-8. This required workers to tunnel through an eight-foot-thick concrete wall to divert the needed water.[34] The first Shasta generator produced power on February 25, 1943, and the second joined it on May 7.[35]

Meanwhile work progressed on the second set of three large units for pits L-4, L-5, and L-6 and construction continued on the east, or right, powerhouse structure. Then, on October 27, 1942, the War Production Board officially cut the priority for generators L-7, L-8, and L-9 and also suspended work on the right powerhouse.[36]

Raver repeatedly urged the War Production Board to reconsider, but the delay between ordering the units and their probable delivery made them an increasingly low priority. The fighting, along with power shortages, had at first accelerated the installation program and the drive to build transmission lines. But the war generated the opposite affect, creating a paradox. Without the war, installation of the generators would have followed a predetermined pattern over a number of years. The war created an immediate demand but it also caused a delay that actually brought in the final generators later than anticipated. Unit L-6 went on line on August 9, 1943, L-5 followed on November 8, 1943, and L-4, completing the generators that contributed to the war effort, added power to BPA's system on February 12, 1944.[37] With six large units, two house units, and two Shasta units, the Grand Coulee powerhouse could produce 818,000 Kw, making it, by the standard of the day, a giant.[38] But no more generators came until after the war ended.

The dam required more than generators. Huge transformers altered the voltage before sending the power to its destinations. The men who assembled the outsized machines, refusing to be outdone by their fellows at coastal shipyards, began "launching" transformers which they draped in American flags while rolling them from the construction site to their permanent positions on the powerhouse transformer deck. To top off the occasion, they broke a bottle of champagne over each new unit.[39] Such symbolic acts perpetuated the myth that Grand Coulee won the war.

Because of the dam's perceived significance in the war effort, even an obscure event at Grand Coulee caught the public's attention. When Fred W. Nolte, Jr., a worker at the dam faced with a clogged drainage pipe, tied a string to the tail of a cat and coaxed the animal to run through the pipe, it made news. Nolte only wanted to thread a plumber's snake through the drain; little did he imagine it would bring him a moment of national fame. Reporters descended on the dam to take pictures of the pipe, the worker, and the cat. When they tried to re-stage the heroic moment, the original cat fled and a stand-in took its place. The whole thing demonstrated the dam's place in the national consciousness, where its fame and reputation outpaced its actual contribution, however noteworthy.[40] It also showed the public's need for some light-hearted diversion.[41]

The rush to utilize available power throughout the Northwest strained, although it did not yet exceed, the region's existing generating capacity. As early as mid-1940, Paul Raver had recommended to Secretary Ickes an integration or pooling of all public and private Pacific Northwest power facilities. Because of long-standing competition and animosity, however, the two seemed unlikely partners and some saw this as a move toward government control of private facilities and part of the push for a Columbia Valley Authority.[42] In May 1941, Kinsey M. Robinson, President of the Washington Water Power Company, wrote an article that appeared in *Electrical West*. He criticized government attacks on private power and claimed that "socialization of the power business in the Northwest," was inevitable:

> There is room in the Northwest for both public and private power, with the huge, federally owned projects generating vast amounts of electricity and transmitting it over existing lines where they are adequate, or over newly built lines if necessary, leaving distribution of that power to private companies or public bodies already in existence.[43]

Robinson's seemingly surprising proposal reflected his fear that a successful CVA might some day take over his company. He suggested instead that the government should limit its electrical supply service to a trunk transmission system while the Public Utility Districts and private companies would sell power and distribute it to consumers.[44]

In January 1942, reversing his position, Raver said he saw no need for the power pool advocated by private interests. He and others viewed the whole idea as a ploy to stall public power's progress. But a few months later Raver reversed himself again and arranged an interconnection with the Washington Water Power Company and the Pacific Power and Light Company.[45] The government, despite continued lack of enthusiasm, agreed to supply the two private producers with power so they could conserve water in the reservoirs behind their dams. Two months later, in late July, the War Production Board ordered the Bonneville Power Administration and eight major electric utility systems in the Northwest, using BPA lines for their interchange, to create the Northwest Power Pool.[46] It was a significant breakthrough toward efficiency that has benefited the region ever since.

Everyone involved during the struggle to build Grand Coulee Dam well remembered how the private power interests had insisted that the dam would glut the market with a huge surplus. It was ironic that during World War II, through the pooling agreement, Grand Coulee Dam saved the private power companies, particularly the Washington Water Power Company, when they did not have enough energy available for their customers.[47] The private power companies missed the point and at the end of the war renewed their campaign against the federal dams by again predicting that they would become white elephants in the wake of an eminent power surplus.[48] They predicted that conditions in the Northwest would return to pre-war levels. Government economists, on the other hand, saw a coming power shortage. Acting on their data, the government began preliminary work on Hungry Horse Dam forty miles from Kalispell, Montana, and later McNary Dam at Umatilla Rapids on the Columbia. Well before that, however, the Bureau of Reclamation re-ordered units seven through nine for Grand Coulee. On July 14, 1945 the Office of War Utilities authorized them to replace the abducted Shasta machines.[49]

Removing the Shasta generators took longer than expected because of difficulty extracting them from the concrete housing them.[50] The government hoped for a brief lull in the Northwest's economy after the war, when it would have time to replace the Shasta machines. But the region boomed more than even the optimists predicted and the loss of the two generators, without the immediate availability of the three large ones, exacerbated a growing regional power shortage that overwhelmed both public and private producers. The need to shut down units L-1 through L-6 for routine maintenance and repairs worsened the problem.[51]

During the fall of 1946 the power load throughout the Northwest became so heavy that in peak periods the BPA system suffered frequency drops.[52] It was not during the war, but after it that power shortages plagued the region. The Bureau expedited plans to finish Grand Coulee's right powerhouse and its first three generators. The flood of workers into the area resembled the construction heydays of the 1930s. Nearly 400 temporary house trailers filled spaces in Mason City and the hillside around the dam.

The seventh generator, L-7, the first installed since the end of the war, began producing power on October 20, 1947.[53] During 1948 unit L-8 on February 12 and L-9 on April 23 added capacity to the still-inadequate system. On January 2, 1948 workers began assembling R-1, the first unit in the right powerhouse.[54] For the next three years, while the Pacific Northwest suffered repeated power shortages,[55] the Bureau of Reclamation and its engineers added generators until it completed all eighteen. The left powerhouse at Grand Coulee set a new power production record during 1948, with the unusually cold winters of 1948-49 and 1949-50 adding to the burden. Running the generators at over-full capacity, requiring them to produce as much as 135,000 Kw despite their rating of only 108,000 Kw, drew down the level of Lake Roosevelt and threatened an eventual water shortage.[56]

President Truman pushed a button in the White House placing into operation unit R-1, the first generator in the right, or east, powerhouse. It officially began production on May 20, 1949.[57] Unlike the left powerhouse machines, equipment in the right powerhouse rotated counter-clockwise. Generators R-2 and R-3 joined the BPA system in July and September of 1949. The second set of three generators in the right powerhouse went on line during 1950.[58]

On May 11, 1950, President Harry S Truman arrived in person to dedicate Grand Coulee Dam.[59] Accompanied by his wife, Bess, his daughter, Margaret, and his Interior Secretary Oscar Chapman, Truman delivered an address in Mead Circle, in the center of what had been the construction town of Mason City. He reminded the crowd of the fight that opponents had waged against Grand Coulee and the claim that nobody but jackrabbits and coyotes would buy the power it produced. He pledged development of a growing, dynamic nation, despite the private power lobby. With that, he christened the dam and "officially" started generator R-4.[60]

The Bonneville Power Administration completed work on new transmission lines between Grand Coulee and the coast in time to accommodate unit R-5, that began work on July 13; R-6 produced its first power for the power pool on October 7.[61] Some looked ahead to the day when Grand Coulee might produce a power surplus. California Representative Clair Engle introduced a bill that called for an interconnection between the regions, but it died. In 1951 linking the Northwest with the Southwest was an idea before its time.[62] Grand Coulee achieved its full complement of eighteen large units and supporting house generators during 1951. On April 13,

June 1, and September 14, units R-7, R-8, and R-9 began turning. Thus the first generator began producing in 1941, at the start of World War II, and the final machine started in 1951, during the Korean conflict.[63]

Workers performed other tasks at Grand Coulee through the late 1940s and early 1950s besides adding generators. Despite the experiments during construction, operation revealed unforeseen design flaws.[64] The force of water sliding down the spillway dug pits in the concrete—an action called cavitation, where water forms vacuums which tear away pieces of material. The water and debris then swirl about in the bucket or curved spillway bottom. In one sense, the bucket is nothing but an immense gutter about thirty feet deep designed to absorb the momentum of the falling spillway water by somersaulting it over on itself. The resulting force dislodged sand, gravel, and even huge boulders from the downstream riverbed and sucked them, against the current, into the bucket. This aggregation of abrasive materials rolled around grinding and lifting up chunks of concrete, leaving big holes. Downstream, the volume of water, increasing and diminishing during the day, eroded the river bank, causing slides and threatening buildings—in a way a continuation of the slide problem that plagued construction in the 1930s.[65] In January 1943 men chipped away spalled concrete on the spillway face and replaced it with fresh material. That year two divers entered the cold water of the river at the base of the dam and inspected the spillway bucket to verify the damage.[66] Frank Banks reported that it was badly eroded with holes up to twenty-eight-feet deep.[67]

By 1948 one of the government design engineers estimated that full repairs might cost as much as $500,000 yearly for ten years or more, with annual maintenance running between $200,000 to $400,000 after that.[68] To remedy the problem, government engineers designed a floating caisson that workers could move from place to place in the bucket, remove water from the area, and make repairs.[69] Named the "Iron Duke," the unique craft had a bottom shaped to fit the bucket's contours. With preparations complete, the government combined the three problems (the spillway bucket, spillway face, and river channel) into one operation and put the package up for bid. On January 14, 1949 the nearly $2 million award went to the Pacific Bridge Company of San Francisco.[70] The company worked through 1949, placing riprap along the banks of the river below the dam, using the floating caisson to repair the spillway bucket, and patching the spillway face. The government negotiated a second contract for over $2.5 million in additional work with the same company in August 1950.[71] Even this did not solve the problem and the riverbank erosion continued into the 1990s.

The dams on the Columbia River generated the base power load for the region. This meant they produced the electricity always needed while other sources supplied power for heavy loads (so-called peaking power) in the mornings and evenings. Keeping the machines running at a constant rate minimized the damage caused by the rise and fall of water in the river below the dam. Consequently, the Bureau of Reclamation rejected suggestions that it use Grand Coulee generators only for peaking power. When the power shortages of the early 1950s ended, however, and when the push to build more dams ran into environmental concerns, the power sold during the hours of heavy use became increasingly valuable. In the late 1960s and early 1970s the Bureau reversed its position, and today the generators at the large dams cover peak loads while thermal units run constantly to fill base load needs.[72] This has, as predicted, increased riverbank erosion.

When Congress authorized Grand Coulee in 1935, it listed flood control as one of the dam's major objectives,[73] and the dam's backers always said that Grand Coulee would end floods downstream. Frank Banks estimated that Grand Coulee's annual flood control benefits would be $5 million.[74] Secretary of the Interior Harold Ickes, at the bid opening for the second construction step on the dam, reiterated "It will serve to reduce flood peaks."[75] Reclamation Commissioner John C. Page repeated the flood control claim a year later while at the same time *Atlantic Monthly* carried an article claiming that "The dam will control floodwaters down the Columbia 450 miles to the sea."[76] In spite of all the flood control hoopla, however, it is significant that when the government allocated the costs of the dam to its various functions, the amount indicated for flood control and navigation combined came to only one million dollars. Despite its apparent large size, Franklin D. Roosevelt Lake is a very small reservoir on a very big river.[77] Lowering water in the lake increases the volume needed to produce power, making the electricity more costly. The Bureau of Reclamation thus resisted efforts to use the dam for aggressive flood control by keeping the lake full, maximizing power production and irrigation potential.

Grand Coulee's inability to control major flooding became apparent in May 1948. An unusual period of warm weather in the mountains, where snow had accumulated 20 to 40 percent deeper than usual, combined with sudden rain storms in April and early May. By coincidence, flows on both the Snake and main stem of the Columbia peaked at the same time. The resulting rapid rise of the Columbia startled even old timers. By May 24 the Columbia reached the previous year's peak, which they thought was high. Flooding then affected three states with more damage predicted, some saying it would exceed the record 1894 flood. The river had reached a flow of 347,000 cubic feet per second (cfs) at Grand Coulee, three and one-half times the average.

On May 29 the dikes that protected the city of Vanport, located midway between Vancouver, Washington, and Portland, Oregon, a temporary city originally built to house shipyard workers during World War II, broke and the Columbia washed away one of Oregon's larger cities. Upstream the Columbia continued to rise and at Rock Island the operators shut down when the difference between the upstream and downstream levels of the river became insignificant. By then the flow of the Columbia at Grand Coulee reached an estimated 573,000 cfs.[78] It peaked on June 12 at 633,000 cfs and to those who lived there, it seemed as if everything in much of Idaho, Washington, and northern Oregon was under water.[79]

At Grand Coulee the high water tested the structure, which survived, but not without some damage. The river banks eroded, driftwood and other debris piled against the back of the dam, and the left powerhouse lost an access road.[80] The action of the water further pitted the face of the dam and deteriorated the spillway bucket.[81]

The Army Corps of Engineers and the Bureau of Reclamation suffered considerable embarrassment. The Bureau argued that Grand Coulee Dam could not have eased the flood but did not actively publicize its findings. In an internal letter, Bureau engineer Sinclair O. Harper wrote, "The ability to control floods at Grand Coulee Dam is not such as to provide material for attractive publicity."[82] Grand Coulee supervisor Alvin F. Darland wrote to the Boise office that using storage space in the Roosevelt reservoir would not be a practical way to reduce downstream flooding.[83] A Bureau memorandum estimated that the modifications at Grand Coulee needed to achieve real flood control would cost at least $3.5 million.[84] Even then, if the full active storage capacity of 5,200,000 acre-feet was available every spring, it would hold only around 10 percent of large floods and less during another flood like 1948.[85]

Like it or not, the Bureau realized that, if only for public relations, it must do something about flood control. As one of the Bureau's regional directors put it,

> In view of the adverse publicity which nonuse of [the] Grand Coulee Reservoir space for flood control resulted in last year, it seems advisable to investigate the possible use of 1,000,000 acre-feet as suggested by Colonel [Theron D.] Weaver. Regardless of the actual benefit that can be achieved, a concerted effort on our part might, at least, have considerable psychological value.[86]

In April 1950 the Bonneville Power Administration, Corps of Engineers, and Bureau agreed to use 900,000 acre-feet of storage at Grand Coulee for flood prevention. This would involve lowering the level of Franklin D. Roosevelt Lake about six inches daily for about ten days each spring, providing space to absorb the high run-off.[87] Under continued pressure over the next two years, the Bureau again considered increasing this to perhaps five million acre feet.[88] Hungry Horse Dam also helped, providing a lake forty miles long with storage capacity of 3,500,000 acre feet, thus decreasing the peak flow downstream.

The flood control plan weathered its first test in 1954. The combined measures reduced the flood crest at Portland by over one foot.[89] The Bureau felt more pressure to use Grand Coulee for flood control. In 1961, Reclamation Commissioner Floyd Dominy responded to Senator Henry M. Jackson,

> Although the Rivers and Harbors Act of 1935 which authorized Grand Coulee Dam specified flood control as one of the project functions, the flood control benefits anticipated at that time were small and incidental to other project operations. Flood control advocates generally believed then that adequate flood control could be achieved solely through the construction of levees. As a result, Grand Coulee Dam was not designed specifically for flood control operation. The disastrous flood of 1948, however, showed conclusively that control of large floods could not

be achieved by levees alone and would require, in addition, reservoir storage. Since that time Grand Coulee Dam has been operated to achieve significant reduction of lower Columbia river floods and is an important element in the plans of the Corps of Engineers to provide comprehensive flood control for the Columbia River.[90]

Dominy continued that further modification at Grand Coulee would be economically unjustified if an agreement with Canada allowed for the construction of storage dams above Grand Coulee. They would solve the flood problem, even out the flow of the Columbia River and, at the same time, allow the downstream dams to generate more power. Until those dams materialized, however, the government credited Grand Coulee with significant flood control benefits. Those ranged from $79,000 in 1960 to a peak of $25 million in 1956, and depended on the degree of the flood in any given year.[91]

Lowering the level of Roosevelt Lake provided a curious side benefit. The outlet tubes drew water from deeper in the lake than the spillway, which carried water off the surface. Water through the generators and that from the tubes was four to eight degrees colder than surface water. The Atomic Energy Commission at Hanford paid the Bureau of Reclamation thousands of dollars to offset the cost of operating the tubes to receive the chilled water used in its cooling operation downstream.[92]

In the 1940s and 1950s the Bureau dealt with power shortages and with the vagaries of the Columbia River's annual floods. In addition, on-site engineers learned how to operate their facility and, in doing so, made one spectacular mistake. In 1951 Congress cut the Bureau of Reclamation's budget, leading to layoffs.[93] The Korean War brought on increased security at the dam, including locked doors and measures to prevent sabotage. The personnel shortage and the restricted access precipitated a near disaster.

The problem centered on the outlet tubes. Those tubes are large tunnels going through the dam from upstream to downstream, with a gate at each end. There are also gates or manholes in the dam's passageways that provide access into the tubes. On March 14, 1952 two workers inadvertently opened the upstream gate on one of the tubes. Unbeknownst to them, that tube had a loose manhole cover and a closed downstream gate. Once in the tube the water, under great pressure, had nowhere to go except into the dam. It threw the 200-pound manhole cover up to the cement ceiling and sent it reeling down the passage. The men fled. Thirty-four degree water poured through the passages at the rate of 70,000 gallons a minute, flooding walkways, rushing into elevator shafts, and short-circuiting sump pumps in the lowest chambers. It took a few minutes for operators in the powerhouses to realize that something was dramatically wrong. Soon water reached the west, or left, powerhouse. Temporary dikes built by workers gave away and the water headed for the turbine pits, flooding L-9 and L-8. Minutes later water reached the right powerhouse. It floated away the lubricating oil that kept the bearings in the turbines from burning up, threatening the machines. Senior operator Roy Peterson kept the machines running as long as possible for the sake of the Northwest Power Pool that, in the early 1950s, needed all the power it could find. For Peterson, it was a gamble. If water damaged the machines it could be months or longer before they returned to service, plunging the Northwest into a monumental power crisis.

Workers heroically tried to reach the switches that operated the upstream gates, hoping they still worked. If the generators failed before they closed the gate, they might lose the entire plant. The locked doors made this more difficult, frustrating workers as they tried to reach the trouble spot. Finally Norman Holmdahl, a well-liked Puget Sound native who had worked at the dam since 1937, managed to withstand the frigid water and its current. He reached the switch, threw it, and miraculously the machinery still worked, despite the water. Generators L-9, L-8, and R-9 were already off line. In all nine turbines were under water with six of them still running!

Twelve generators still operated at Grand Coulee, though precariously. Around the Pacific Northwest the word traveled fast. In Tacoma and Seattle public and private power producers activated steam plants reserved for emergencies. The aluminum companies, which always asked for three days notice before losing power, had three minutes. Radio announcers urged housewives to postpone dinner or not turn on the family stove or oven at all that afternoon. As generating capacity dropped below the ability of the entire Northwest system to provide power, the electricity the remaining generators produced fell under the usual sixty cycles. Lights dimmed and clocks slowed throughout the region.[94]

Norm Holmdahl became a hero, although his effort went largely unsung.[95] With the water now shut off, workers drained away over ten million gallons that had relentlessly savaged the inside of the dam and powerhouses. They quickly pumped out the turbine pits and returned one generator to service by 4:30 that afternoon. At night, after the power demand dropped, engineers inspected the generators and discovered that, despite the water and the gamble that kept most of them running in the face of catastrophe, none suffered appreciable damage. On March 15 all the generators went back on line.[96] If nothing else, the flood demonstrated the critical role played by Grand Coulee in the Northwest. Understandably the Bureau of Reclamation tightened its procedures and vowed that no similar error would ever threaten any of its facilities.

With the lesson of the 1952 incident firmly in mind, and the real or imagined threat of the cold war hovering in their consciousness, Northwest residents ruminated about the security of Grand Coulee and other Columbia River dams. "What if Grand Coulee Hit by H. Bomb?" asked a June 1954 *Wenatchee Daily World* headline.[97] That August the government held a civil defense exercise, dubbed "Operation Floodout," investigating the results of a speculative devastating strike against the dam. It projected inundation of the main industrial areas of Portland along with considerable downstream damage.[98] The fear of an attack on the dam continued for a few years. In 1957 the *Wenatchee Daily World* carried on its front page an artist's rendering of Grand Coulee as it might appear after being hit with an atomic bomb.[99] When authorities found explosives in the tourist center near the dam, right after a ceremony involving 25,000 people, some felt the fears of an attack justified.[100] In 1959, Admiral Hyman G. Rickover, while in Seattle, specifically named Grand Coulee Dam as a nuclear bomb target. Only in the 1990s did such fears subside. It is amazing, however, considering the number of dams that have failed in this century, from the St. Francis Dam in California to the more recent Teton Dam disaster in Idaho, that nobody expresses concern over the safety of the Columbia River structures.

The people around Grand Coulee Dam realized that the end of powerhouse construction in 1951 meant a drop in local population and they turned to recreation as a substitute. The Bureau estimated that nearly 300,000 visited the dam each year through the 1950s. As during construction, it conducted talks at the vista houses. A diesel locomotive pulled flat cars, fitted with bleacher seats, from the vista house to the left powerhouse, taking tourists on a brief jaunt through the project. To encourage people to stay over night, the Bureau added a twenty-minute display of colored lights playing on the face of the dam. Installation of the equipment cost over $150,000 and the mechanisms finally went into place late in the summer of 1957.[101] The light show officially began during the 1958 season and Fred Seaton, Secretary of the Interior, threw the switch that first illuminated the dam.

The communities at and around the dam developed annual celebrations connected with Grand Coulee. Grand Coulee residents linked their festivities to the colored lighting. Their annual "Colorama" lasted well into the 1970s when it diminished, and finally died away entirely. Soap Lake celebrated its annual "Suds 'n' Sun" festival. In 1963 its residents joined to honor the thirtieth anniversary of the dam with an unusually large festival. The only disappointment was the inability of President John F. Kennedy, who had declined an invitation, to attend the September 9 ceremony. But large groups of vacationers have not come and the communities have not prospered. The industrial aspect of the agricultural/industrial empire that Rufus Woods sought for eastern Washington did not develop as the power traveled elsewhere. The area around the dam, except during periods of construction, is almost as unpopulated as ever.

As for the dam itself, by the end of the 1960s it had weathered its break-in period, survived an onslaught by nature, and although it did not win World War II, was a valuable resource critical to the region if not the immediate vicinity. Now the largest power plant in the world, it waited only for the addition of more generators to increase its capacity and realize its full potential.

Perhaps even more than the power, the war, and all the aluminum, one thing during those first years helped immortalize Grand Coulee. In May 1941 the Bonneville Power Administration hired folk singer Woody Guthrie for thirty days. In that brief period he produced twenty-six songs about the Pacific Northwest, with most concerning the Columbia River and its dams. In one entitled "The Grand Coulee Dam," Guthrie wrote, "Uncle Sam took up the challenge in the year of '33 for the farmers and the workers and for all humanity—From the rising of the river to the setting of the sun, the Coulee is the biggest thing that man has ever done."[102] And in "Roll on, Columbia," Guthrie repeated, to the tune of "Goodnight Irene," "And far up the river is Grand Coulee Dam, The Mightiest thing ever built by a man, To run the great factories and water the land, It's roll on, Columbia, roll on."[103]

May 15, 1948: turning on the first water to enter the Columbia Basin Project.
*Courtesy Bureau of Reclamation, Columbia Basin Project Office.*

# Chapter Sixteen

# Irrigation Comes to the Columbia Basin

It has always seemed to me that projects should be examined
and authorized primarily on the basis of the greatest good to
the greatest number of people. I understand the urge to get
Federal money for expenditure in every congressional district
of the United States but I believe, as I have often recommended
to the Congress, that instead of taking a bite here and a bite
there, there should be a well-considered and well-rounded plan
for projects to be undertaken in a definite order of human and
national preference and desirability instead of putting them only
on a local geographical basis.

*Franklin D. Roosevelt, 1939*[1]

It is easy to forget that the government built Grand Coulee Dam to
irrigate the Columbia Basin, the nation's largest reclamation project.
In 1933, when the New Deal assumed the project, it took the vision
of the agricultural/industrial empire and added the dimension of
planning.[2] President Roosevelt encouraged planning at the federal
level and appointed advisors and cabinet members with similar
views, including Raymond Moley, Adolf A. Berle, Milburn Lincoln
Wilson, and Rexford Guy Tugwell.[3] In its early stages, the New
Deal planners wished to create organized rural communities, and
they intended to plan from the top down with guidelines coming
directly from leaders at the federal level.[4]

Their goal was what historian Richard Lowitt called the
"Planned Promised Land."[5] Free from the vagaries of rainfall—because of guaranteed water delivery—and with its economy controlled by the government, the project would accommodate
displaced dust bowl refugees on thousands of subsistence farms.[6]
But by 1948-52, when the first acres went under the ditch, changes

in national politics and in the economy, combined with pressure from farmers and area business people forced planners to modify project guidelines. In the process, New Deal hopes for population redistribution, promotion of the family farm, and large-scale economic and social planning died.[7] The size of project farms grew and the trend toward large-scale and even corporate enterprises began.

Those changes took time. In the beginning the planners insisted on careful preparation and investigations to guarantee that the Columbia Basin would produce the desired outcome. Preliminary work began in 1935, well before the start of World War II, when Congress provided $250,000 for economic surveys.[8] These studies, which continued through 1942, included topographic mapping to identify irrigable acreage, a search of ownership records, and detailed soil classifications. Additional appropriations in 1937 and 1938 brought the survey costs to well over a million dollars and at times the work involved nearly 200 people.[9]

The surveyors also searched for potential storage reservoirs, dam sites, and canal routes.[10] The largest study looked at the ancient Grand Coulee, where the Bureau of Reclamation anticipated creating a large equalizing reservoir.[11] Here they would store water pumped up from behind Grand Coulee Dam before it moved into the project's distribution system. The surveys were the most comprehensive preliminary investigations undertaken by the government on any reclamation project.[12] They covered nearly two million acres, 1,095,000 of them able to receive water, and divided that land into six classes indicating which to and which not to irrigate.[13] Record searches showed that individuals and banks owned 85 percent of the project, the federal government 5 percent, the State of Washington 5 percent, and the Northern Pacific Railroad retained 5 percent—property left from its nineteenth-century land grants.[14] Most of the area's 10,000 residents lived in the south, near Pasco.

After 1933, when construction started at Grand Coulee Dam and it became clear that some day the government would irrigate the basin, land prices from Ephrata south to Pasco rose rapidly and rumors of speculation soon reached the national capital. Rufus Woods quoted Interior Secretary Harold Ickes as threatening to abandon the project if inflated values interfered with government plans.[15] Then the Interior Department sponsored a law to control the situation. Passed on May 27, 1937, the Anti-Speculation Act limited each project farm to forty acres, regardless of soil characteristics. A husband and wife together could own eighty acres. After the government appraised the land based on its value before irrigation, the present owners could sell excess holdings only at or below that price. The law required the landowners to form an irrigation or reclamation district that would then sign a contract with the government providing for repayment of construction charges.[16] No previous reclamation project had such restrictive land ownership regulations. In fact, within the next three years the government exempted both the Colorado-Big Thompson Project and the Truckee Project in Nevada from even the traditional, and more generous, 160-acre limit, realizing that farmers needed larger plots to succeed and that the Bureau of Reclamation often allowed farmers to hold larger parcels despite the law. Ickes's advisors warned him that the Bureau usually failed to regulate land acreage limits and that it had a poor record on land redistribution.[17] The Secretary still insisted on the strict rules, vowing that the Columbia Basin Project would become the successful model that others would later follow.

In 1939 a new Reclamation Act, designed largely by geographer Harlan H. Barrows of the University of Chicago, granted irrigators on new projects a ten-year free period before starting the traditional forty-year repayment schedule.[18] The act added that some costs could be charged off to flood control and navigation and based the farmers' obligation on their ability to repay rather than actual construction costs.[19] The law directed the government to develop new reclamation projects in an orderly way by dividing them into blocks, each with a set number of units. It would then irrigate those blocks following a predetermined pattern on a specific schedule. This law embodied the heart of the New Deal planning concept,[20] and it applied directly to the Columbia Basin Project. Within its guidelines the government moved to build the "Planned Promised Land."

The first step required that project landowners form an irrigation district. On July 29, 1937 over 1,000 farmers gathered in Ephrata to initiate that process.[21] But forming a district proved more difficult than anyone had imagined. Although the Bureau of Reclamation urged quick action, rivalries among farmers living in different sections of the project complicated the process. Members of the old Quincy Valley Irrigation District, which still represented over 400,000 acres, or about one-third of the project, approached Frank Banks for assurances that the government would develop their land first. The other areas sought similar guarantees.[22] Residents on the west side of the district feared that Spokane interests would arrange to irrigate the eastern area first so that they could control the project.[23] The same jealousies that had plagued backers of the dam remained.

The Bureau conducted organizational meetings through the rest of 1937. However, James O'Sullivan and others had already concluded that area rivalries made its "one district" concept untenable.[24] Besides the competition and distrust on the west side, east side wheat farmers, unwilling to sell excess land as required by the Anti-Speculation Law, now wondered if they even wanted irrigation, further complicating the discussions.[25] Nevertheless, the Bureau insisted on one district. Landowners around Quincy organized, requesting that O'Sullivan help them form a separate district. Commissioner John C. Page warned that their attitude could jeopardize the whole Columbia Basin Project.[26] But at the same time he added that the government had little sympathy for the large landowners on the east side and that it would act first on the west side. The Bureau wanted to develop the land, and its officials, including Commissioner Page, told factions whatever they wanted to hear to achieve that end.

A few weeks later Page weakened and agreed to accept two districts, but no more than two. O'Sullivan then urged the farmers to move quickly, fearing that New Deal money might run out, resulting in long delays. The Bureau of Reclamation, irritated by the bickering, hinted it could begin irrigation as soon as 1942 if the landowners immediately formed districts. In October 1938 members of a now formally structured Quincy-Columbia Basin Irrigation District, organized by O'Sullivan, filed a petition with the Grant County board of commissioners asking for a vote to legalize their district.[27]

On February 18, 1939, Grant County officials held an election to ratify the agreement. Organizers arranged for special low-cost excursion trains from Spokane to Soap Lake and from Seattle to Quincy so that the hundreds of absentee landowners could easily travel to the polls. Under Washington state law, the 500 out-of-state owners could not vote. Successful district formation required two-thirds approval by the participating landowners; 709 voted in favor and 34 opposed, the first step toward fulfilling the requirements and acquiring irrigation.[28]

Landowners on the project's east side then began organizing a second district, but their talks went badly.[29] People in the south expressed concern that the east would usurp early water deliveries, delaying for years irrigation in their area. When he realized that the east might divide into two areas, Commissioner Page suggested pumping water from the Columbia River near Pasco to expedite irrigation in the south. Despite this, his effort to hold the east side together failed.[30] Franklin County commissioners announced in September that they wanted a separate south district.[31]

By the end of October the county commissioners had established boundaries for the East-Columbia Basin Irrigation District and the South-Columbia Basin Irrigation District. Both proposed districts held successful authorization elections on December 9, 1939 and again, mostly absentee landowners participated.[32] To make things tidy, members of the Quincy Irrigation District met in October and dissolved their older organization.[33]

With the districts formed, the government could legally develop Columbia Basin Project irrigation. But federal reclamation experiences, particularly in the Upper Rio Grande Basin and along the Mississippi River, led New Deal planners to advocate further social and economic studies before construction. The Washington State Planning Council and the Pacific Northwest Regional Planning Commission also recommended more detailed investigations.

On May 8, 1939, Reclamation Commissioner Page took Harlan Barrows to lunch. Barrows, active in Tennessee Valley Authority planning, was a member of the Mississippi Valley Committee and was instrumental in the 1939 reclamation law that affected the Columbia Basin. Page told Barrows that the Columbia Basin Project was like a blank sheet of paper where the government could create a project as nearly perfect as planning could devise.[34] It was an opportunity for the academic to apply his theories and create the model reclamation project. To that end Barrows agreed to supervise what became the Columbia Basin Joint Investigations.[35]

The Joint Investigations explored twenty-eight potential problems that fell into sixteen categories.[36] Government analysts enlisted federal, state, and local agencies, including many in the private sector, such as railroads and chambers of commerce, each pursuing questions about which they had particular interest.[37] By the time the work concluded four years later, over 300 people, representing some forty agencies, had participated in the process.[38] Without question, taken together with the earlier surveys, the end product was the most comprehensive economic, social, and technical planning program ever undertaken before starting an irrigation project.[39]

By 1943 the studies were complete, and through 1945 the government published all twenty-eight. The outcome of the work envisioned the orderly development of all sections of the project, guaranteeing a balance of trade and services for residents. The studies determined farm unit sizes, the potential income per acre, and the ability of farmers to repay irrigation costs. But by then world events, including World War II, had already revitalized the national economy and altered some of the assumptions that guided the work. For example, dust bowl refugees of the 1930s had found employment, but the specter of war veterans needing farms in the post-war years loomed large.

"After the war, the threat of terrific economic dislocation and unemployment will be relatively greater on the West coast than anywhere else in the country," warned an article in *New Republic*.[40] The authors anticipated a post-war depression and urged development of the Columbia Basin as a public works project providing employment and settlement opportunities for war veterans "on a huge scale."[41] Meanwhile war needs drew attention away from domestic projects and the rapidly dwindling group of planners never adjusted to these changing conditions. As a Department of Agriculture economist later lamented, "We were planning *for* a group about whom we knew very little, and were not planning *with* them."[42] The people who farmed the Columbia Basin in the 1950s were not the people pictured by the planners in the late 1930s.

In the spring of 1943 the War Department appropriated approximately 9,000 acres of irrigable land near Moses Lake for use as an air base.[43] Further withdrawals came later, both there and on property connected to the atomic project at Hanford. This military

Pouring concrete at the east powerhouse, 1942. The "elephant's trunk" below the bucket channeled concrete to the exact site where needed. *Courtesy Bureau of Reclamation.*

Above: Grand Coulee street scene, 1938. *Courtesy Eastern Washington State Historical Society, no. L87-1.13899-38.*
Below: Inside the Silver Dollar on Grand Coulee's B Street, *c.* 1939. *Courtesy Bureau of Reclamation.*

Above: Consolidated Builders, Inc. store in Mason City, 1942. Below: Bank and post office in Mason City, 1935. *Both courtesy Bureau of Reclamation.*

The last concrete pour at the dam, November 12, 1942. *Courtesy Wenatchee World.*

The floating bunkhouse and boats used to clear the reservoir behind Grand Coulee Dam, 1942. *Courtesy Bureau of Reclamation.*

Sheep crossing the dam on their way to summer range, 1943. *Courtesy Bureau of Reclamation.*

A turbine pit under swirling water during the flood of
1952. *Courtesy Bureau of Reclamation.*

Frank Banks and Franklin Roosevelt at the dam. *Courtesy Wenatchee
World.*

John C. Page. *Courtesy Bureau of Reclamation.*

Michael W. Strauss. *Courtesy Bureau of Reclamation.*

Floyd E. Dominy. *Courtesy Bureau of Reclamation.*

Harlan H. Barrows. *Courtesy Department of Special Collections, University of Chicago Library.*

Donald Dunn, center, watches Bureau of Reclamation Commissioner Michael Strauss open the gate allowing water onto the land Dunn won in the "Farm in a Day" contest, 1952. *Courtesy Wenatchee World.*

President Harry Truman visits a display in Ephrata showcasing the produce of the Columbia Basin Project, 1952. *Courtesy Bureau of Reclamation.*

A concrete-laying machine lines the West Canal. *Courtesy Wenatchee World.*

Princesses of the Washington State Apple Blossom Festival pour water gathered from 48 states into the main canal, symbolizing the contribution of all the states to the Columbia Basin Project, 1951. *Courtesy Bureau of Reclamation.*

Installing one of the pipes that carry water up the hill from the pump plant behind the dam to the equalizing reservoir in the Grand Coulee, 1947. *Courtesy
Bureau of Reclamation.*

Blasting the east end of the original dam to make way for the third powerhouse, 1969. *Courtesy Bureau of Reclamation.*

Senator Henry M. Jackson at ground-breaking ceremonies for the Second Bacon
Siphon and Tunnel, 1976. *Courtesy Bureau of Reclamation*.

Water from the Main Canal pours over cliff into Billy Clapp Lake, forming Summer Falls. Glade Walker photograph. *Courtesy Bureau of Reclamation.*

activity brought an added and unanticipated urban population to the Columbia Basin area, further changing its characteristics. The Joint Investigations were largely outdated, inappropriate, and flawed before the government even printed them.

Nevertheless, the Columbia Basin Joint Investigations provided the basis for the Columbia Basin Project Act which Congress passed in 1943. Sponsored by the Bureau of Reclamation and the Interior Department, the new act replaced the Anti-Speculation Act of 1937.[44] It outlined a four-point plan giving the government authority to buy private property, to resell that land in sizes based on soil quality, and to help settlers with federal loans.[45] Representatives of the three irrigation districts generally backed the new law but voiced apprehension about the power granted the federal government, fearing that Secretary Ickes, or one of his successors, might create a system of "agricultural peonage" in the Columbia Basin.[46] In April the Interior Department submitted an altered bill with changes that reflected those concerns. The House of Representatives Interior Committee made changes further reducing the power of the Interior Department and sent the bill to the full House where it passed. Now called the Columbia Basin Project Act, the Senate approved it and President Roosevelt signed it on March 10, 1943.[47]

The new law stipulated that the government could not deliver water until the irrigation districts and the Bureau of Reclamation signed contracts. The government could establish townsites on project lands and the multipurpose nature of the project allowed power revenues to subsidize irrigation.[48] Most significantly, farm units could range from ten to 160 acres depending on land quality, and owners of record before 1937 could retain up to 160 acres regardless of land quality. For five years after the first delivery of water, the farm owner must not sell his land for more than the appraised value before improvements. This law, passed before any water arrived, started the trend toward increasingly larger land holdings, and that trend has continued.

The new law allowed farmers to withdraw their land from the project with the understanding that then they would receive no water. World War II brought higher wheat prices and, coincidentally, the start of a wet weather cycle that dampened both the land and farmer enthusiasm for irrigation. As wheat prices soared, the farmers saw no reason to sell excess lands, particularly with farming dry wheat now profitable, perhaps more so than a future speculative venture into irrigation on a small plot.[49] By 1946 east side farmers withdrew over 300,000 acres—nearly a third of the project total.[50]

The land withdrawals were a result of factors unique to the Columbia Basin Project. Not all the land was always arid; farmers had successfully raised grain on the east side since the 1880s. Only during dry cycles or when grain prices dropped was their interest in irrigation heightened, and even then they produced a paying crop if they owned enough acres. Unless conditions suited their purposes, they saw no reason to work with the Bureau of Reclamation and a few refused to comply with stringent land ownership restrictions under any conditions. They wanted cheap subsidized irrigation and the right to keep all the land they owned.

The land withdrawals made east side canal construction impractical.[51] Bureau planners, however, took the position that eventually the government would irrigate most of the land and while they postponed indefinitely construction of east side facilities, they nevertheless drew blueprints for those works.[52] But this period was a watershed in the history of the Columbia Basin Project. After 1946 the Bureau of Reclamation no longer considered the project's immediate full development. Instead of building the project all at once, the land withdrawals resulted in the piecemeal construction of irrigation facilities stretched out over a number of decades.

The loss of almost one-third of the project's land traumatized Bureau officials. Although they put the best face on things, they now understood that in the Columbia Basin, where many of the farmers could get along without them, they could not make assumptions or act with impunity. Despite all the planning, forces over which the Bureau had little control would determine, at least in part, the direction of the project.

There were other changes. Nationally, planning fell into increasing disfavor.[53] In 1943, Congress abolished the National Resources Planning Board (NRPB) which had been the main force behind the national planning effort.[54] This meant that subsequent planning for the Columbia Basin Project occurred locally, although perhaps not always by local people alone.[55] It reduced the chances of easily making corrections in project guidelines. The Columbia Basin Project was now trapped with an outdated master plan, and no mechanism existed to remedy the problem.

Determined to push ahead and realizing that competition for post-war dollars might be fierce, irrigation promoters, especially representatives of the three Columbia Basin irrigation districts, asked the Washington legislature to form a new Columbia Basin Commission.[56] They hoped it would fill the void left by the loss of the national planning board. In the post-war years they wanted a viable state agency to promote Columbia Basin reclamation. The legislature acted in April 1943 and formed the third Columbia Basin Commission that, over the next two decades, promoted the project and lobbied the federal government for funds.[57]

The government needed to negotiate repayment contracts with the three irrigation districts. Commissioner Page had ordered preliminary work on the contracts early in 1940. Before the war interrupted construction plans, Page had urged fast action on the contracts because he anticipated delivery of water as soon as 1942 or 1943.[58] On September 30, 1940 he went to Ephrata and presented the directors of the three districts with a preliminary draft contract.[59] The Quincy-Columbia board of directors voted to accept it, but the other two districts demurred. East-Columbia District members objected to any charges prior to water delivery, and the forced sale of excess lands rankled others. The war postponed further negotiations.

On March 17, 1944 the Bureau resumed meetings in Ephrata with district directors. During the interim, Bureau officials had redrawn the contracts, making many items more specific. The Bureau flatly stated that the cost per acre for farmers would be $85, spread over a forty-year period, starting eleven years after initial water delivery.[60] The new contracts appeared satisfactory. In all, the farmers would, over the fifty-one-year repayment period, return to the government $85,465,000, or about 25 percent of the irrigation costs.[61] James O'Sullivan organized an education campaign to inform landowners of the terms.[62] Many farmers were apprehensive, and some felt that the charges were high, complaining that the sale of Grand Coulee power should carry more of the burden. In March, Interior Secretary Ickes signed the contracts and issued a statement which implied that the government might lower the $85 per-acre charge at some later date.[63]

The Washington legislature passed a bill allowing absentee property holders to participate in the elections through which the districts would ratify the contracts. Any of the 8,000 landowners who had not withdrawn were then eligible to vote. The directors of all three districts approved the documents, as did their constituents in the election held on July 21.[64] On August 31, 1945, during a ceremony held at Grand Coulee Dam, they signed the contracts that became the official documents hiring the Bureau of Reclamation to build the Columbia Basin irrigation project.[65]

With the war ending, and the repayment contracts signed, Ickes worked with the Truman administration's budget directors to plan annual allocations for project construction.[66] He first requested $25 million, stating that the project would provide farms for thousands of war veterans.[67] That request became part of a larger and older struggle between the administration and Congress over funding Western reclamation—the same debate that had plagued allocations for the dam in the 1930s when opponents, largely in the Eastern states, referred to federal projects as the "reclamation racket."[68] In 1945 the Bureau of Reclamation reduced its request to $8 million to start intensive Columbia Basin construction. Post-war inflation caused project cost estimates to rise rapidly. Figured at around $300 million before the war, by the end of 1945 they reached $583 million. At this point, Harry W. Bashore, tired of the battles, resigned as Reclamation Commissioner.[69]

President Truman replaced Bashore with Michael W. Straus and in 1946, when Harold Ickes left, Julius Krug became Secretary of the Interior. Krug backed expansion of federal power policy and reclamation, but he was cool to population resettlement. Straus, uninterested in social and economic problems or planning, concentrated on building, and under his leadership the Bureau expanded rapidly.[70] For the Columbia Basin Project, these changes broke the last link with long-term planning and initiated a period of almost twenty years during which government leaders emphasized only construction.

Annual congressional debates continued into 1948 when legislators also considered reducing irrigation's power subsidy. The Washington state delegation, backed by the Columbia Basin Commission, successfully fought that effort. The Columbia River flood of May 1948 helped convince Congress that it needed to harness the river, if not irrigate adjacent land, and the appropriation for 1949 was the largest on record despite another contentious debate.[71]

In the 1948 national election, reclamation was an issue. Republican vice presidential candidate Earl Warren assailed Democrats and the Bureau of Reclamation for their uncontrolled spending. Presidential candidate Thomas Dewey advocated shared control of federal projects. Truman continued calling the Republicans "Saboteurs of the West."[72] After the Truman victory in November, appropriations for the Columbia Basin, while still not automatic, came more easily in 1949 and 1950. But by 1949 the cost estimate of the project had reached $780 million, and, to many in Congress, its increases seemed endless,[73] a far cry from the record at the dam where the contractors stayed within budget and finished work ahead of schedule. This began the era of hefty cost overruns and rapidly accelerating price tags on large projects, especially those financed by the government.

**

The Columbia Basin Project, exclusive of Grand Coulee Dam, is a collection of canals, dams, reservoirs, laterals, wasteways, and ditches. The aggregate, built over more than two decades, from before 1946 until after 1966, was an accomplishment larger in size, more complicated in engineering, and more costly than Grand Coulee Dam, the project's key feature.[74]

First construction centered on the equalizing reservoir in the ancient Grand Coulee. Engineers checked to guarantee that it would hold water and they located sites for dams at both its north and south ends. At the end of the war, when construction seemed eminent, Frank Banks, who had moved from Grand Coulee to become the Bureau's regional director in Boise, Idaho, resigned that job and became supervising engineer for the Columbia Basin Project.[75] Banks was not only integral in the construction of the dam, but also the reclamation.

On Monday, January 14, 1946, 400 farmers, officials, and business owners gathered in Ephrata for a "kick-off" dinner marking the "start of construction."[76] The whole basin area responded to the increased activity; Moses Lake called itself the "Busiest town in the State of Washington."[77] Home building, road construction, and other improvements invigorated by the project, stretched from Pasco in the south to Ephrata and Quincy in the north. Work started on the main canal between the Grand Coulee pump plant and the equalizing reservoir, and the first section of the West Canal. In October the Bureau awarded more contracts for work on the Potholes Dam, Long Lake Dam, and the East Low Canal. At the end of 1946, North Central Washington experienced its greatest construction boom and influx of workers since the building of Grand Coulee Dam.

In 1946 the Bureau called for bids on the dam at the south end of the equalizing reservoir. During the summer of 1949 farmers harvested one last wheat crop in the Grand Coulee, and then sixty families vacated the area while others abandoned the small town of Osborne. By 1951 the State of Washington completed relocation of the highway that connected the town of Grand Coulee with Grand Coulee Dam.[78] That year, the twenty-seven-mile-long reservoir was ready to receive water.

One of the biggest parts of the project was the Potholes Dam, south of Moses Lake. The dam itself, when finished, was over three miles long and cost nearly $10 million.[79] By January 1947 a combination of three contractors worked to move over nine million cubic yards of earth and rock that eventually formed what was then the fourth longest and fourth largest dam in the United States.[80] In February 1948 the Moses Lake Chamber of Commerce recommended that the Potholes Dam be renamed O'Sullivan Dam in honor of James O'Sullivan and the work he did promoting Grand Coulee Dam and the Columbia Basin Project.[81] Senator Warren Magnuson introduced a bill in the Congress to that effect, and it passed on June 1. President Truman signed it on June 29, 1948.[82]

Monday, September 27, 1948, was "James O'Sullivan Day" in the Columbia Basin. Ephrata residents named O'Sullivan their mayor-for-a-day. Senator Magnuson, Frank Banks, and other dignitaries hosted him that night at a formal dinner. Secretary of the Interior Julius Krug came to rename and dedicate the large dam that was not yet complete.[83] The honor to O'Sullivan was timely for he fell ill that day and died February 15, 1949.[84] It was fitting that the dam be named for O'Sullivan. Although he had backed Grand Coulee Dam and fought for it with an unusual dedication, his greatest substantive contribution actually came in organizing the irrigation districts and preparing the way for the delivery of water. O'Sullivan Dam facilitated that process.

In the fall of 1950, Frank Banks retired after forty-four years of government service. He died December 14, 1957. His contribution to Grand Coulee Dam and the irrigation facilities had been formidable, and residents throughout North Central Washington sincerely respected him. With the reservoir behind Grand Coulee Dam named for Franklin Roosevelt, the reservoir behind Chief Joseph Dam for Rufus Woods who died in 1950, and the Potholes dam after James O'Sullivan, the *Wenatchee Daily World* urged calling the equalizing reservoir after Banks.[85] The United States Board of Geographic Names sanctioned the change and the State of Washington solemnized the event with two plaques installed on September 5, 1958.[86]

By the time Banks left the Bureau, it had completed the Soap Lake Siphon, the West Canal, the South Dam at the Grand Coulee, and O'Sullivan Dam. At that time the Bureau had work underway on $100 million worth of contracts. As the finished project took shape, the initial delivery of water appeared imminent. In April 1951 work started on a two-mile-long tunnel through the Frenchman Hills.

Crucial to the project were the huge pumps that would raise water from the lake behind Grand Coulee Dam into the equalizing reservoir. The California Institute of Technology had begun designing the pumps in 1938.[87] Their studies described twelve of the largest pumps ever built, and the Bureau of Reclamation called for bids on the first six.[88] In 1948 the Bureau let a contract on the pump plant itself. Installation of the pumps started in 1950. When the engineers ran preliminary tests on the first two units, they encountered unexpected vibrations and, while further studies and alterations diminished the problem, they never eliminated it.[89] When operating, two pumps required the full output of one of Grand Coulee's eighteen large generators.[90]

In 1946 the Atomic Energy Commission (AEC) began buying land around the Wahluke Slope in the southern part of the project. By 1947 it had acquired 88,000 acres that it included in the Hanford Reservation's permanent "Central Control Zone."[91] Over half of the land was irrigable. Further, the AEC created a secondary zone of 173,000 acres including 107,000 acres of prime project land.[92] The Bureau of Reclamation announced that it would never accept such a large loss from its project, and local business people and the three irrigation district boards expressed similar sentiments. They held meetings and pressured the AEC to reconsider.

In 1949, AEC Commissioner David Lilienthal said that his office would pursue health and safety measures that would allow release of some secondary zone land, but nothing happened.[93] In 1951, Senator Henry M. Jackson urged the AEC to reopen about 100,000 acres. Finally, in 1958, the AEC released all land outside the original control zone and in 1967 it added 39,000 acres from the original control zone.[94] It was a strange situation. Eventually people shunned nuclear developments, wanting to be as far from them as possible or to have them built elsewhere—something known as the "Not In My Backyard" phenomenon. But at that time on the Columbia Basin Project the irrigation districts and residents fought the government for the right to take the risk and live closer to the atomic reactors.[95]

As the Bureau of Reclamation completed the physical components of the project, it also readied farm units. It anticipated a population increase in the area from 11,000 in 1946 to 300,000 by the late 1950s. Some of those would live on 10,000 to 15,000 new farms, but the rest would fill the expanding cities.[96] Any remaining thought of relocating dust bowl refugees or poor veterans vanished when the Bureau announced in 1948 that a family must have $7,500 in cash to move onto a unit, and perhaps well over $20,000 more to develop it.[97]

In planning farm units, Bureau engineers frequently drew boundary lines along land contours. This was an innovation, as previous developments ignored geography and followed traditional rectangular patterns.[98] By 1949 the Bureau had platted nearly 600 units on some 41,000 acres. Although private owners held most of the land, the Bureau emphasized that they must sell excess property, anything over 160 acres, before water delivery, and in 1946 the government itself began a $10 million land purchase program to acquire property for resale later. Reclamation law stated that the government would deliver water only to "conformed" units; that is, the units it platted.

To determine who owned what land, reclamation legislation required that property holders file "recordable contracts" with the government. Those holding more than fifty-five acres of Class One lands, more than 100 acres of Class Two lands, or more than 160 acres of Class Three lands, unless they had owned the property before 1937, had to sell the excess at previously assessed dry land prices. Purchasers had thirty days from the date of any sale to produce affidavits indicating how much they paid per acre.[99] By 1947, 4,000 landowners had filed recordable contracts, but 2,000 others had not.[100] The Bureau set deadlines, extended them, and warned landowners that without the completed contracts, they would not receive water. When the final deadlines passed, most of those who did not return their contracts lived on the project's east side.[101] Again and again the Bureau learned that those farmers who could live without irrigation would not bow to its regulations just to obtain water.

Demand for project lands appeared high. In 1946 the Bureau received nearly 2,000 inquires from veterans who wanted to buy any available unit. The Department of the Interior issued a booklet that described the project and detailed availability of land and requirements for settlement, stressing that it would take between $20,000 and $35,000 to start such a farm.[102] The authors speculated that dairy farming and grape or sugar beet growing would dominate the project. For the serious, the Bureau, in conjunction with Washington State College, issued a larger 134-page booklet describing ways to build a house, how to borrow money, what crops grew best, what fertilizers to use, and other suggestions on ways to be a successful settler.[103]

To guide prospective settlers, the government set up development farms. There, pumps drew water from deep wells to irrigate land where the Bureau planted various crops and tried different techniques to see what performed best on what soils. The first two farms were near Pasco and Moses Lake.[104] With help from Washington State College's Agricultural Experiment Station, they began operation in 1947, leased to renters who worked together with officials conducting various studies. In 1948 the Bureau established a third farm near Winchester, later a fourth near Burke, and finally a fifth near Othello.[105] The development farm at Moses Lake proved that sugar beets grew well and a company based in Utah and Idaho started a processing factory in Moses Lake that opened in October 1953.[106]

The first irrigated land was in the south, farthest from Grand Coulee Dam, rather than at the project's north end. Not content to wait for water delivery through the long canals, the Bureau of Reclamation expedited the opening of blocks near Pasco. Engineers designed a pump plant, powered by electricity from Grand Coulee, to lift water directly out of the Columbia River and deliver it to eighty farms.[107] In the spring of 1948 veterans who could prove at least two years experience working on a farm and who had the necessary money drew lots for available government units.[108] The Northern Pacific Railroad, one of the larger landowners in the area, sold more units from its holdings.[109]

At 10:15 in the morning on May 15, 1948, after a ceremony at the site of the discharge pipes 167 feet above the level of the Columbia River, government officials and members of the Pasco Chamber of Commerce started the pumps, and water began flowing through the main canal. In just under an hour it reached the farm unit of O. C. Dillum, a Navy veteran who owned eighty-five acres twelve miles northwest of Pasco. Disappointed that President Truman turned down their invitation to attend the celebration, the Pasco Chamber of Commerce carried on with a large celebration in town. But an untimely electrical storm and heavy rain drove the crowd into nearby buildings and dampened most of the festivities.[110] Two more blocks followed and the three were the first in the Columbia Basin Project to receive water. It was four years later before the first water arrived from behind Grand Coulee Dam.[111] Nevertheless, thirty years after Billy Clapp proposed construction of the great dam, some water finally flowed on Columbia Basin Project lands.

In 1951 the government advertised thirty units in the blocks that would open in 1952 and it received over 900 applications, almost all from veterans.[112] Officials held a drawing on March 15, 1951 in Moses Lake. The winners traveled to the project to select their units, but after looking at the land, five of the thirty backed out, either disliking what they saw or, when faced with the reality of farming, changing their minds. More applicants, however, continued to over-subscribe subsequent lotteries.[113]

The Bureau of Reclamation experimented with sprinklers, small "part-time units," and town building. The first two were successful and the latter met with mixed results. Irrigation technology changed and sprinklers supplanted "rills" (water channels dug into the ground).[114] In time farmers adopted the new methods. Part-time units were immediately successful. Near various project towns, the Bureau set aside small farm plots, some as tiny as one-half acre. The idea was to give those who worked in an urban setting a chance to own and live on a plot large enough for garden vegetables, perhaps even producing a surplus to sell.[115] At the same time, the Bureau did not anticipate the large number of farmers on full-sized units who chose to live in cities and commute to their land. These people wanted both farming and the amenities of urban life. The planners had never imagined farmers who did not want to live on their farms.

The Columbia Basin Project Act of 1943 allowed the Bureau of Reclamation to establish new townsites within the project. The effort met with limited success. The first attempt was in 1950, at the proposed village of Burke, inside the Quincy-Columbia District. When the Bureau released plans for Burke, a storm of public protest caught it by surprise. "It was evident that insufficient attention had been given to informing the irrigation districts and project leaders of townsite plans," commented the Bureau in its annual project history for 1951.[116] Bureau planners had not foreseen that existing towns would worry about the new settlements taking away their business. These established city people, not farmers, protested the Burke development. In January 1952 the Bureau withdrew plans for Burke.[117] The incident again demonstrated the authority business people and professionals retained over project development and the inability of the Bureau of Reclamation to do whatever it wished. Just as landowners had spoiled the one district ideal, so business people thwarted the townsite plan.

In 1957 the Bureau did better with the town of George, which it did establish in the Quincy-Columbia District. While 3,000 people, including Governor Albert D. Rosellini, attended the site dedication, the *Wenatchee Daily World* dubbed the place "the town that is not."[118] And today, George, Washington, remains little more than a name on the map, a wide spot along the road, and the home of a unique postmark sought annually on February 22.[119]

In May 1951 the giant pumps at Grand Coulee Dam began lifting water 300 feet up into the main canal and from there to the equalizing reservoir.[120] Despite tightened security because of the Korean War, the Bureau of Reclamation and project boosters planned a celebration to mark the occasion. On June 15, Secretary of the Interior Oscar L. Chapman sent a signal from Washington,

D.C., that "officially" started the first pump, although it had already moved 70,000 acre feet of water. Fifty princesses from the Washington State Apple Blossom Festival, dressed in formal gowns, poured water from fifty jugs into the canal. Dubbed the "Water-of-All States Ceremony," the organizers had gathered the water from each of the forty-eight states and the territories of Hawaii and Alaska as a publicity stunt to advertise the project nationally. Bureau of Reclamation press releases, ever sensitive to the complaint that the project, which cost millions of federal dollars, benefited only a few in Washington state, declared that the act symbolized the Columbia Basin Project's contribution to the national wealth.[121]

The "gimmick," as a preliminary report issued by ceremony organizers called it, worked.[122] Most officials returned their jugs not only filled with water but decorated to identify their state. The governor of Pennsylvania got publicity when he had a girl wade into the Susquehanna River to fill his state's jug. In all, it was a colorful show, and it spotlighted the initial flow of water into the network of canals and ditches that formed the Columbia Basin Project.

The ceremony in 1951 was small compared to the celebration in 1952. In October 1951 over 600 volunteers began organizing the "Water Festival," advertising it as a "month of fun" for tourists and residents alike.[123] The highlight was the "Farm-In-A-Day" promotion. The idea centered on finding the most deserving veteran of either World War II or the Korean War and then providing him and his family with a complete farm and farmhouse, free of any costs, and all built within a twenty-four-hour period.

The Veterans of Foreign Wars conducted the search for the "most worthy veteran"—a man who truly merited what celebration organizers estimated would be a $50,000 farm.[124] The winner was thirty-year old Donald D. Dunn, a veteran of World War II, who farmed in Kansas until 1951 when the Cottonwood River flooded and ruined him.[125] Dunn, with a wife and two children, was the perfect candidate, and project authorities feted them royally.[126]

At one minute after midnight on May 29, 1952, as fireworks signaled work to begin, over 300 volunteers, guided by floodlights, began leveling an eighty-acre farm unit three miles north of Moses Lake. Airmen from the Moses Lake Air Force Base worked on a seven-room house that, by six in the morning, already had a roof and side walls. A stiff wind and blowing dust hampered the effort, but everyone kept at the job. Late that night the finished farm was ready for inhabitants and water.[127]

The Water Festival, sponsored by an independent corporation of business people known as the Columbia Basin Celebration, Incorporated, ran for eleven days, from May 22 through June 1, 1952. Each project town took part: Soap Lake had a "Cavalcade," Quincy ran "Pioneer Days," Ephrata put on a "Little World's Fair," Coulee City conducted a "Frontier Days Celebration," and Pasco and the Tri-Cities added a parade and an "Aqua-Rama." Postal authorities issued a commemorative stamp to mark the occasion. They canceled their plans to unveil it on top of Grand Coulee Dam due to security problems stemming from the Korean War. The west vista house stood in as substitute. Donald D. Dunn received the first stamp.

Everyone agreed that the celebration was a great success and, at last, water from behind the dam flowed onto the land. As the biennial report of the Washington State Columbia Basin Commission put it:

> The long awaited land opening on the Columbia Basin, the greatest agricultural opening in all history, began in 1952, after 19 years of construction, anticipation and prediction as to what would come when this vast area in the heart of the State of Washington would be opened to settlement.[128]

But 1952 also brought political change. In November Republican Arthur B. Langlie was elected Washington's governor, and the nation sent Dwight David Eisenhower to the White House. Although Langlie claimed to back the project, he resented its federal controls.[129] Eisenhower, voicing similar sentiments, said,

> The whole hog method is not the way to develop Western resources. . . . Nor can it be done by the states alone. Nor can it be done by free enterprise alone. We need river basin development to the highest degree, but not at the expense of accepting super government in which the people of a region have no voice . . . we want this to be done through partnership . . . bringing in the federal government, not as your dictator, but as your friendly partner.[130]

Eisenhower's so-called "Partnership Program" limited the government to performing only services the states could not do for themselves and then executing them in close cooperation with local authorities and private enterprise. Douglas McKay of Oregon, Eisenhower's Secretary of the Interior, declared that the Columbia Basin Project must stand on its own merit and could not be a pork-barrel job.[131] In so saying he implied the project was, in the past, just that. With reason, project backers worried about future appropriations, and they found themselves in good company. To the horror of Westerners, few new reclamation projects started under the Eisenhower administration, and the struggle to fund those in progress grew as prolonged and acrimonious as those in the 1930s.

Eisenhower announced in his State of the Union message that he intended to complete "sound planned projects," but then he directed McKay to reorganize the Bureau of Reclamation and McKay announced his intention to fire its Commissioner, Michael Straus. Not waiting, Straus resigned on February 6, 1953, and Eisenhower appointed Wilbur A. Dexheimer in his place.[132] This further demoralized Bureau employees already chafing under attacks by Republican Senator Joseph McCarthy of Wisconsin. In speeches delivered to the Congress, McCarthy pictured reclamation advocates as purveyors of subversive social and economic ideas. Although these were the same sentiments that had helped the Bureau stifle plans for a Columbia Valley Authority, they also threatened its large reclamation projects. The Bureau retrenched and tried to save what it could.[133]

In 1953 Republicans in Congress slashed the reclamation budget and took $1.5 million away from the Columbia Basin Project. The Bureau of Reclamation laid off 118 employees in Washington and slowed construction.[134] In 1954 the House of Representatives again cut the reclamation budget, but the Senate restored the money, leaving the Columbia Basin Project with almost all the $12 million requested by Dexheimer. Congress repeated the process in 1955. Faced with mounting crop surpluses the government initiated its expensive "Soil Bank" program to help ease the problem. It paid farmers not to plant crops on land that would produce unwanted excesses. In light of this plan the Eisenhower administration found spending to bring more acreage into production a troublesome contradiction.[135]

Surprisingly, however, despite the attacks and debates over funding, from 1952 through 1960 the project experienced its fastest and most consistent growth. During those years, 437,172 acres came under the ditch—an average of about 48,500 annually.[136] The Eisenhower administration, despite its antipathy toward reclamation, did not appreciably slow development of the Columbia Basin Project. Three things account for this. The Bureau of Reclamation had learned during the funding battles of the 1930s and now it always requested far more money than it needed, knowing that Congress would cut its allocations. Western congressmen exerted tremendous pressure on the administration and were willing to trade votes for other pet projects to insure backing for theirs—if not to start new ones, at least to keep the existing ones going. And finally, reclamation still retained its position as a fashionable conservation procedure. Despite reservations about bringing more land into production during times of surplus, nobody doubted that irrigation stopped erosion and prevented precious topsoil from blowing away.

Costs for various aspects of the project continued their rise. By 1953 the government outlay required to build laterals to carry water away from the main irrigation canals reached $84 per acre, up from $75 in 1948.[137] In 1953 government analysts estimated it would cost nearly $500 million just to complete the remaining irrigation aspect of the project alone. For each acre of newly irrigated land, the federal government then spent $470. In addition, for each acre domestic water cost $34, farm development reached $258, county and state roads added $36, schools came to $34, and a few smaller items brought the total to $839. Combined costs to all government agencies for a seventy-five-acre farm exceeded $60,000.[138]

The settlers who benefited from this were not the people envisioned by the planners. Despite the hope for population redistribution, 81 percent of project farmers were Pacific Northwest residents, 53 percent from within the state of Washington and over 30 percent from around the project. They were not young men looking for a start; their median age was 39.7 years—younger, however, than the median age for farmers nationally. Just over half of the settlers were veterans, but only half purchased units through the Bureau of Reclamation's drawing system, the rest buying directly from previous owners with excess land.[139] They were not poor, either. Over half came to the project with assets exceeding $20,000, and the median for the total was almost $15,000. Three-fourths took out loans with the Farmer's Home Administration.[140] They were better educated than the typical American farmer and one-third lived in cities, usually no more than twenty to thirty miles from their land. The largest religious denomination was the Church of Jesus Christ of Latter Day Saints, whose members comprised 15 percent of the total. Most of them came with irrigation experience and moved due to pressure on the land in Utah.[141] The *Wenatchee Daily World* noted the high expectations among settlers by pointing to the large number of new farm houses with television antennas.[142]

From the date that they settled their farms, the new owners had eleven years before they began forty years of repayments to cover construction costs. They made those payments on a sliding scale, with charges being low in the beginning, increasing during the final years. The amount owed depended on land quality, but the average was the $85 per acre paid over forty years as agreed upon in the 1945 contracts.[143] In addition, from the start the farmers paid an annual surcharge covering day-to-day operation and maintenance expenses.

As an irrigation dividend, some recreational opportunities were enhanced. The Washington State Department of Game planted fish in the various lakes. The equalizing reservoir in the Grand Coulee became the site of three parks that the state operated and improved with recreational facilities. Ponds and small lakes that developed from the runoff of irrigation water drew water fowl in significant numbers. In 1955 the United States Fish and Wildlife Service bought 32,000 acres in the Potholes and Crab Creek area north and west of Othello and formed the Columbia National Wildlife Refuge.[144]

As farmers moved on to each new irrigation block, the population of the adjoining areas rose. North Central Washington found itself in the midst of "irrigation fever," reported the *Wenatchee Daily World* in 1957.[145] In 1953 alone, Quincy gained 229 residents. The same year three new schools opened in Moses Lake, which doubled its population between 1952 and 1956. In 1950 about 25,000 people lived inside the project area and the number rose to 48,000 by 1955, peaking at 80,000 in the 1970s and dropping after that.[146] But many came due to the military activity at Moses Lake and the atomic works at Hanford. Even with them, however, it was hardly the 300,000 predicted by the planners.

In 1955 the 1,816 farms[147] within the project (including those on dry as well as irrigated land) produced a total gross income of

$17 million, double the figure in 1953. Three years later it rose to $27 million, and a study done by the Bureau showed that irrigated land produced ten times the value of adjacent dry land.[148] While the total gross income rose in the wake of increased irrigation, the amount per acre dropped due to a steady decline in farm prices.[149]

The project's infancy ended around 1958, although it was then only about half complete. Ten years after the first water reached the early Pasco units, those farmers ended their developmental period and began the repayment schedule. While construction work continued adding new units, that process slowed noticeably.

Bureau publications offered optimistic forecasts that recalled the visions that promoters had painted years before:

> The Columbia Basin Project will provide employment for a large number of the Nation's fast expanding population. the [*sic*] climate is healthful, the level of living is high, and production of a wide variety of agricultural commodities is possible. . . . The power generated can be utilized to greatly expand the employment base and thus round out the economy of the region. These things are being accomplished without cost to the nation except for the loan of funds for development of irrigation works.[150]

Despite the positive tone, the late 1950s ended the project's rapid growth period. In its place, long-simmering problems emerged that slowed and eventually halted expansion. The project had already departed markedly from the original blueprints and, over the next ten years, solutions to complicated problems brought more changes. Prior to construction, the Columbia Basin Project received the most detailed scrutiny and planning of any single federal reclamation effort. But as the project grew and conditions changed, the larger goals set in the 1930s proved antiquated and unrealistic. Long-term planning is fragile at best, especially with projects built over a decade or more. In their desire to cope with new circumstances and to make a profit, people forget the idealism that initiated the effort. They seek pragmatic and self-serving changes. The events on the Columbia Basin Project from 1933 through 1958 support that thesis, and the changes from 1955 to 1965 provide even more evidence. Those years moved the project even farther from the early vision.[151]

James O'Sullivan. *Courtesy Bureau of Reclamation.*

# Chapter Seventeen
# Water, Water Everywhere

And the people of Israel said to him, "we will go up by the highway; and if we drink of your water, I and my cattle, then I will
pay for it; let me only pass through on foot, nothing more."

*Numbers 20:19*

The best laid schemes o' mice and men Gang aft a-gley.

*Robert Burns*

The years 1958 through 1963 were a period of crisis for the Columbia Basin Project. After starting irrigation in 1952, Bureau of Reclamation engineers soon realized that much of the project required
unanticipated and expensive work to drain away excess water. The
situation worsened with the entry of each new block. To pay the
rapidly rising costs, the Bureau asked farmers to renegotiate their
1945 contracts and increase the amount they paid. Busy starting
their families and facing both rapid inflation and a drop in farm
prices (what economists called the "cost-price squeeze"), farmers
hardly welcomed suggestions that they also pay higher water
charges.[1]

What farmers most wanted was an end to restrictions on land
ownership. With improved technology and the economies of size,
they reasoned that working more land with modern, efficient methods would increase their incomes. Consequently they put their time
and energy into asking Congress to change the law while avoiding
repayment negotiations with the Bureau of Reclamation. The resulting conflict between the Bureau and its project settlers eventually halted project construction with work only half done. It again
showed that the Bureau did not control the project or its residents.
It also demonstrated that, despite the extensive studies done

before construction, the government had not asked all of the right questions or fully understood the geology or geography of the area it sought to irrigate. Tinkering with nature on such a large scale is a venture fraught with surprises—some quite unpleasant.

The trouble started during the ten-year experimental period, before the repayments began. The first indication of drainage problems came at Soap Lake in 1952 and 1953. A series of small lakes, including Park Lake, Sun Lakes, Blue Lake, Lake Lenore, and finally Soap Lake, which is five miles northeast of Ephrata, dot the bottom of the Grand Coulee below Dry Falls. Because Soap Lake forms the final step in the series, and because it contains no outlet, it has, over the years, become rich in minerals that give its water a bitter, salty taste. Winds frequently whip the surface into masses of white foam that give the lake its name. Since the early 1900s, area residents touted the waters for their medicinal powers. Consequently, a small industry developed that catered to people who traveled there either to drink the water or spend some time soaking in it.[2]

When the Bureau of Reclamation built Columbia Basin Project canals, dams, and reservoirs it anticipated some leakage and drainage problems.[3] But it was unprepared for Soap Lake's rapid rise. At first, Bureau engineers attributed this to increased rainfall. Throughout the ensuing controversy they maintained this attitude and Soap Lake residents admitted that the lake went through cyclical fluctuations.[4] Whatever the cause, the lake quickly threatened the town's water and sewer system, public swimming pool, beach, and private property. By January 1953, Bureau engineers had installed four pumps that, working constantly, slowed but did not stop the rising waters.

By early 1954 the lake was four feet above its pre-project level and water filled the basements of some homes.[5] Worse yet, because of the inflow and the pumping, the mineral content of the lake dropped from 30,000 parts per million to 22,000. Residents protested the dilution of the waters on which many depended for their livelihood. They remained unhappy when the Bureau announced that while it had stabilized the level of the lake, it could not guarantee maintenance of its high mineral content. The lake might eventually become just another body of fresh water.[6] Even though the Bureau had already spent $630,000 installing pumps to remove excess water both from Soap Lake and Lake Lenore, the problem worsened.[7]

In December 1955, Soap Lake businessmen charged the Bureau of Reclamation with "foot dragging" and demanded an independent investigation. They flatly rejected a contract, offered them by the Bureau, that reimbursed them if they pumped water out of the lake themselves and released the government from any liability resulting from decreased salinity. The Bureau then dug two more interception wells. Washington Senator Warren Magnuson charged that the steps taken by the Bureau to preserve Soap Lake were inadequate.[8] He and Senator Henry M. Jackson introduced legislation to "save Soap Lake."[9] Although the Bureau argued that preserving the mineral content of the lake was not the government's obligation, Congress appropriated $233,800 to stop the fresh water encroachment.[10] With the money the Bureau built more interception wells, and by the middle of 1958 it stabilized Soap Lake's salinity. Although the Bureau repeated that freshening the lake might eventually be inevitable, Soap Lake has remained about the same since 1958.[11]

Soap Lake was only a prelude to the drainage dilemma.[12] When farmers introduced forty to sixty inches of water to land where rainfall was six to ten inches annually, striking changes were inevitable.[13] Nobody, however, foresaw how fast those changes would occur or the amount of land they would involve. World War II prevented completion of ground water and geological studies originally planned for the Columbia Basin. When the military drafted many of the Bureau's men, it forced the government to rely on preliminary drainage projections done in the 1930s with inadequate methods.[14] The limited studies failed to detect the often-shallow depths at which solid rock underlay project land. That rock kept water from draining away and caused problems.

Soon water appeared elsewhere, creating soggy lowlands or even small lakes. In the spring of 1953 the Bureau warned settlers to limit water use to avoid crop damage and wetland development. One year after first delivery, the Bureau began building wasteways and drainage ditches, and in some areas the cost of those facilities exceeded the initial irrigation price.[15] By the summer of 1954 new lakes dotted the project and wildlife officials stocked some with fish while most drew pheasants and waterfowl. In 1955 the Bureau

started sinking over forty wells to facilitate drainage, and it estimated that such costs might run to $40 million or more for the
whole project.[16]

From 1956 through 1958, seepage washed out small sections
of roads. Project residents blamed poor planning of the wasteways
(channels designed to carry off the excess water) and ignorance of
geologic conditions.[17] A survey in 1956 showed one-third of the
project's farmers dissatisfied with operations, and almost half of
them cited poor drainage among the chief reasons.[18] In the Quincy-
Columbia Irrigation District, settlers pointed to flooded basements
and to the water table, which in places rose over 150 feet in just
four years.[19] Landowners with affected property filed hundreds of
thousands of dollars in damage claims against the government.[20]
The Bureau of Reclamation promised to solve the problem but it
asked for patience, pointing out that it was simultaneously building
not one but two independent systems, one to bring in the water
and another to take it away. The Bureau reluctantly admitted that it
had not fully understood the region's geology, and its annual project
history for 1956 commented, "Experience to date indicates that this
problem will be more difficult and costly than early estimates, which
were preliminary in nature."[21]

In 1957, 1,400 acres were too wet to farm. In that year Reclamation officials spent $600,000 on drainage work, bringing the over-
all total to $2.5 million.[22] Adding the pumping at Soap Lake and
other similar works, the figure more than doubled. The Bureau
offered to resettle those with wet lands who had bought their units
from the government, but for the majority, who bought from private owners, the Bureau would only cancel water charges. By 1961
the Bureau had built thousands of shallow and deep wells and over
700 miles of drains and wasteways. Engineers tested experimental
sealers that they poured into canals and laterals trying to stem the
loss of water. Yet in that year nearly 5,000 acres remained out of
production due to high water.[23]

The drainage work grew increasingly expensive, and that became a critical problem in itself. When Congress ratified the repayment contracts with the three districts in 1945, it included a ceiling
of $8,176,000 for drainage work, based on government estimates
made in 1940.[24] By 1957 the Bureau had already spent $5.8 mil-
lion.[25] Under existing legislation, there was no way the Bureau could

pay for the drainage already needed, let alone any that would come as it opened new blocks. Officials of the three irrigation districts annually assessed their water users small fees to operate and maintain the facilities. Once the Bureau passed the $8 million ceiling, it would have to add the costs, now estimated at more than $1 million yearly for many years to come, to the annual fees. Everyone realized that farmers faced potentially oppressive charges—and this before the scheduled repayment period had even begun![26]

Columbia Basin Project farmers already had enough financial trouble. Increased demand for farm produce and the high prices farmers received during World War II encouraged unprecedented production nationally, peaking with the record output of 1948. Accelerated mechanization, greater use of fertilizer, improved conservation practices, better plant varieties, and superior insecticides—all developed in the 1930s but repressed during the Depression—suddenly revolutionized American farms.[27] Farmers now needed more equipment, which meant higher start-up costs. In those good times, farmers willingly borrowed to buy machines and land. That trend lay behind the great interest in settling the Columbia Basin Project in the late 1940s and early 1950s and it accounted, in part, for the number of applicants seeking available units.

Furthermore, farmers in the post-war years were increasingly unwilling to endure rustic conditions; they wanted the comforts and amenities of their urban cousins. This was particularly true of the people who populated the Columbia Basin Project; with their higher levels of education, their television sets, and their automobiles, they wanted to live comfortably and send their children to college.[28] The studies that planned the project anticipated none of this, having anticipated subsistence dust-bowl refugees.

The law of supply and demand eventually reversed the favorable economic situation that temporarily made small-plot farming seem desirable and profitable. Although the market welcomed increased production during the war, starting in the late 1940s European farms resumed operation while American farms dramatically increased output. World and national shortages ended, and prices dropped. At the same time the federal government suspended economic controls, unleashing rapid inflation that continued into the 1950s. It was the misfortune of Columbia Basin farmers that they began settling project land just as this cost-price squeeze started to hurt.[29]

An unusual case, but one that illustrates aspects of the problem, was that of Donald D. Dunn, who received the free eighty-acre farm with a home built in one day. In 1952 he grossed $18,152 with potatoes as his main crop. In March 1955, after three seasons, he sold the farm for $75,000 and moved to Colorado.[30] Dunn claimed that tourists pestered him and that he had trouble securing loans. Money, he argued, was available only for developing farms, making him and his "instant" farm ineligible.[31] The Bureau of Reclamation countered that Dunn managed the unit badly, especially by renting too much additional land.[32] Dunn replied that the only way to succeed was with more land; the government-configured units were too small. The owner who followed Dunn also sold within two years. The next farmer combined two more units, making a total of 455 acres, and stayed into the 1970s.[33]

Dunn was not the only settler to leave. Of 725 units sold by the government from 1952 through 1956, 16 percent changed owners during that period.[34] One-quarter of the settlers dropped out in the first four years.[35] Other farmers rented additional units or bought them directly from private owners, including those who left. The longer a farmer lived on the project, the more likely he was to operate multiple units. If they bought more units, they often registered them under the names of family members to avoid contravening land ownership limitations.[36]

In 1955 the Washington State Columbia Basin Commission asked Congress to raise the ownership limitations on Columbia Basin lands.[37] The Commission organized petition drives among farmers and landowners, backing a bill that would at least bring project lands into line with other reclamation areas, most of which used the traditional 160-acre limit.[38] In 1956, Senators Magnuson and Jackson introduced the legislation but the Bureau of Reclamation opposed it, stating that existing farm sizes were adequate. But Assistant Commissioner Floyd Dominy revealed the Bureau's deeper motive when he said that any changes in the law must also include higher repayment charges, something the farmers flatly rejected.[39] This conflict between the farmers for more land and the aim of the Bureau to charge more money set up a conflict that drug on for years. The immediate result was that Congress adjourned without taking action on the Magnuson-Jackson bill.[40]

Between January 3 and 5, 1957, Assistant Secretary of the Interior Fred G. Aandahl visited the project and held hearings at Moses Lake, Quincy, and Mesa.[41] In all, over 1,000 people attended, and 125 testified. Aandahl heard repeatedly that modern farm machinery could easily handle more than one unit, that income from one unit was not enough to support a family at current living standards, that one-unit operators usually had one or more family members working away from their farms to supplement their incomes, and that diversified farming and good soil management required more land.[42] A minority of farmers grumbled that a change would lead to large corporate farms and drive them from their land.[43] The Bureau argued that average farm size on other similar projects was even smaller than in the Columbia Basin.[44] Clearly impressed by the side urging change, Aandahl urged Congress to loosen the land ownership limitations.[45]

The House of Representatives held hearings on the Magnuson-Jackson bill, and this time the Bureau of Reclamation made no formal objections, although it quietly opposed the measure.[46] An article in the *Wall Street Journal* backed the legislation, stating that the Columbia Basin Project had become an "ailing utopia," existing under stagnant rules while economic and social conditions had changed.[47] "The basic trouble is that the plan has stood still while time and the technology of farming have moved on," the article added.[48]

On September 2, 1957, Congress approved the law, and President Eisenhower signed it three days later.[49] The changes allowed any operator to hold multiple units up to 160 acres. A husband and wife together could own up to 320 acres. Leasing of more land remained legal; however renters now needed approval from the Bureau of Reclamation. In reality, the law only recognized what, through renting, was common practice on the project.[50] The 1957 law resulted in larger project farms; the officially estimated 1958 average of 84.1 acres rose to 107.3 acres by 1973.[51] The Bureau of Reclamation noted that many farmers also increased their profits and reasoned that now it was time to discuss increasing the 1945 repayment contracts and resolve the drainage problem. The new law cost the Bureau its leverage with the farmers, however, and achieving a revised contract, in light of changing economic conditions, proved increasingly difficult.

In 1943 the government estimated the cost of the finished Columbia Basin Project at $487 million. In 1954 the projection reached $740 million and continued to rise.[52] The 1945 contracts committed farmers to repay $85 per acre over forty years, with charges starting eleven years after first delivery of water. That would have returned about 25 percent of irrigation costs, with Pacific Northwest power ratepayers covering the rest and the federal taxpayer underwriting the interest.

The same inflation that caught farmers in a cost-price squeeze also raised project expenses. By 1959 the Bureau of Reclamation concluded that the existing $281 million ceiling set years earlier by Congress on all project costs would now cover irrigating only 455,000 of the ultimate 1,029,000 acres.[53] Furthermore, by 1959, drainage expenses had almost exhausted the $8 million previously allocated. In 1953, when Project Manager Philip R. Nalder told the irrigation boards of the three districts that drainage would eventually cost not $8 million but perhaps $40 million or more, the farmers, especially those not yet receiving water, flatly rejected his talk of increased payments or contract changes.[54] They recalled government claims made in the 1930s and 1940s that guaranteed low-cost irrigation at set prices, with power sales picking up all remaining costs.[55] The repayment contracts of 1945 formalized the idea that farmers would repay a fixed cost and not a percentage of actual expenses.[56] While the farmers welcomed altering the rules to allow them more land, they bitterly opposed changes increasing their financial obligation.

In April 1957, Reclamation Commissioner Wilbur A. Dexheimer sent letters to the three irrigation districts proposing a new contract that extended the repayment period by around ten years to cover drainage costs, but did not raise the annual payments.[57] He pointed out that, unless the farmers settled the matter, work on the project would stop in 1963 with less than half of the proposed project land irrigated. Nothing came of the initiative and the Bureau complained that the board members intentionally kept the Commissioner's letter quiet so that water users remained uninformed.[58] At this point, any remaining trust between the Bureau of Reclamation, project residents, and project leaders rapidly deteriorated.

Nevertheless, the Bureau tried to persuade project residents to renegotiate the contracts. Through the 1958 irrigation season, it mailed postcards to water users every two weeks detailing system operation costs and reporting how much water they had used. Assistant Reclamation Commissioner Dominy, while visiting the project in September, stated his feeling that district board members did not accurately reflect the views of the farmers, and he announced that the Bureau would take its proposals directly to the water users.[59] After discussions with Dominy, the Quincy-Columbia and the South-Columbia district boards agreed to re-open negotiations with the Bureau, with the East-Columbia District joining later.[60]

A support group formed by farmers in the South District sided with Assistant Commissioner Dominy about the boards not representing the farmers, but they argued that the boards had not stood firmly enough against the Bureau. They added their feeling that the Bureau of Reclamation, having failed on the acreage limitation issue, now used drainage costs as a club to force renegotiation of the repayment contract.[61] As tension increased, three key issues emerged: how much Grand Coulee power should subsidize irrigation; how efficiently the Bureau had built the project; and the method for determining the farmers' ability to repay construction costs.

In 1945 the $85 per-acre charge would have covered about 25 percent of the estimated project cost, down from 50 percent in 1935. By 1952 inflation had further reduced that to around 16 percent, and by 1959 some calculated that it reached 11 percent.[62] The farmers found no problem with this. When discussions turned to the mounting drainage costs, farmers said the Bureau should not require them to pay for its miscalculations and they pointed to dry drainage ditches in some areas while rising water tables ruined hundreds of acres elsewhere as evidence. They said the Bureau badly used the $8 million allocated for drainage and that it did so deliberately to enhance its negotiating position, an outrageous charge which showed the depth of ill feeling on the part of the farmers toward the Bureau.[63]

The Bureau and farmers also disagreed over the farmers' ability to repay. Precedent and law based the water users' charges on their financial well being and not project costs. In 1958 a study by the Farmer's Home Administration indicated that average farm income rose nearly $2,000 from 1955 to 1957, increasing their ability to repay. The farmers countered that crop returns in 1958 were only 60 percent of 1952 levels, reducing their ability to repay.[64] When negotiations began in January 1959, the Bureau of Reclamation refused to include any evaluation of Grand Coulee power revenues and Harold Nelson, the Bureau's regional director, limited the talks to determining the water users "reasonable ability to pay."[65] Each side manipulated the data to its own advantage and the discussions stalemated.

In May, after months without progress, the Bureau submitted three formulas to the irrigation districts. The first called for charges of $133.50 per acre over fifty years on a scale that graduated from $1.50 the first year to $2.70 the final year. The second recommended charges of $125 per acre over forty-five years with higher annual charges. The third also recommended $125 per acre, but over fifty years with limits set on the total available for drainage.[66] After lengthy discussions, the directors of the three irrigation boards accepted the second option, which allowed a total expenditure of $44,500,000 for drainage, although Reclamation officials refused to guarantee that this was enough.[67]

The directors of the three irrigation districts lobbied their memberships to ratify the documents, but on July 29 the East-Columbia Basin Irrigation District, the first of the three to vote, soundly turned down the new contract.[68] While 327 voted yes, 897 voted no.[69] Stunned by its miscalculation and realizing that approval required all three districts, the Bureau canceled the Quincy and South district elections scheduled for August. In an editorial the *Tri-City Herald* snarled:

> In our opinion renegotiation isn't really dead because it never was alive. When you looked into the subject you discovered two worlds—one the hazy world of the Bureau of Reclamation and the Irrigation District commissioners, and the other the grim world of the farmer.
>
> How the feelings of the farmers could have been so badly misjudged is completely beyond us. It was even circulated that opposition to the new contract was confined to a "small vocal group in the South District." That myth was exploded yesterday.[70]

Leaders in the Quincy District stated that their district would work for a new contract with or without the other two and it was not surprising that this section of the project, which most needed irrigation to raise any crops, was the one that to this point still cooperated with the Bureau. The South District remained more aloof

while the East District, where dryland farming continued successfully, and where the fewest acres received irrigation, raised the
greatest obstacles.

Rather than profit from its better rapport with the Quincy farmers, however, the Bureau of Reclamation did two things that soured
relations even with them and drove them into a greater alliance
with the South and East. It announced that it would mothball the
project so that costs would not exceed contract limits, and it stunned
farmers with a projection that 1960 operation and maintenance
charges, including drainage work, would climb by nearly $1.65 an
acre. Those charges, estimated by the Bureau of Reclamation at
around $2.65 an acre in 1953, already averaged $5.75 in 1957. The
increase appalled farmers and, added to quadrupled state and federal tax hikes over the previous ten years, outraged them.[71]

Floyd Dominy became Commissioner of Reclamation on May
1, 1959. That November he urged farmers on the Columbia Basin
Project to vote again on the new contracts because he said the government had nothing better to offer. Still exercising its heavy hand,
the Bureau determined to have its way. It did not count on the
strength of farmer resistance or the depth of their feelings. These
immediately surfaced. On Thursday evening, November 19, around
200 project residents met in Mesa with directors from all three districts. Walter Le Page, a school teacher turned farmer and an outspoken critic of the Bureau, demanded that Congress investigate
the project. He and others praised Senator Magnuson who urged
farmers not to pay the high charges and who recommended two
contracts, the old one for land already irrigated and a new one for
land still dry.[72] Finally, at just after midnight, the farmers voted to
organize a "water users' strike" and refuse to pay the new operation and maintenance charges.[73] It was a direct challenge to Bureau authority.

The Bureau of Reclamation said that without payment it would
not deliver water during 1960.[74] Caught in the middle, the district
directors, who had urged farmers to accept the new contract, were
furious at Magnuson. Earl Gregory, Chairman of the Quincy board,
told the press that the senator first told them that changes in the
contracts were necessary and not to bother him. Now, Gregory
accused Magnuson of using the issue as a "political football."[75]

In December, 125 water users met in Othello for four hours and demanded that power revenues pay the drainage assessment and finance the East High Canal. They reaffirmed that they would not pay the fees, and they asked that Congress investigate the project. William "Billy" Clapp, the lawyer credited with the idea of Grand Coulee Dam back in 1918, received applause when he said, "Don't put any more burden on the back of the G.I.s."[76] At other meetings around the district, angry farmers suggested that a $500,000 reserve fund, held by the Bureau, be used to offset the 1960 drainage charges. When Commissioner Dominy rejected the suggestion, the Columbia Basin Land Owners Association, one of the farmers' pressure groups, sent telegrams to President Eisenhower asking him to remove Dominy because the Commissioner exhibited "ruthlessness and inconsideration."[77]

From February 9 through 11, Commissioner Dominy traveled across the Columbia Basin Project, meeting with the district boards in Pasco, Mesa, Warden, and Quincy. Dominy seemingly took a personal interest in the project and was then in the process of replacing Philip R. Nalder, who had recently resigned as project manager. The *Wenatchee Daily World* printed the rumor that Dominy "deported" Nalder to Afghanistan because he was too easy on farmers.[78] Apparently Dominy then went to eastern Washington to see things for himself. What he found showed the intensity of feeling among the farmers and the failure of the Bureau to deal effectively with it.

In a three-hour meeting in Pasco with about 200 present, Dominy fielded angry questions. When one farmer shouted, "Don't try to get money out of us because we haven't got it," the others cheered. The Commissioner remained pleasant despite repeated cries of "four-flusher," and in the end he gave them some advice. He told them to go directly to Congress with their problem.[79]

At the Mesa meeting over 300 attended. They kept Dominy until almost midnight while every speaker berated the Bureau of Reclamation. One settler called the project, "an extortion game." Dominy countered, "If the contract I'm offering you is worse than the one you got [*sic*], I'm a monkey's uncle."[80] He told the farmers they had closed minds and were unwilling to listen to facts. At Quincy, Dominy had a better time of it. Again he spoke frankly, advising the farmers to approach Congress. There was criticism, but in reaction to some of his comments the Commissioner received applause.

Project residents took Dominy seriously and approached their congressmen. On February 15 senators Jackson and Magnuson introduced a resolution asking Congress to grant a two-year moratorium on Columbia Basin Project annual drainage charges and to investigate the project for one year.[81] With the farmers refusing to pay the assessed fees and the Bureau threatening to cut off delivery of water, the senators commented, "In effect the Bureau of Reclamation has set a time-bomb to go off in the Columbia Basin Project this Spring." The senators also criticized the Department of the Interior for its inability to provide Congress with accurate and up-to-date project data.[82]

The Bureau of Reclamation backed down somewhat and announced that it would probably deliver water to the three irrigation districts in 1960. Sounding a conciliatory note it added that pending legislation, if passed, would change conditions.[83] As the Senate Committee on Interior and Insular Affairs reviewed the repayment problem, senators Jackson and Magnuson added an amendment to their bill guaranteeing water deliveries in 1960 and 1961. The Senate Interior Subcommittee on Irrigation and Reclamation approved the bill, as amended, despite objections from the Bureau. Everything went smoothly as the bill passed the Senate.

In the House of Representatives, congressmen cut the 1961 appropriation for the Columbia Basin Project to its lowest level since the war, and Representative Walter Edward Rogers, a Democrat from Texas, announced that he was in no hurry to conduct hearings on the Magnuson-Jackson moratorium bill. Buoyed by that and by a new repayment capacity study, done partly by the Department of Agricultural Economics at Washington State University that showed the farmers in a more favorable economic position than they admitted, the Bureau of Reclamation hardened its position. Commissioner Dominy demanded that the districts either pay the charges or put the money in escrow before a May 1 deadline.[84] The Quincy-Columbia District acquiesced while the South and East districts thought it over. Finally all three irrigation districts together placed $620,000 in escrow accounts and then watched the House of Representatives.[85]

In the middle of May, the House Interior and Insular Affairs Committee held hearings on the Magnuson-Jackson bill. From the outset, the exchanges lacked a conciliatory tone. Committee members noted the extensive use of power revenues to offset irrigation charges. They also remarked that drainage was traditionally the responsibility of irrigation districts.[86] Nineteen farmers from the Columbia Basin Project addressed the committee in favor of the bill. One of them, Earl Weber of Quincy, complained, "Veterans are going broke."[87] The Bureau sent officials who testified against the bill. On the final day matters worsened when Pennsylvania Representative John Phillips Saylor, a Republican, responded to Walter Le Page's statement that the "process of negotiation stinks" by stating that such comments reflected on the integrity of the House committee. At the conclusion, the House Committee frankly told the farmers to "Go home and work out your problems with the Bureau of Reclamation."[88] At the same time, the House Appropriations Committee withdrew funding from the project and called a stop to all new construction.

In July 1960, Commissioner Dominy accompanied subcommittee chairman Walter Rogers and three other members of that subcommittee on a three-day project tour. They held what became a six-hour meeting in a crowded gymnasium in Moses Lake. Over 300 farmers attended, and the hot weather did nothing to cool tempers on either side. In response to unfriendly shouting from the audience, one of the congressmen retorted that he was angry too, that he had traveled to Washington state to help and had done so at personal expense and inconvenience. Most of the thirty people who testified did so calmly, and the exchanges were civil. However, when an Ephrata farmer told Representative J. T. Rutherford, Democrat of Texas, that power revenues from Grand Coulee Dam were as important to the Columbia Basin Project as offshore oil was to his home state, Rutherford exploded, "I told you what you'd get—nothing. Don't start throwing barbs at me buddy boy."[89] When the four Congressmen departed the next day, two declared against the drainage moratorium bill, one said he favored it, and one remained undecided.

The House and Senate conference returned some funding to the project. Then on September 1, Congress adjourned. When the three irrigation districts placed their fees in escrow, they agreed to turn them over to the Bureau of Reclamation if Congress failed to act on the moratorium bill. When the session ended, the bill died, and the districts released the money without protest.[90]

The next blow to the farmers fell a few weeks later. The Bureau of Reclamation announced that it managed to hold drainage charges to $1.65 per acre in 1960, but in 1961 they would be $2.30. This, added to the normal operation and maintenance cost of $6.38, raised the total to an average of $8.68.[91] Farmers became outraged again, insisting that the Bureau levied the drainage charge as a bludgeon to force their acquiescence.

Although the Quincy District had begun new negotiations with the Bureau in July 1960, the presidential election that fall slowed the process. On November 4, a few days before the election, Senator John Fitzgerald Kennedy, presidential candidate for the Democratic Party, sent a widely publicized letter to Washington Governor Albert D. Rosellini:

> It would be the goal of my administration, if I am elected President, to seek an equitable solution to the drainage and repayment problem. We must carry forward the program and ideals established by the administrations of Franklin D. Roosevelt and Harry S. Truman for the development of Grand Coulee Dam and the great Columbia Basin Irrigation Project.[92]

The press reported that Kennedy favored the moratorium proposed by Jackson and Magnuson, and all negotiations stopped for the election.[93]

On November 8, 1960, Republican presidential candidate Richard M. Nixon carried all of North Central Washington.[94] Kennedy, however, narrowly won the election nationally. Within a few weeks, irrigation district officials called on him to fulfill his promise and back the moratorium. "Some Republican farmers are saying now, 'Let's see what the Democrats can do,'" Walter Le Page wrote in December to Senator Magnuson.[95]

Both Governor Rosellini and the three irrigation district boards formally asked Kennedy to proclaim a moratorium on drainage payments for 1961. Instead, Kennedy turned the matter over to his new Secretary of the Interior, Stewart L. Udall, who handed it to Under-Secretary James K. Carr.[96] Through that spring, Carr met in Washington, D. C., with board members from the irrigation districts, with members of the Columbia Basin Commission, and with officials from the Bureau of Reclamation. After those discussions, he appointed a board of three consultants and told them to recommend an equitable solution.[97]

On August 1, 1961 the three-man board issued its report. It recommended retention of the $85 per-acre charge, added to $46.60 over a new ten-year period to cover drainage. That ten-year period would follow the original forty-year repayment period, thus lengthening the total commitment of farmers from forty to fifty years and from $85 to $131.60.[98] It suggested turning over operation and maintenance to the three irrigation districts, and it urged renegotiation of new contracts to raise the construction ceiling, allowing completion of the project.[99]

By the middle of September the Bureau of Reclamation and the three irrigation districts began talks.[100] Out of the discussions, two contracts emerged for each district: a very long and detailed document and an abbreviated version known as the "short form." The long form described all the new benefits. This meant that rejection by Congress would negate the effort. The short form established a contract but left the controversial benefits for Congress to include or reject.[101] The Quincy District approved the short form.[102] The South and East districts decided not to gamble and worked toward the long form.

While Senators Jackson and Magnuson agreed that power revenues should pay more of the costs, they pointed out that the House of Representatives would reject any such bargain.[103] They were pragmatic enough to realize the limits they faced in Congress and grew irritated at farmers who did not appreciate the constraints under which they labored. The House and Senate passed the measure accepting the new contracts and President Kennedy signed it on October 1, 1962.[104] On January 28, 1963, in a special ceremony at Ephrata, the principals signed the last of the three new contracts.

During the negotiations, reclamation officials and members of the irrigation district boards urged letting the districts themselves assume responsibility for project operation and maintenance. From 1962 through 1968, meetings between Bureau administrators and district representatives discussed the issue.[105] One sticking point was whether the districts would operate jointly or separately.[106] Old antagonisms the districts held toward each other, and irrigation facilities that did not follow district boundaries, complicated the matter. The discussions eventually turned into negotiations, and by 1965 the district directors reached tentative agreements. The districts finally chose to manage the facilities separately, but they agreed to coordinate their operations closely.[107]

It took ten years from first mention of the proposal in 1958 until the boards signed a contract for the joint takeover on December 18, 1968.[108] The agreement authorized the largest transfer of facilities in United States reclamation history and became official on January 25, 1969.[109] Under the terms of the contract, over 200 employees of the Bureau of Reclamation left government service and went to work for the water users. The Bureau agreed to deliver water to each district, but from that point on it was up to the districts to manage and pay for distribution. Slowly the Bureau was learning to let the farmers deal with costs directly.

The new repayment contracts and the transfer of operation and maintenance resolved the problems and tensions that spoiled the long-awaited Columbia Basin Project. At the same time, the new laws ended the notion that the project would facilitate the return of veterans to civilian life. The new law also repealed anti-speculation limits, and land values rose quickly, as did farm size. By 1968 some land reached $500 and more per acre. With the changes, the way opened legally for large-scale farming, either through single ownership or even corporate enterprise, to control the project.[110]

During the ten-year debate over drainage and contract renegotiation, farmers on the project showed little appreciation for the subsidy they received from power ratepayers. If the government required the farmers to repay the same percentage of project costs in 1960 that they accepted in 1945, the charges they faced would have dwarfed anything ever mentioned by the Bureau.[111] But the farmers simply concluded that hydroelectric power purchasers should pay the bills regardless of how high they were.

While the Bureau of Reclamation tried to continue project development and solve the finance problems, the irrigators created adverse publicity and thereby embarrassed their already frustrated benefactor. There is evidence that farmers exaggerated the economic hardships they faced. For example, the State of Washington continued selling land it owned during the late 1950s and early 1960s. Free from anti-speculation restrictions, due to a legal technicality, its land went for three times the appraised value.[112] It is doubtful that the land would have commanded such a price if buyers did not anticipate adequate returns.

Because of its visions of planning and resettlement, the New Deal and the Joint Investigations saddled the Columbia Basin Project with limitations more stringent than those on other projects. In 1970 region one director, Harold T. Nelson, speaking for the Bureau of Reclamation, addressed the twentieth annual Irrigation Operators' Conference in Pasco, Washington. Among his remarks, Nelson said:

> The Columbia Basin Project is quite different from the project designed by planners in the 1930s. . . . A number of the well-intentioned 28 studies completely missed their mark. Subsequent readjustments in legislation and policies in the last 30 years certainly demonstrate that all planning must be flexible and subject to change. . . . Probably the greatest error was the assumption of size of farm unit, which was natural enough during the depression years when first desires were considered to be for food, shelter, clothing, and economic security with independence. Those policies resulted in initial layout of farm units from 45 to 60 acres each in size, and the proposed establishment of several small project towns. As we know, this Utopia failed to materialize for several reasons. . . . Today the average farm size in the Basin is 240 acres. Not one of the planned government-sponsored towns ever came into being. As late as 1945, the population of the project at full development was estimated to be 400,000 persons or about 400 per 1,000 acres. Today, the project has a population density of 190 per 1,000 acres, which is less than one-half of the original projection.[113]

The utopian goals set by the New Deal for the Columbia Basin Project failed. The vision of small farms with minimal amenities held no appeal for post-World War II settlers. The Truman administration speeded the process, even before the first units received water. Afraid of a post-war depression, it nationally subordinated the family farm ideal in favor of commercial agriculture and economic growth.[114] Understanding this, the Bureau of Reclamation, recognizing that renting was legal, allowed farmers to do it well beyond the spirit of the law. Whether it was ever realistic or not, the "Planned Promised Land" failed before it started.

For their part, farmers voiced some valid arguments. In the 1920s and 1930s, the Bureau of Reclamation and local project backers in eastern Washington repeatedly promised that the revenues from Grand Coulee would pay most of the irrigation costs. From the beginning, everyone understood that without such a subsidy the farmers faced prohibitive charges. Repeatedly the businessmen and professionals who lobbied the government to build the project depended on cheap water as the foundation for their agricultural/industrial empire. Those early promoters never coped with the difficulties that technicians faced when they actually introduced water onto the land. Nor did they anticipate the economic changes that sent costs soaring. But the farmers remembered the promises and, with self-serving justification, they expected the government to honor them.

The Bureau of Reclamation, in its push to irrigate the project quickly, moved too fast. A slower pace might have allowed for adjustments that avoided costly problems. In its handling of project farmers, the Bureau showed a lack of tact and understanding that aggravated the situation. Over the decade of the controversy, neither side was entirely right.

Because the landowners on the east side of the project could farm without irrigation, they retained their independence and willingly confronted the Bureau of Reclamation. Those on the west side, particularly around Quincy, could not operate without water and found disagreement with the government more difficult. The south fell somewhere in between. But even the Quincy District proved troublesome. In the end, the Bureau lost what power and influence it had on the project. Rather than endure continued confrontation with farmers, it willingly relinquished operation and management. This meant that while the Bureau abdicated full control of the project, future problems were not entirely its responsibility. After 1963 the administrative bureaucracy of the Bureau of Reclamation on the Columbia Basin Project diminished appreciably. That trend has continued ever since.

Finally, after the renegotiation controversy, the Washington legislature and the national Congress lost any remaining enthusiasm for the Columbia Basin Project. In 1964 Washington state disbanded the Columbia Basin Commission, and it went out of existence on June 30, 1967.[115] At the national level, the House of Representatives decreased project funding; since 1968 only around 40,000 more acres have received irrigation.[116] Work on the project, for the most part, stopped about half-way. After the early 1960s, the goal of Columbia Basin Project backers and of the Bureau of Reclamation has been to regain the momentum lost during the renegotiation conflict and to finish the second half of the job.

Irrigation on the Columbia Basin Project, 1973. *Courtesy Bureau of Reclamation.*

# Chapter Eighteen
# To Build, Or Not To Build

Descendants of Central Washington's farmers who scrabbled
for a living during the Depression now tabulate balance sheets
on computers, invest in $30,000 circle systems to irrigate 3,000-
acre spreads and relax with television shows from their satel-
lite dishes.

*Seattle Post-Intelligencer, 1985*[1]

In 1963, when the Bureau of Reclamation and the three irrigation
districts completed their negotiations on new repayment contracts,
the Columbia Basin Project was about half complete. After the mid-
1960s, the Bureau of Reclamation and most Columbia Basin Project
residents, along with urban business people and professionals in
the area, worked toward one overriding goal. They wanted the government to finish irrigating the project's remaining dry land, and
for thirty years they moved toward that end. Yet by the mid-1990s
little had changed; just over 550,000 of the ultimate 1,029,000 acres
had received water and only 47,318 acres had entered the project
since 1968. Repercussions from the repayment controversy, continued high drainage costs, and changing national priorities combined to slow, and then stop the flow of federal dollars that sustained
construction after World War II.

At a chamber of commerce banquet in Ephrata early in 1964,
Reclamation Commissioner Floyd Dominy assured those present
that strong local support would persuade Congress to complete
the project, and he predicted adding 15,000 acres to the system
annually.[2] That fall support groups formed the Columbia Basin
Development League—a reincarnation of the old Columbia Basin
Irrigation League or the Columbia River Development League.
Membership lists for the organization show that it, like its predecessors, contained mostly urban businessmen and professionals.[3]

At its first annual meeting Secretary of the Interior Stewart Udall spoke about the anticipated third powerhouse at Grand Coulee Dam and the proposed Pacific Northwest/Pacific Southwest power intertie that would link the two regions. Development League members wondered if the emphasis on power would distract attention from reclamation.[4] Their immediate target, which Udall did not mention, was a feature known as the Second Bacon Siphon and Tunnel.

All the irrigation water pumped from behind Grand Coulee Dam flowed through a main canal and then into Banks Lake. From there it traveled south in another canal, then through a 10,000 foot-long tunnel and a 1,000 foot-long siphon, together called the Bacon Siphon and Tunnel. With about half of the project complete, those facilities operated at full capacity during the irrigation season. To service more land the system needed a second siphon and tunnel.

In 1965, Congress funded a third powerhouse at Grand Coulee Dam, and while senators Warren Magnuson and Henry Jackson assured project residents that this would not diminish irrigation money, escalating American involvement in the Vietnam War led to deep reclamation budget cuts. Incredibly, despite increased resistance to funding, in 1965 the still-functioning Columbia Basin Commission recommended tripling the project's size, increasing it to three million acres by adding 600,000 acres in the region from Odessa to Ritzville, 250,000 acres north of Ephrata, and over a million acres in the area from Waterville to Davenport.[5] Bureau engineers paid serious attention to the suggestions, named the East Banks Project, the West Banks Project, and the East Project—or together, the Central Washington Project. By the middle 1970s, however, the Bureau of Reclamation, generating a feeling of *déjà vu*, abandoned all the plans as impractical because of their enormous expense.[6] It was the 1920s all over again with supporters ignoring costs and asking the federal government to spend not millions, but now billions of dollars to irrigate ever more land.

Liberalized land ownership policies, suspension of anti-speculation limits, and a number of unusually dry years did revitalize interest among east-side farmers on the existing project. Many recommited land withdrawn in the 1940s and now worked for project completion.[7] The Columbia Basin Development League and the Columbia Basin Commission persuaded legislators to increase the

1966 appropriation by $2.5 million, thus allowing irrigation on two additional blocks.[8] But the next year legislators, "disenchanted" with irrigation in the Pacific Northwest, cut the funding dramatically.[9] A number of factors complicated the already difficult process of finessing dollars out of Congress. In developing the project, the lowest areas received irrigation first. The higher sections required expensive pumping facilities and more complicated canal works. And the relative prosperity of the middle 1960s reduced the number of people seeking farm settlement opportunities.[10] Every year made it harder for backers to justify the expenditures necessary to enlarge the project.

In 1967 the Lyndon Johnson administration noted that the government had already spent $315 million irrigating the Columbia Basin and estimates for finishing the project now exceeded one billion dollars. The projected cost just to remedy existing drainage problems rose to over $100 million. The Bureau realized increasingly less for each dollar spent, and inflation made this even worse.

And there were growing complaints from project settlers. In the south some landowners complained that the water they received, recycled after use in the north, was contaminated by herbicides, pesticides, dissolved salts, and weed seed.[11] The Bureau tested the water and denied finding significant pollution. The drainage problem, however, was undeniable. "We ought to have the government declare us a disaster area," shouted one angry farmer at a 1967 meeting in Basin City. "Why spend money to bring in new land when we have top land, already producing crops that is being flooded out of existence by too much water," he added.[12] The Bureau installed more expensive pumps, especially in the Quincy area, and continued searching for solutions.[13]

"I'm through! I've had it! I'm gonna sue the bureau!" insisted farmer Charles Smith when he spoke in 1968 to members of the East Columbia Basin board of directors.[14] He claimed that the rising water had reduced his 136 acres to only seventy that remained usable. "They've been giving me the run-around up at Ephrata for the last four years. They've promised to exchange the land for dry land down here—but they won't do anything about it."[15]

From the frequent farmer complaints it appeared as if the Bureau only reluctantly drained land. That, however, was not always the case. First the Bureau determined if the outcome justified the expenditure. But in some cases it manipulated the figures so they looked favorable. For example, in examining a number of questionable blocks, Commissioner Dominy wrote,

> We have agreed that we should defer indefinitely the development of these high cost drainage areas, but we do not feel it is necessary to get into the economic justification of constructing irrigation works for areas having a reasonable drainage cost. If we do this, we are certainly going to get into cases where the ratio of benefits to costs will be less than one to one, unless we either use the 'sunk cost' approach, or in the case of Wahluke Slope areas, reappraise our benefits recognizing the fruit potential of the area.[16]

When a farmer near Basin City found thirty acres of his unit flooded, he asked the Bureau for help. It turned him down because the work was "not economical." The farmer appealed directly to Commissioner Dominy, who ordered his staff to calculate benefits for the entire farm rather than the wet land alone. "After reviewing the data used in determining the drain spacing for the 30 acres," noted a Bureau employee, "it appeared that a slight adjustment could be made to bring the assumed water use more in line with that delivered to these lands."[17] In other words, by "adjusting the data," they made the benefits outweigh the costs. For all land drained through 1973, the average cost exceeded $100 per acre. The Bureau had then drained over 66,000 acres and it estimated that nearly 200,000 acres more of the 517,537-acre total still needed work.[18]

For the farmers on the Columbia Basin Project, it was fortunate that the renegotiation crisis occurred when it did. Had it come a few years later, when the Bureau of Reclamation better understood the full cost of drainage or the inflation rate during the late 1960s and 1970s, the government would undoubtedly have required the farmers to pay considerably more.[19] In summing up the situation in 1973, Assistant Reclamation Commissioner William E. Warne made this understatement:

> Adequate preparations were not made in advance for drainage, and high water tables were experienced quickly in some areas. Water pollution proved to be a problem rather soon. The project manager worked at programs to correct these problems, but not without some controversy.[20]

Still, partisans exerted pressure on the Bureau to irrigate more land. But without the second Bacon Siphon and Tunnel and additional pumps at Grand Coulee Dam, no more land could receive water. When Reclamation Commissioner Dominy retired in October 1969, President Richard Nixon replaced him with Ellis L. Armstrong, who announced significant reductions in Reclamation spending. "In another decade or so," Dominy moaned as he left office, "our farmers will be hard pressed to grow enough food and fiber to satisfy the needs of our own people."[21] In saying that, he almost exactly repeated the people who backed the Columbia Basin Project in the 1920s. Despite such gloomy forecasts, the Nixon administration appeared unmoved and remained determined to cut the Bureau's budget. In North Central Washington, any hope for extensive new work on the Columbia Basin Project dimmed.

The future was not entirely bleak, however. In an unexpected way power again came to the rescue. The Bureau of Reclamation's long-range plan called for twelve pumps to lift water out of Franklin D. Roosevelt Lake. Initially the Bureau installed only six. With improvements in technology, government engineers considered pump-generators for the remaining places. Operated as pumps during periods of low demand, when the cost of power was cheapest, the pump-generators would lift the water into the reservoir (Banks Lake) in the Grand Coulee. When power demand increased, and its value rose during the morning and evening hours, the same water would flow back through the units creating electricity.[22] The pump-generators would help relieve a growing regional power shortage. Using the justification that they would aid in producing electricity, not in pumping water, backers sought money for the six new machines. Congress agreed.

In 1971 the government called for bids on the first two pump-generators. Nohab Incorporated, a company in Sweden, built the units and installed them in 1973. Able to generate 50,000 kilowatts each, they cost about 35 percent more than pump units alone.[23] The Bureau installed the last four pump-generators in 1979. The new units necessitated enlarging the feeder canal from the pump station to Banks Lake to control surges created by the new units and to allow them access to the top two feet of water in the reservoir.[24] Now there were sufficient pumps in place to irrigate the entire project.

Promoters turned to the second Bacon Siphon and Tunnel. Washington governor Daniel Evans emphasized their importance when, in the fall of 1968, he ordered a temporary stop to well drilling in the Odessa area. There farmers increasingly tapped the water table, which had dropped over nine feet in two years.[25] To provide a more reliable irrigation source and compensate for the loss, Evans hinted that the state might help fund Columbia Basin Project expansion. Then Congress appropriated over $1.5 million for the siphon and tunnel, but the Budget Bureau cut the funds. Congress returned the money in 1969 and the Bureau of Reclamation announced that it would go ahead with construction. In March 1969 it called for bids, but in April it canceled them when the new Nixon administration reconfigured the Johnson budget.[26] Senator Jackson warned that Nixon's people seriously considered a "no-new-starts" policy that would postpone the siphon and tunnel indefinitely.[27]

Upset, members of the Columbia Basin Development League successfully appealed to Congress for the $4 million that Nixon cut. Again the Bureau of Reclamation called for bids. But in January 1971, President Nixon impounded the money.[28] In response, the Bureau put all remaining money into drainage facilities. Again recalling the past, the Columbia Basin Development League argued that finishing the project would provide thousands of jobs, easing unemployment, and the government should build the $30 million siphon and tunnel for that reason alone. The League added trustees from Spokane, Portland, and the Puget Sound area to broaden its base of support.[29] In doing that and in its propaganda it repeated the tactics used in the 1920s.

In the Spring of 1974, Secretary of Agriculture Earl L. Butz visited the Columbia Basin Project and openly dismissed economic justification for completing a project where the final cost approached $2,000 an acre.[30] Nevertheless, the Nixon administration indicated a willingness to reconsider funding if state and other public agencies shared the costs. Through the rest of 1974 the Washington legislature considered the overture. In December 1975 it allocated the necessary $15 million, and the governor signed the measure.[31] Optimistic, Governor Evans stated that the construction would be a "toll bridge" which would pay its own way in the long run.[32]

The Bureau of Reclamation announced that it would quickly issue a contract for the siphon and tunnel, the Office of Management and Budget released the one million dollars previously appropriated, and the Gerald Ford administration recommended an additional $2.5 million for the project in 1976.[33] A ruling by the Supreme Court that stated that impounding congressionally allocated funds was unconstitutional undoubtedly helped the Office of Management and Budget make its decision.[34]

The state's $15 million came with a demand that the Bureau build the siphon and tunnel large enough to serve the entire second half of the project.[35] The Bureau agreed, provided that the water users pay fees linked to their ability and not just a set amount. As early as 1975 members of the Columbia Basin Development League had announced that construction of the second Bacon Siphon and Tunnel might mean higher assessments for landowners and even a new repayment contract.[36] But when the Bureau of Reclamation proposed a new repayment price of $1,560 per acre over fifty years, a twelve-fold increase over the existing $131.60, members of the three irrigation districts rebelled.[37] The Quincy board announced that it would not participate in any discussions.[38] In its study of the repayment capacity of project residents, the Bureau calculated that the farmers could easily afford annual charges of $50 per acre.[39] This was less than the cost to those on adjacent land who dug wells and pumped their own water. But the three boards rejected the figures and asked the Bureau to recalculate them.

Professor Norman K. Whittlesey, agricultural economist from Washington State University, complicated the negotiations when he issued a report showing that economic conditions did not justify the existing project, let alone expensive additions. Whittlesey stated that the energy cost to residents of the Pacific Northwest for increased irrigation was $100 per irrigated acre, and something he called "social overhead" increased that to nearly $2,000. Others, not the direct beneficiaries, bore those costs, Whittlesey wrote.[40] Based on his calculations, he advised against siphon and tunnel construction.[41]

The boards of all three irrigation districts ignored Whittlesey and urged Governor Evans and the Bureau of Reclamation to start work on the siphon and tunnel while at the same time negotiating with farmers. They feared, based on previous experience, that those discussions might take two years or even more. Meanwhile, the key to construction was the state's $15 million and a condition tied to it by the legislature that it be used within two years. Senator Jackson argued that the government might not let power ratepayers underwrite any additional project costs, and he felt that a new contract with farmers should come first. The Office of Management and Budget sided with the senator and demanded that new repayment contracts be negotiated and signed before any work started.[42] Everyone involved realized that getting the farmers to pay higher costs after construction would be almost impossible.

To break the impasse with the irrigation districts, the Bureau proposed a two-step plan. First, it suggested negotiation of a master water-service contract specifying the charge on each acre-foot of water delivered from the new facilities. Interested parties could then buy water at that price, guaranteeing an income for the government and satisfying its demand that farmers underwrite at least part of the costs. Then the farmers could negotiate a new repayment contract later. To sweeten the deal, the government said it would deduct the state's $15 million from the total cost, lowering the bill to the water users. This also meant that new project farmers would pay considerably more than those already receiving water.

Farmers in the Quincy and East districts, anxious for work to start and not seeing any immediate cost to themselves, agreed. The South Columbia Basin Irrigation District balked and, defying the Bureau, approved independent water-service contracts for landowners within its boundaries.[43] Further, the South District refused to sign the new master water-service contract.

In August, Senator Jackson, while in Ephrata, told the press that the Bureau had approved the master water-service contract, and that Secretary of the Interior Thomas S. Kleppe had signed it. The senator said that the East District had voted to accept it and that the Quincy District appeared ready to do so. He added that construction of the siphon and tunnel would proceed on the strength of support from two out of three districts.[44] That June the government opened bids for the second siphon and tunnel.[45]

The Guy F. Atkinson Company submitted the lowest estimate at just over $32 million.[46] On October 9, 1976, Interior Secretary Kleppe spoke at groundbreaking ceremonies initiating construction on the second Bacon Siphon and Tunnel. Flanked by senators Magnuson and Jackson, and Reclamation Commissioner Gilbert Stamm, he praised the work and predicted increased productivity and prosperity for eastern Washington.[47]

The *Wenatchee World* reported that Quincy and East district farmers agreed to the master water-service contract because they believed that the new Democratic Carter administration favored irrigation and would be their ally. What President James Earl Carter really believed, and did, shocked them. In March 1977 he ordered a re-evaluation of water projects nationally, including the Columbia Basin Project and the second Bacon Siphon and Tunnel. The new President found support for charges, frequently made in the East, that irrigation subsidized by the government produced huge crop surpluses. He then listed nineteen unnecessary reclamation schemes, including the Columbia Basin Project. It took all the power and influence that Senator Jackson and, especially, Senator Magnuson could summon to rescue the siphon and tunnel and maintain their funding.[48]

Events on the project hardly helped Magnuson's cause. Despite partial acceptance of the master water-service contract, negotiations on a new repayment contract stalled. While most farmers insisted on retaining the $131.60 per acre figure from the 1963 contracts, the Bureau had raised its estimates of the amount they should repay from $1,550 to $1,700 over fifty years.[49] By the fall of 1978 the Quincy board left the talks and said it would not return, even if it meant an end to the development.[50] Further, the three irrigation districts squabbled among themselves about what area to develop first after completion of the new siphon and tunnel. The Bureau suddenly saw itself building expensive facilities that nobody would either use or pay for. The *Seattle Post-Intelligencer* asked if the Bacon construction was a "Bargain or [a] Rip-off?"[51] Using material provided largely by Professor Whittlesey,[52] the paper suggested that in a period of energy shortages, crop surpluses, and budget deficits, the tunnel was unnecessary. Furthermore, estimates by the Bureau of Reclamation stated that the completed project now would cost over $1.5 billion. The Bureau had calculated about $1.50 in benefits for every dollar spent on the project. But the article pointed out that this was at 3 percent interest. At 5 percent, however, the ratio changed to eighty-two cents in benefits for every dollar spent, and 5 percent was then well below actual interest rates.[53] None of this included hidden costs. For example, the power used to pump the water out of the reservoir behind Grand Coulee Dam was electricity lost to commercial sale, and the irrigation water itself generated no power at downstream dams, causing lost revenue.[54]

Amazingly, and largely because of Magnuson's and Jackson's influence, Congress, with the Carter administration acquiescing, continued funding the siphon and tunnel—spending $17 million in 1977 and $4 million more in 1978. By September 1977 the contractor had cut through the full length of the 9,950-foot tunnel and by early 1978 the work was nearly a year ahead of schedule.[55]

Meanwhile, negotiation of the repayment contract was behind schedule. Farmers already receiving water who refused to pay more, and those who might receive it in the future, caught the Bureau between them. The latter group asked if it was right that they might pay perhaps ten times more than their neighbors. On the other side, the Bureau experienced increasing scrutiny from a Carter administration that wanted all farmers to pay much more of the bill.

Late in 1979 an Energy and Water Development Appropriation Bill passed Congress and President Carter signed it. This provided final funding for the Bacon Siphon and Tunnel, bringing combined state and federal contributions to $32 million.[56] The contractor finished the tunnel's concrete lining and the siphon neared completion. On March 22, 1980 water flowed through the structures.[57] Because the government and the three irrigation districts had failed to negotiate a new repayment contract, however, the Bureau kept additional water out of the existing irrigation system.

Both sides now waited for the newly elected President, Ronald Wilson Reagan, to reveal his intentions. The first impression was that Reagan favored irrigation. In the spring of 1981, with that thought in mind, project manager James Cole sounded optimistic when he announced the resumption of talks.[58] The South District refused to cooperate with the Bureau[59] but talks with the Quincy and East districts proceeded apace. Because of increased well drilling, and the resulting depletion of the underground water table, East District residents were considerably more amenable to discussions than in the past. Further, representatives from the Bureau had learned to talk to farmers in terms of yearly costs for each acre receiving water rather than stressing the total figure for the full fifty-year obligation.

Taken in this way, and compared with the price of irrigating adjacent land with water pumped from wells, the dollar amounts looked reasonable.[60]

By the fall of 1982 the two sides, minus the South District, achieved a settlement. The supplemental master water-service contract signaled a major change in the government's approach to project development. It allowed the East and Quincy districts to subcontract with landowners not yet served by the government. They then could pump and transport irrigation water from the existing canals and facilities already built by the Bureau but maintained by the districts. To receive that water, the contracting farmers had to install their own delivery systems.[61] In that decision, the Bureau turned the cost problems directly over to the farmers. The result, called "piece-mealing" by some, had severe critics who argued that it was the government's responsibility to complete the project as promised fifty years earlier. But most found the solution satisfactory. It allowed farmers to take or reject water as they chose. The income from water used promised the government some return on the investment in the new siphon and tunnel. The compromise was immediately successful as farmers subscribed the full 10,000-acre allocation in the first phase.[62] In the ten years between 1979 and 1989, about 12,000 to 15,000 acres, inside and outside the project combined, acquired irrigation through this process, including land using contract water from government facilities and land receiving water from wells.[63]

With a firm hand on their own destiny and an eye to cutting costs, the districts pursued a number of ideas. In 1978 the government approved plans that they install generating plants which they, not the Bureau, would operate.[64] These would use water flowing through the various main canals, which drop around 1,000 feet over the length of the project. Engineers estimated that during the irrigation season, up to 100 megawatts of potential power might yield $200 million in earnings over forty years.

In another venture, however, the basin farmers failed. The problem involved sugar beets. Project planners speculated that sugar beets would be one of the area's more successful crops. But at the same time, due to surpluses and declining prices, the government set limits on the number of acres farmers across the country could plant in sugar beets.[65] Despite those constraints, over 100 farmers in the Quincy District alone raised sugar beets, and in response, the Utah and Idaho Sugar Company built refineries at Toppenish and Moses Lake, with the latter being one of the nation's largest.[66] In December 1960, responding to pressure from farmers, Congress eliminated the acreage restrictions, allowing unlimited sugar beet planting. Disaster struck about ten years later. In the middle of the 1970s, due largely to increased sugar prices, the nation's soft drink industry switched to cheaper corn sweeteners.[67] In late 1978, in a surprise announcement, the Utah and Idaho Sugar Company closed both the Moses Lake and Toppenish factories.[68] Through that winter, basin farmers appealed for help to Senator Jackson, and pondered what to do. Conferences held with company representatives in Salt Lake City offered no solutions, and Jackson admitted his inability to find any remedy.[69]

The best suggestion was for farmers to form a cooperative, then buy and operate the Moses Lake refinery. The farmers went ahead with preliminary organization of what they hoped to call the Washington Sugar Company, and then negotiated a selling price with the Utah company. In March they announced that they needed a capital base of $9 million. To guarantee this, each farmer had to commit somewhere between $100 and $200 for each acre they intended to plant in sugar beets until together they underwrote the $9 million figure.[70] Although the sugar company extended its deadlines, cooperative promoters failed to secure enough backing. And so the sugar beet refineries remained closed and farmers switched 42,000 acres to other crops.[71] The plants had employed over 1,000 workers with an annual payroll of nearly $9 million. Throughout the Columbia Basin, the sudden loss of the sugar beet refineries affected most people in some way.[72] The episode showed the need to diversify and draw a variety of industries to the basin—the failed dream of Rufus Woods. It also showed the inability of farmers to work together and, whether they admitted it or not, it reaffirmed their reliance on the government.[73]

The 1980s brought congressional action on a different but lingering problem. The preliminaries began in 1972 when a United States District Court in San Diego held that despite the practice of the Bureau of Reclamation, sections of the reclamation law of 1902 remained valid. The court specifically pointed to a clause that said those receiving government water had to live within fifty miles of the project.[74] Four years later, in 1976, the Ninth Circuit Court of Appeals in San Francisco held that the 1902 law limited government-subsidized irrigation water to owners of under 160 acres.[75] Armed with these decisions, the Carter administration started to enforce the law.[76] In testimony before a Department of the Interior panel investigating the ramifications of enforcement, farmers from projects throughout the West argued that adherence would bring chaos. The Bureau of Reclamation, reversing its position of the 1950s, added its statistics that showed little or no economic viability in farms under 160 acres.[77]

Some legislators wanted to raise ownership limits to a total of 960 acres while others wanted to allow 960 acres including rented land. They also debated the fifty-mile rule. The Carter administration disliked any changes and the bill stalled until after the 1980 election.[78] When Ronald Reagan won, the new reclamation bill picked up support. As governor of California some years earlier, Reagan convened a special task force that recommended raising land ownership limits on federal projects to at least 640 acres. Most who knew him felt that he might back even more generous changes.[79] He did.

In the final deliberations the Senate allowed 1,280 acres for farms receiving subsidized water, but the House of Representatives cut the figure to 960 acres hoping this would still insure a family farm preference. The Senate concurred. Both houses agreed to end the fifty-mile rule. On October 12, 1982, President Reagan signed the measure.[80] Considering the amount of renting among Columbia Basin Project farmers, the law only recognized and legalized common practice and did not bring any sudden or major changes. Once and for all, however, it did kill the notion of the project as fostering small family farms.

The Reclamation Reform Act of 1982 was the last piece of major national legislation directly affecting the Columbia Basin Project. Two issues remained: the availability of water and whether to irrigate the second half of the project. On the first question, the debate centered on whether there was enough water in the Columbia River to cover all uses: irrigation, power, navigation, fish, and reclamation. In May 1971, W. H. Keating, acting for Reclamation Commissioner Ellis Armstrong, wrote Timothy Atkenson of the Council on Environmental Quality that,

> There is sufficient water available [in the Columbia Basin Project] for full multipurpose development of the land and water resources. Some 2,500,000 acre-feet of water is [*sic*] now pumped annually from the Columbia River and when the project is completed this annual withdrawal will have doubled. This amount of withdrawal is not controlling since the minimum recorded annual flow of the river at Grand Coulee is about 51 million acre feet. Also some 1,225,000 acre-feet of return flow will be available in the Columbia River system.[81]

This optimism continued into the 1990s among representatives of the Bureau of Reclamation who maintained that there was sufficient water to irrigate all the Columbia Basin Project while at the same time satisfying the other demands put on the river.[82] In 1976 a representative of the Bureau declared that the Columbia River could irrigate seven million more acres without significantly cutting power generation.[83] He did not mention fish.

In the middle to late 1970s a number of unusually dry years exacerbated potential water shortages. This, combined with a power shortage, which the 1973 Middle East oil embargo worsened, led to increased concern about use of Columbia River water.[84] On the Columbia Basin Project, the Bureau and water users realized that the irrigation pumps were the second largest single user of electric power in the Northwest, just after the aluminum industry. Irrigating the Columbia Basin Project consumes nearly three billion kilowatt hours annually. A study done for the Northwest Agricultural Development Project in 1979 showed a rise in use of the Columbia River for both irrigation and power generation that put the two into direct conflict.[85] In 1981 the *Tri-City Herald* quoted a study done at Washington State University as stating, "'The economics of the irrigation energy [*sic*] tradeoff is not encouraging.'"[86] The authors complained that current practice did not require that the value of electricity lost to downstream generation because of water withdrawals for irrigation be charged against the costs of the reclamation project.

Besides power generation, fish and wildlife needs drew further on the river's finite resources. To sustain runs of anadromous fish, the government increasingly required the Bureau of Reclamation and especially the Army Corps of Engineers, the agency with overall authority to regulate the river, to spill or release water sufficient to maintain adequately high levels. At a Washington State

Department of Ecology Commission meeting in Wenatchee in 1979, Quincy-area farmers expressed fears that the need to spill such water in dry years could cripple irrigation.[87] It was becoming increasingly clear that every use of the river came at some cost to every other potential or existing use.

Against this background, backers of the Columbia Basin Project became concerned about the supply of water for present irrigation and future project expansion. They knew that the state guarantee of the allocation necessary for the second half of the project expired in December 1989. At that point, Washington state law required that water go to the "highest beneficial use."[88] A new law, then under consideration, contained sections that would end the water reserve held for the second half of the project.[89] Irrigators worried that if they did not acquire the water rights, someone else would.[90]

In 1984 the *Tri-City Herald* urged farmers to pressure their congressmen and declared, "Renewal of water rights to irrigate the second half of the Columbia Basin Project no longer can be taken for granted."[91] In 1987, responding to pressure from Columbia Basin Project backers, the Washington legislature enacted a law that reserved the water right for the irrigation project. Further, it indefinitely guaranteed that allocation without the need for any further renewals until the government either completes or abandons the project.[92] With more recent discussions surrounding salmon preservation, this priority may fall under increased scrutiny. It will be moot, however, if the project is not enlarged.

The question of finishing the project is undecided. Should the government go ahead, a more complicated question is, who will pay the bill, and how? The Bureau and project backers have debated this since the middle 1960s. The most formidable obstacle is a clause in the Flood Control Act of 1936 which states that on any public reclamation project the benefits must exceed the costs.[93] To determine the existing benefit-to-cost ratio, in 1969 the Columbia Basin Development League signed a contract with Battelle-Northwest Laboratories. That study showed definite economic gains from increased irrigation.

But construction estimates rose and in 1979 the *Wenatchee World* asked, "Is the project, expected to cost over $2 billion to complete, needed?"[94] When pushed on how increased irrigation would

affect Pacific Northwest power rates, Bureau of Reclamation representatives said that the figures remained inconclusive.[95] Agricultural economists at Washington State University speculated a
significant hydropower loss, and further stated that its replacement
with thermal power could increase rates throughout the region 43
percent above 1980 levels.[96] The second half of the project could
cost Northwest ratepayers over $2 billion in fifty years.[97] An earlier
study, also done at Washington State University, speculated that
the additional irrigation would return only twenty-five cents for
every dollar invested, and that it would further depress already
low farm prices.[98] While project backers relied on arguments remi-
niscent of those used in the 1920s, so did opponents who summoned
visions of astronomical costs, surplus crops, and, in this case, terri-
bly high-cost, rather than ruinously low-cost, power.

Early in 1984, Washington state legislators considered a bill
authorizing a $100 million bond issue to initiate work on the sec-
ond half of the Columbia Basin Project.[99] Opponents and backers
engaged in hot debates over the proposal's merits. Proponents argued that this would show the federal government good faith and
help garner the money needed to complete the project. By February the measure had passed the state Senate. Norman K. Whittlesey
dominated the hearings subsequently held by the House of Representatives. He testified against the project and accused local people
in the area, sprinkler dealers, bankers, and professionals, of backing construction only because they hoped to benefit financially.
Whittlesey asserted that they did not care if the nation really needed
new agricultural land. He could have said the same about Spokane
and Wenatchee interests in the 1920s. The House voted against
the bill. The Columbia Basin Development League blamed
Whittlesey directly for the failure. "All the things he said were
wrong," complained Chan Bailey of Othello, executive secretary of
the League.[100]

With the state nearly out of the picture, at least for the moment, the future of the Columbia Basin rested with the federal government. Existing law required an environmental impact statement
for any large new construction project. To show some good faith,
the state contributed $100,000 toward that study with the hope that
it would help draw federal money.[101] The Bureau of Reclamation
began work on the study in January 1984.

The government paid the engineering and consulting firm of CH2M Hill of Bellevue, Washington, just under $800,000 to conduct the study.[102] But in January 1986 the General Accounting Office (GAO) released its evaluation of the Columbia Basin Project. It concluded that completion of the Columbia Basin Project would show a benefit to cost ratio of 1.4-to-1 if it included interest rates figured at 3 percent. Even at that rate, the ratio was negative if it also considered the value of lost hydroelectric potential downstream. With changes bringing interest rates into line with current figures and using corrected accounting procedures, the GAO calculated the benefit to cost ratio of the second half of the project at .2-to-1.[103] The GAO further recommended that the Bureau, in its forthcoming Draft Environmental Impact Study (DEIS), show figures that represented realistic interest rates and current accounting practices.[104]

The Bureau put the DEIS on hold.[105] It then established an Interagency Water Conservation Steering Committee that it charged with recommending water conservation proposals for the DEIS. During the deliberations, the Bureau reduced the proposed project alternatives to three. The first called for full development of the remaining land, irrigating 172,900 acres over the initial fourteen years with the rest to follow. The second alternative, which the Bureau preferred, recommended irrigating 87,000 acres next to the East Low Canal and enlarging existing facilities, with further development deferred. The third alternative advised no additional irrigation.[106]

Project residents and interested parties received copies of the DEIS early in September 1989. In looking at the ratio of benefits to costs, it rated the first alternative at 1.8-to-1 based on an interest rate of 3 percent. It admitted that, at higher rates, the costs outweighed the benefits.[107] For the second alternative, the study showed that with downstream power losses added, and with all interest figures included at the highest level then generally paid by the government, the benefits could be as low as .027-to-1.[108]

In other areas, the DEIS projected that the loss of power potential resulting from project completion would equal the removal of Hungry Horse Dam, and the estimated annual cost to Pacific Northwest electricity users in the year 2036 would be, on the average, three dollars per household.[109] Finally, in looking at the environment, the most significant finding in the study, also a speculative and certainly a highly subjective one, claimed that failure to complete the project would result in greater harm than finishing the work. The losses to the water table due to continued pumping, especially on the east side, would exceed any damage done by expanding irrigation canals and other facilities.

For the government to finish the project, Washington state must show a willingness to finance some of it and Congress must underwrite costs that increase with each new estimate. Finally, the farmers must be willing again to negotiate a new contract and repay considerably more of the costs involved. Should they consent to just 10 percent of the figure currently projected, they are looking at two or three thousand dollars an acre over fifty years—in addition to daily operation and maintenance charges. The tendency is for farmers to feel that no price is too high for water as long as someone else pays the bill, and they are certainly not concerned about the cost/benefit ratio if they make money.

The present conclusion is that the agricultural aspect of the Columbia Basin Project never paid for itself and will not under almost any foreseeable economic conditions. Without the link to power and its direct subsidy, plus additional subsidies from the government, such as its assumption of interest charges, the project could not exist. In looking at developing the second half, state and federal legislators must consider that reality.

The economics of the Columbia Basin Project involve an intricate mix of hydroelectric power, irrigation, recreation, wildlife, navigation, conservation, agricultural surpluses, and Native American rights. Furthermore, these aspects of the project do not exist in a vacuum. Economics and politics on the local, state, national, and international levels all affect them.[110] The interplay between these factors changes from year to year, sometimes dramatically, making predictions about the future no better now than they were in the past. Further, the 1990s have brought increased attention to the salmon as a changed factor in the equation.

Those making the decision about proposed Columbia Basin Project expansion must also realize how much the project has changed internally over the last forty or even the last twenty years. Although small- to medium-sized family-run farms still dominate, they are significantly larger than they were and collectively they are home to increasingly fewer people. That trend will undoubtedly continue. Some corporate farming of a sort has developed; in places, a few families combined their lands and resources to form a larger business. Critics see this as a move away from the true family farm, but others see it as inevitable to take advantage of technological advances. So far, no large impersonal corporate entity exists on project lands.[111] The potential, however, is there.

Over the last twenty years, project farmers have hired growing numbers of salaried employees. Most were Hispanic, and by 1970 nearly 10 percent of Grant County spoke Spanish, while in Moses Lake the figure was twice as high.[112] The *Wenatchee World* has recently found it profitable to issue *El Mundo*, a Spanish version of its newspaper.[113]

Finally, a significant change occurred in the 1960s when the three irrigation districts assumed responsibility for day-to-day operation and management of all project facilities other than the main canals and Grand Coulee Dam. With the negotiations surrounding the second Bacon Siphon and Tunnel, the three districts gained the right to sell water that they had obtained from government irrigation works. Under the resulting revised master water-service contract it is increasingly possible for farmers to buy water when they wish and to build the distribution apparatus themselves. As the role of the Bureau of Reclamation diminishes, the position and authority of the irrigation districts grows. They will certainly have an increasingly important part in determining the project's future direction.

When asked in 1989 if he thought the government would irrigate the last 500,000 acres, Columbia Basin Project Manager James Cole said "yes." "But," he added, "the real question is, when?"[114] At the moment, the Bureau of Reclamation recommends only limited development of the second half, and argues that at present more is unnecessary. "It would be a shame to wait until you need a crash program to develop the land," Cole continued. "That is when the government makes its biggest errors."[115]

June 23, 1971: beginning construction of the third powerhouse. *Courtesy Bureau of Reclamation.*

# Chapter Nineteen
# The Third Powerhouse

Everything about this third power plant installation is so
large as to be nearly unbelievable.

*Reclamation Commissioner Ellis L. Armstrong*[1]

The Columbia River does not hold a uniform amount of water. Spring
floods or heavy rainfall dramatically expand its volume; cold winters bring snow and ice in the mountains reducing runoff and diminishing the river to a fraction of its peak. Because of the vagaries
of nature, it is impossible to predict exactly how much water will be
available at the Grand Coulee powerhouse at any given moment or
from year to year.

Only water passing through the turbines in the bowels of the
Grand Coulee powerhouses, which in turn spin the giant generators, actually produces electricity. Excess water, above that held in
the reservoir behind the dam, passes over its spillway providing a
spectacular display of wasted energy—falling water that generates
no power. Conversely, during low water periods the river lacks
enough water to operate all the generators, aggravating the regional
energy shortage. That power crunch notwithstanding, having water escape without producing power, or allowing expensive machines to sit unused, reduced the efficiency of the Grand Coulee
plant. The solution lay in finding a way to even out the flow of the
river, making it consistent and dependable from month to month.
Such a feat would not only increase the power production potential
of Grand Coulee's existing eighteen generators, but also make possible more units. With proper control, the river could turn machines
in a third and perhaps even a fourth powerhouse. The increase
would return Grand Coulee to first place among the world's power
producing stations, a position that it lost to the Russians in the 1950s.

It would also allow more efficient use of downstream dams, further increasing the amount of electricity available.

In 1944 the United States government began the $42 million Hungry Horse Dam, near Kalispell, Montana, which provided significant upstream storage.[2] Later in the 1940s the government went ahead with facilities at Albeni Falls. But these alone were not enough to even the river's flow. Franklin D. Roosevelt Lake, Grand Coulee Dam's reservoir, touches the international border. The storage needed to guarantee a constant water supply would, for the most part, come only with new dams and large reservoirs in Canada, and that required complicated agreements between the two governments. Determining who would receive what percentage of the ensuing benefits proved troublesome, and forging the treaties took years of unprecedented diplomatic maneuvering.

The immediate post-war region-wide power shortage quickened interest in upstream storage, and the Bureau of Reclamation explored alternatives. In March 1944 the United States government referred the storage issue to the International Joint Commission (IJC) which, in turn, created the International Columbia River Engineering Board, composed of two members from each country. The board set up an engineering committee to collect data and analyze the situation.[3]

From the outset the planners realized that increased reservoir storage in Canada would produce massive benefits in the United States. Through its boards and committees the IJC struggled to determine the exact value of those benefits and the fairest way of crediting Canada with a reasonable share. This was more difficult than imagined and almost two decades transpired before both sides agreed to the details.[4] The extensive technical studies, often slowed by insufficient funding, dragged on until 1959.[5] The flood of May 1948, and its widespread damage, led directly to accelerated action by the IJC and other planning bodies. The IJC's "Second Progress Report to the Federal Inter-Agency River Basin Committee" showed increased Canadian willingness to accommodate storage in return for electricity or financial compensation.[6] The revised "308 Report," produced jointly by the Bonneville Power Administration, the Army Corps of Engineers, and the Bureau of Reclamation, described 20,000,000 acre feet of potential storage in the United States and in Canada that would help alleviate downstream flooding.[7] For its part, the United States proposed a multipurpose project for the Kootenai River near Libby, Montana, that would impound four million acre feet in a reservoir extending well into Canada.

Ex-Senator Clarence C. Dill, now a member of Washington state's Columbia Basin Commission, made direct contacts in Canada and arranged a meeting with British Columbia officials early in 1949.[8] He received a favorable hearing and everyone agreed to study the possibility of storing even more water in reservoirs behind dams. At a subsequent meeting in Ephrata, Washington, the Columbia Basin Commission passed a resolution urging the creation of large storage dams in the Arrow Lakes area, just north of the international border near Castlegar. The Harry Truman administration signaled its approval and urged fast action on the matter.[9] The International Columbia River Engineering Board's preliminary report to the IJC in April 1949 said that a dam at Arrow Lakes would take five years to build, that it would enhance downstream generating plants, and that alone it would not end downstream flooding.[10]

Through 1950 Congress considered the $271 million Libby Dam, then proposed in an omnibus Rivers and Harbors-Flood Control Authorization bill.[11] With congressional approval, the United States asked the International Joint Commission to sanction Libby Dam, complete with the forty-two-mile-long reservoir extending into Canada. The IJC conducted four public hearings, but in 1953 Canada denied the application and the United States withdrew it.[12]

Some British Columbia officials saw resolution of the Libby Dam question as a crucial test. If they successfully struck an agreement, it would show Americans that Canada was willing to provide reservoir space with reasonable sharing of downstream benefits as the best compensation.[13] Others in Canada, however, enjoyed their unique position and did not want to rush any decisions. They saw that for once they controlled something that their giant neighbor wanted and could not obtain without their approval. Having the upper hand, they stalled the negotiations.

In 1952 the election of a Social Credit Party government in British Columbia and a Republican President in the United States dramatically changed the predisposition of the negotiating parties. Before 1952, those authorities in Canada who enjoyed aggravating United States' interests or, as historian Neil A. Swainson put it, "plucking the eagle's feathers," held important political offices.[14]

Social Credit people criticized them, and Social Credit's rise to power seemed to presage a quick and happy end to negotiations. But the Dwight Eisenhower administration, with its "No New Starts" directives and its "Partnership Policy," which together stopped progress on any project not already underway and said that the government would work with private industry rather than alone on any further undertakings, dampened American enthusiasm for federal projects. The Republicans, who emphasized private development or, at most, public and private joint-development, specifically discouraged the push for upstream storage.[15] Such increased capacity would have benefited the federal dams that directly competed with private installations. By the time this position softened some years later, John G. Diefenbaker, a Conservative, was Canada's Prime Minister. Diefenbaker steered toward an independent foreign policy not unfriendly toward the United States, but far less amiable than the Social Credit position. The switches on both sides resulted in confusion and waning enthusiasm for a treaty.

Meanwhile, in 1954 the United States resubmitted the Libby proposal.[16] The Army Corps of Engineers redesigned the dam, moving it several miles upstream, reducing the area of Canadian land it would flood. But both sides came forward with tougher proposals, including a Canadian demand for increased benefits. This stalled the negotiations. In the background, Canadian government representatives hinted at a new plan that appalled their American counterparts. They pointed out that under existing international treaty provisions, Canada could do what it pleased with the water in the Columbia River on its territory. General Andrew G. N. McNaughton, International Joint Commission Canadian Section Chairman, suggested diverting up to 67 percent of the Columbia's flow at Ravelstoke into the Fraser River to the north, consequently reducing the volume of water in the river downstream. To underscore its intent, the Canadian Parliament appropriated money to study ways to divert both the Kootenay and the Columbia.[17] How seriously the Canadians really took this suggestion is debatable. It is certain that with this action and other comments they signaled displeasure over the IJC negotiations and their desire for increased benefits from storage that supplied more water only to American generators. Perhaps the river diversion plan was nothing more than a ploy or bluff to achieve that end.[18]

Other ideas surfaced and died. In 1954 the Kaiser Aluminum and Chemical Corporation proposed a $30 million dam at Castlegar to increase Arrow Lakes storage capacity, creating more downstream power for aluminum production. Kaiser signed a contract with the British Columbia government, but the Canadian Parliament refused to approve it. This angered British Columbia residents, and their leaders resolved that the federal government in Ottawa would never again assert control over Columbia River usage and development.[19] This accelerating feud between British Columbia and the Canadian Parliament complicated the situation. For example, the Puget Sound Utilities Council had no luck talking to British Columbia about a proposed Mica Creek Dam, near where the Columbia River reaches its farthest northern point, about 200 miles above the international border. The Council offered to pay $250 million toward construction, giving the dam to British Columbia after completion in return for access to the additional 1,790 Kilowats (Kw) generated downstream.[20] While British Columbia favored the idea, the Canadian government held out for a larger chunk of the power. Consequently, nothing came of that proposal. Then, in 1955, the Canadian government passed its International River Improvements Act that prohibited construction of any dams in Canada without a federal license, stopping negotiations between United States' interests and the British Columbia government. The action further infuriated political leaders in British Columbia who again saw Ottawa usurping their authority. At that point British Columbia planners looked briefly at developing the Peace River to satisfy all provincial power needs, making any development on the Columbia River superfluous and thus undercutting Ottawa. The outcome of the Canadian brouhaha held the key to any further development at Grand Coulee.

Through 1955 and 1956, General McNaughton pushed his diversion idea, hoping for more favorable negotiations and more American concessions.[21] A 1956 article in the Canadian magazine *Maclean's* expressed Canadian fears about the Columbia. Once the United States bought the power, the author concluded, Canada would lose it forever. "The plug" he continued, "cannot be pulled." He suggested that Canadians still had control of the river, that Canada needed it, that Americans wanted it, and that the fight over it could become "the biggest hassle in our history."[22]

In 1958 Eisenhower met with Diefenbaker in Ottawa. They agreed to do something about Columbia River negotiations and they scheduled a series of high level discussions, beginning in February 1960, that bypassed the IJC. The Americans still wanted to build Libby Dam while the Canadians preferred storage on the Kootenay. Both sides compromised. They would build three dams in Canada: High Arrow, Duncan Lake, and Mica (for 15.5 million acre-feet of storage) along with Libby Dam.[23] The five million acre-feet at Libby would bring the total to over twenty million acre feet—more than enough to contain floods the size of the 1894 or 1948 disasters. The United States would pay one-half of the value of resulting flood control benefits, or about $65 million, to Canada.[24] The ten-year financial agreement, representing a triumph over difficult obstacles, created a $458 million joint program and included the promise that Canada would not divert Columbia River water for at least sixty years.[25]

In October, President Eisenhower announced the agreement for cooperative river development. On January 17, 1961, Canadian Prime Minister Diefenbaker and Eisenhower, in one of his last actions before leaving office, stood together in Washington, D.C., where they signed the Columbia River Treaty. Congress ratified it with a ninety to one vote. John F. Kennedy's Secretary of the Interior, Stewart Lee Udall, praised it and declared it the equivalent to a new Grand Coulee Dam.[26] At last everything appeared to progress smoothly.

But it was not to be just yet. In Canada the jealousies over control that simmered between the federal government in Ottawa and provincial leaders in British Columbia undid the treaty.[27] The Canadian federal government had negotiated without first settling its internal difficulties. The provincial government immediately asserted its own goals. It expressed concern over the recovery of the Canadian share of jointly produced power and rejected the use of American money for construction of storage projects. It took the two Canadian political units two years to resolve their disagreements. At its heart the matter turned on the issue of who controlled the valuable resources in the province and who determined the future of Western Canada's economy.[28] The resolution involved considerable haggling and time.

Despite American prodding, the debate raged in Canada through 1963. That April a national election brought Lester B. Pearson and the Liberty Party to power. Pearson, who campaigned on a policy of commitment to action after what he called a period of hesitancy, stated that the Columbia River Treaty should be renegotiated. The *Vancouver Sun* agreed, printing editorials calling the existing agreement a "sell-out" and insisting that the United States would realize most of the benefits.[29] The new government agreed, leading to resolution of the rift between Ottawa and British Columbia.

Diplomats resumed negotiations in December 1963. Out of those meetings came another quick agreement. Actually, little changed from the previous negotiations, although W. A. C. Bennett, British Columbia's Premier, claimed that the new arrangements would save British Columbia hundreds of millions of dollars. Canada agreed to complete Duncan Dam by 1968, Arrow Lakes (later named High Keenleyside) in 1969, and Mica Creek Dam in 1973, together providing 20.5 million acre-feet of storage. The United States would largely finance construction of the Canadian projects in return for the energy produced. British Columbia would proceed with the Peace River development and it would sell all of its share of the new electricity produced in the United States back to the United States over the next thirty years.[30] The Chelan, Douglas, and Grant county Public Utility Districts led in forming the Columbia Storage Power Exchange, a non-profit corporation that issued the revenue bonds needed to buy the Canadian power, or entitlement. Bonds, sold by that new organization in August through September 1964, raised the $254 million to pay B. C. Hydro the initial money to start construction.[31] Canada also gained $64.4 million in flood control benefits.

The Canadian House of Commons ratified the treaty on June 10 and the American Congress quickly followed. On September 16, 1964, President Lyndon Johnson and Prime Minister Pearson signed the Columbia River Treaty and Protocol at Peace Arch in Blaine, Washington, almost twenty years after both governments had started the negotiations.[32] It was a fitting location to end one of the most bitter debates ever waged between the two countries.[33]

The Columbia River Treaty yielded two immediate results in the United States. First, it guaranteed the additional power needed to justify the long-awaited Pacific Northwest/Pacific Southwest Intertie. The power intertie allowed the BPA to sell Canada's share of the Columbia River Treaty power in the Southwest during those periods when it was an unwanted surplus in the Northwest.[34] Second, as the BPA stated, the treaty made "essential the construction of a third powerhouse at Grand Coulee Dam."[35] The close link between the two developments resulted in the government planning and promoting them at the same time.

The Pacific Southwest and the Pacific Northwest are heavy power users during alternate seasons. The Southwest draws large amounts of electricity in the summer, while Northwest usage peaks in the winter. As early as 1919 different agencies proposed joining the two areas and in 1939 the BPA's first administrator, J. D. Ross, advocated a direct current transmission system as the best way to do so.[36] The urgency for the intertie came more from the South than from the North. Technological limits and politics prevented the logical interconnection, but when it finally came it raised the efficiency of existing generating facilities in both areas. This took time, however, because Northwesterners, jealous of the water in their rivers and their cheap electric power, were afraid that Californians would somehow find a way to appropriate both.[37] In April 1961, as scientists perfected long distance transmission and with the Columbia River Treaty an apparent reality, BPA administrator Charles F. Luce suggested building two 500,000-volt lines from the Columbia River to Bakersfield. With Northwest concerns in mind, Congress adopted legislation early in 1963 that guaranteed its first right to power generated at federal plants on the Columbia. Only the surplus, if any, would be available for shipment to California.

On June 25, 1964, Secretary of the Interior Stewart Udall sent Congress a plan describing four principal lines linking eleven western states. Private investment would provide $417 million out of the $697 million total needed to build the system.[38] In August the Senate and House of Representatives agreed to the plan and appropriated $27 million.[39] While in Portland, Oregon, on September 17, 1964, President Lyndon Johnson spoke at the "Intertie Victory Breakfast," praising both the success of that venture and the Columbia River Treaty that he had just signed the day before at Peace Arch in Blaine, Washington. Physical construction of the intertie began in 1966 and extended through 1970.

When the eighteen Grand Coulee generators began turning on September 14, 1951, the press described the next powerhouse that would increase Grand Coulee's capacity.[40] The *Wenatchee Daily*

*World* announced that Bureau of Reclamation engineers already had preliminary surveys for the new powerhouse, that it would sit on the west side of the river and generate 750,000 Kw, but that it had to wait for upstream storage.[41] In 1952, Congress appropriated $125,000 for a third powerhouse feasibility investigation and from mid-1952 through 1953 engineers inspected the area around the existing dam, mapping geological foundations and identifying optimum locations.[42] They recommended two sites, one on each side of the river.

During those years the United States and the Soviet Union orchestrated complex cold war rivalries, one of which brought Grand Coulee Dam a propaganda indignity. When the Russians built the Kuibyshev Power Plant on the Volga River in 1955, it topped Grand Coulee's 1,974,000 Kw capacity by churning out 2,300,000 Kw. Senators Henry Jackson and Richard L. Neuberger moaned that by 1957, three Volga River dams in Russia exceeded Grand Coulee's power potential and in 1958, with the Bratsk project on the Angara River, the number rose to four.[43] In the wake of the 1957 Sputnik I embarrassment, it appeared that the American effort in power plant size also ranked second. Worse yet, the Russians built generators significantly larger than those operating at Grand Coulee. The grand dam on the Columbia was dangerously close to losing all of her first-place claims.

In 1962, after completion of the Columbia River Treaty, Senator Jackson announced that the Bureau of Reclamation would request a $390 to $400 million congressional appropriation for a third powerhouse on the east side of the existing dam. The structure would hold twelve 300,000 Kw units; four coming at first and the rest following as needed.[44] Early in February 1965, when Interior Secretary Udall approved the Bureau of Reclamation's tentative recommendations for the new powerhouse, the *Wenatchee Daily World* noted happily that they would make Grand Coulee again the "world's largest power plant."[45] The Senate's Interior Committee held hearings on April 13 and 14, 1965. With Jackson himself, as committee chairman, pointing out that the Soviets had projects bigger than Grand Coulee, the powerhouse had little opposition and the committee reported it favorably on June 10, followed by the whole Senate on June 16, 1965.

Additional problems caused a delay of almost a year. The government linked the third powerhouse, and the Northwest/Southwest intertie, to the basin account plan. This idea tried to broaden the Columbia Basin Project power subsidy to include all federal reclamation in the area and pay for the greater whole with money from all federal dams. Some objected to the idea fearing that the principal, once applied to one region, would in time become a national concept. Money from the Northwest could then finance such things as diversion of the Columbia River to California. Eventually, however, Representative Wayne Norviel Aspinall, a Democrat from Colorado who put forward the basin account proposal, and Thomas Stephen Foley, Democrat from Washington who wrote the powerhouse measure, convinced enough congressmen that the benefits outweighed the potential abuses. Senator Jackson agreed to the plan after concluding that it was what the BPA had been doing since 1963, so it made little difference.[46] Final legislative approval of the third powerhouse immediately followed acceptance of the basin account plan.[47]

In a White House rose garden ceremony, with representatives from Canada watching, Johnson signed the third powerhouse appropriation bill on June 14, 1966. The act brought considerable symmetry to the Grand Coulee story. Johnson considered himself something of an heir to Franklin Roosevelt's New Deal that began the dam. Furthermore, Johnson loved big things, and the notion that Grand Coulee, enlarged through the efforts of his administration, would again be the biggest power producer in the world, appealed to him.[48] The President immediately announced that he would request a $3 million appropriation to start construction, with first power to come in 1973.

But Johnson had misgivings about the basin account amendment and he cautioned that he would propose a bill to limit the amount of revenue that could subsidize irrigation. Reclamation Commissioner Floyd Dominy agreed that the government should not allow reclamation to appreciably increase Northwest power rates, and he guessed that a $30 million cap on the power subsidy through the year 2026 would be about right.[49]

With congressional approval and the President's signature, action on the third powerhouse moved quickly. Over 200 people attended a meeting in July called by Governor Daniel Evans and held in the Grand Coulee city hall. There the mayors of surrounding towns formed a steering committee that listened to local concerns about the anticipated influx of up to 10,000 people, perhaps 3,000 of them workers, expected during construction.[50]

Interior Secretary Udall announced that the new powerhouse would produce 9.2 million Kw using not 300,000 but 600,000 Kw monster generators.[51] Two months later the Soviet Embassy shocked the Interior Department when, in a telephone call, its representative asked if powerhouse equipment contracts would be open to international bidding. The Soviets pointed out that Russian engineers had already produced 500,000 Kw generators and they wanted to demonstrate their accomplishments by doing the same at Grand Coulee.[52] It is an understatement to say that the inquiry panicked the bureaucrats. If the Russians won the bid and built the units, they could still claim at least a part in the world's largest power plant. That would cloud Grand Coulee's glory when it reclaimed the title as world's biggest. In propaganda terms, the Soviets seized a wonderful opportunity to embarrass the United States.

Udall felt that the United States should build at least the first three big units and that it would be better if it made all of them. Many in Congress strongly agreed. President Johnson and the State Department, on the other hand, argued that allowing the Soviets to bid on at least half the units would be part of an "enlarged partnership" that could ease the cold war. On May 19 the Interior Department announced, however, that it would accept only domestic bids for the project, ending the debate.[53]

Meanwhile, the first contractors began earth removal and other preliminary jobs at the dam. The Bureau of Reclamation informed residents of Coulee Dam that it would remove much of their city to make room for the new structure. Around the dam, new employees occupied hastily built houses and reclaimed buildings that lingered from previous construction periods. The influx recalled the heydays of the 1930s except that this time no Great Depression drove homeless men and destitute families to the dam site. The new breed arrived with mobile homes and around Electric City they formed a settlement of trailer houses, temporary but far more comfortable than the tarpaper shacks and makeshift structures that poverty brought to the same place three decades before. Like the old days, however, the population increased rapidly. In one year

Coulee Dam's student count rose by 102 and in Grand Coulee, up on the hillside above the dam, seventy-six more entered its already crowded schools.[54]

The schedule of events in the construction of the new powerhouse required careful attention to timing geared to the seasonal rise and fall of the Columbia. The planners basically relied on the scheme used thirty years before. Hence, it was a deliberately critical moment when the dam workers called a strike on April 1, 1968 against all three contractors then on the job. In a conflict over pay, they asked for a package increase of $1.11 an hour, including benefits. Denver planners designing the powerhouse concluded that they had to complete a cofferdam before the spring high water or else delay construction by one year. The five-day strike, partly successful for the unions, ended on April 5, 1968.[55]

Workers placed the first cofferdam cells on April 18. As the operators in the existing powerhouses lowered the level of the reservoir, the cofferdam extended across part of the east end of the back of Grand Coulee itself, enclosing a section on the right side of the lake, giving access to the east end of the old dam. This temporary structure allowed work to continue in areas that were normally under water.

Designers in Denver orchestrated the new structure's elaborate plans. Using the Bureau's Engineering and Research Center, they determined the proper configuration and appearance for the new building, the shape of the forebay channel behind the new structure, and the contours for the turbulent tailrace areas, where water coming from all three powerhouses would mix and churn during its flow downstream. An internationally known New York architectural firm, Marcel Breuer and Associates, executed the drawings for the building.[56] Their challenge: to make the new twenty-story structure, the size of four city blocks, complement the appearance of the original dam and yet take advantage of stylistic changes that had occurred during the intervening years. They abandoned the art deco and massive lines that tied Grand Coulee to the 1930s in favor of a more contemporary streamlined appearance. This allowed the two structures to sit comfortably side by side without being completely in harmony. As Reclamation Commissioner Floyd Dominy announced, the imposing and dramatic architectural concepts are designed to establish Grand Coulee Dam and its Third Powerplant complex in Washington State as one of the great showplaces of American resource and ingenuity.[57]

The plan included removing a section of the old dam's east end. A new structure, called the forebay dam, then stretched from the original dam to the north along the east side of the river. Behind it water eventually flooded into the forebay and from there into giant arching penstocks that carried it down to the base of the powerhouse where it turned the turbines. The design showed the powerhouse sitting in front of and forming part of the forebay dam, pouring out its water into the area just downstream from the existing right powerhouse. The finished building held twelve generators. The immediate plan, however, called for only half of the final structure and space for six generators.

To accommodate tourists, the third powerhouse plans included what the Bureau of Reclamation called a "novel and forward looking" visitation deck where the public could view the generators below. To reach the deck, people would travel in a glass-enclosed elevator 475 feet down along a spine merging with one of the arched penstocks.[58] The Bureau went to extraordinary trouble and expense to include these dramatic extras because it wanted to make the facility accessible and a desirable place for people to visit, and to showcase the world's largest power complex with the world's largest generators.

All of this seemed rather spectacular, but not enough for Clarence Dill. In a curious campaign the ex-senator, then in his late eighties, who realized that the new storage and powerhouse would end the cascade of water that often spilled over the dam, urged the Bureau to consider giant decorative fountains along the front of the dam and powerhouse. He called the concept "Fountains Unlimited," and at his own expense he printed an eight-page brochure that described and pictured the concept. When the Bureau rejected his idea, Dill badgered Congress with it for three years. Finally Reclamation Commissioner Ellis Armstrong wrote, "It appears to us there is general agreement that the most preferred display is water cascading down the face of the dam."[59] In a letter to Dill, Armstrong concluded, "The consensus is that these [fountains] would be artificial solutions which would appear to be contrived and strained and not in keeping with the purpose and beauty of Grand Coulee Dam."[60] This finally ended the discussion of Dill's fountains.

With the structure's design decided, the contractors began removing blocks 92 and 93 that formed a 250-foot-long section of the original structure. The gap became the access into the forebay behind the new powerhouse. As one contractor excavated rock and earth to create that forebay, others blasted away the east end of the nearly thirty-year-old dam. In mid-October they set off the first dynamite charges that began reducing the large section of concrete to small pieces of rubble. This continued through November when cold weather shut down the project. Once back to work in mid-January, the rhythm of the dynamiting continued. The old concrete was amazingly hard and it took longer than anticipated to break up the end of the dam.[61] On Friday, February 28, 1969, in one dramatic explosion, the final enormous 11,000-ton piece broke lose as workers neared the end of the task. The way was then clear for water to flow past the old dam into the new, yet-to-be-built forebay behind the third powerhouse.[62]

Despite the hard concrete and the cold weather, the plan moved ahead according to schedule. In the spring of 1969 engineers drew down the level of Roosevelt Lake exposing enough area to allow excavation of the upstream link with the new forebay.[63] With the water so low, Kettle Falls reappeared for the first time in twenty-eight years. For six weeks people came to see a sight hidden for nearly three decades.[64]

Nineteen large pumps kept the excavation area dry. The men raced to complete the work before the spring runoff swelled the Columbia and forced the operators to refill the lake. By mid-May they finished digging, a new cofferdam stood in place, and the water crept up behind the structures. Stretching north from the shortened end of the old dam, a new wall 390 feet high—the forebay dam—paralleled the river.

Everything went according to schedule until September 1969 when the Richard Nixon administration announced budget cuts designed to curtail 75 percent of all government construction. By October the work force at Grand Coulee dropped from 700 to 350, with more men leaving daily. In December, Senators Magnuson and Jackson confirmed rumors that the administrative budget adjustments would indefinitely postpone the third powerhouse.

The Nixon budget crunch came just as the Bureau of Reclamation prepared to take bids on the powerhouse building. Then early in February the new Nixon budget set aside $68 million for the third powerhouse and the government exempted Grand Coulee construction from public works spending reductions due to the "urgent schedule needed to meet Pacific Northwest power needs and to fulfill the Canada treaty commitments."[65]

Back on schedule, the Bureau opened the bids on February 11, 1970. Four firms, Vinnell Corporation of Alhambra, California, Dravo Corporation of Pittsburgh, Pennsylvania, Lockheed Shipbuilding and Construction Company of Seattle, and Mannix Construction Company of Calgary, Alberta, Canada, joined together and submitted the lowest of the four offers. They estimated the job at $112,525,612. The largest contract ever awarded by the Bureau of Reclamation, it covered both the powerhouse and the forebay dam.[66] Although it exceeded Bureau of Reclamation expectations by over $3 million, the government indicated satisfaction with the amount and on February 26, Secretary of the Interior Walter Hickel formally awarded the job to the four companies. The contractor immediately moved men into the area. The atmosphere of the swelling communities around the dam began to resemble the heydays of the late 1930s.

On October 21, 1970 workers poured the first official four-cubic-yard load of concrete. Senator Jackson, Reclamation Commissioner Ellis Armstrong, Representatives Thomas Foley and Catherine May, and Clarence Dill attended the ceremonies, along with a crowd of around 2,500.[67] In a unique ritual unprecedented at Grand Coulee, workers threw pennies onto the bedrock. Then the first load of wet concrete buried them. When asked by a reporter why the penny tossing, the men said that they did it so that someday they could tell people that "their money was there, in the dam."[68]

By January 1971, 700 men worked at the powerhouse site and Vinnell people expected 2,200 by mid-1972. Unlike work on the original dam, both the contractor and the Bureau of Reclamation hired women. Twenty-three-year-old Toby Ann Levy served as the Bureau's first female civil engineer on a heavy construction job.

She inspected for the engineering department and was one of three women the government employed there in 1970.[69]

The Bureau of Reclamation had planned to lower Roosevelt Lake again during the early spring of 1972 to remove the temporary cofferdam. But a series of aggravating labor delays, strikes, and jurisdictional tiffs postponed that operation until December 1972. This delay required raising the cofferdam ten feet to withstand the flood period in the summer of 1972. Fortunately the high water stayed within normal limits and the plan went smoothly.

Powerhouse construction did not go smoothly. The contractor had spent half of the money by January 1973, but the project was well behind schedule. Workers joined the forebay dam to the east end of the main dam, and then in May they lowered Roosevelt Lake eighty-five feet to complete excavating the area where the two units met. This time the new Mica Dam in Canada absorbed some of the runoff, making the situation easier at Grand Coulee. But Vinnell, Dravo, Mannix, and Lockheed frequently disagreed among themselves, and ongoing strikes complicated their already strained relationships. Finally, in May 1973, Vinnell withdrew as "contract sponsor" after billing the government $33 million in construction claims beyond contract specifications. Dravo assumed prime responsibility. Trying to maintain optimism, the Bureau of Reclamation rescheduled the first anticipated production of power in the new plant from 1974 to 1975.[70]

The last forebay dam concrete went into place during the summer of 1973. More strikes in August hindered progress, briefly idling 1,200 workers, but in December, as planned, the level of Roosevelt Lake began dropping 120 feet. Dravo removed the cofferdam and when the reservoir refilled, water ran into the forebay behind the new powerhouse. With the last difficulty overcome, everyone expected that the third powerhouse would generate power within the next few months.

Then an event half-way around the world interrupted progress at Grand Coulee. As a result of the Yom Kippur War in the Middle East in October 1973, Arab nations interrupted oil delivery to the United States. This led to an immediate fuel shortage, which in turn caused a power shortage regionally and nationally. Long lines of cars waited at gas stations and fossil fuel-using generating plants cut production. The governors of both Oregon and Washington asked businesses and private citizens to eliminate their traditional holiday lighting displays to conserve electricity, most of which now came from the straining hydroelectric dams. The Bureau of Reclamation found itself in a dilemma. If it continued lowering Roosevelt Lake, it reduced the power-generating capacity of Grand Coulee Dam. On the other hand, raising the lake would postpone by another year the power available from the new generating station.[71] The government proceeded with the drawdown and unused water spilled through Grand Coulee's outlet tubes while the Northwest faced the threat of brownouts.

On March 21, Kettle Falls re-emerged. Through the first week in May, Native Americans and other interested tourists and spectators made a final visit to what had been one of the richest salmon fishing spots in North America.[72] Downstream, the contractor amazingly worked ahead of schedule, removing the final 700,000 cubic feet of rock, and clearing the way for water to fill the forebay behind the third powerhouse. At 8:34 a. m. on April 25, 1974 men broke the earth dike between the lowered Lake Roosevelt and the forebay, allowing the first water to flow into the new area.[73] On May 11 they closed all the gates and began refilling Roosevelt Lake.

With the power shortage abated and the last drawdown completed the project advanced smoothly until an iron workers' strike shut it down during the summer of 1974. That action was the last significant delay, however, and through the rest of 1974 and into 1975 the work proceeded apace. Finally, the contractor completed the powerhouse building on December 3, 1975, exactly one year and two days late.[74]

The powerhouse was ready for generators. Westinghouse Corporation and General Electric had both inquired early in the summer of 1968 about the specifications for the giant generators, but General Electric withdrew after expressing concerns about government demands that they be made almost entirely in the United States.[75] While the Willamette Iron and Steel Company in Portland, Oregon, began work on the first three turbines, Westinghouse pondered the restrictions that might also eliminate it from providing the generators.[76]

The difficulty resulted from the Soviet interest in 1967. When the government blocked foreign bidding, Interior Secretary Udall invoked what he called a "Buy American" policy intended to eliminate foreign competitors from federal projects. The new guidelines aimed at allowing American manufacturers to "demonstrate their inventive capability in design and manufacturing."[77] The fine print permitted up to 49 percent of the components to come from abroad. Even so, American companies relied heavily on equipment from Japan, Canada, and Sweden, and Westinghouse wondered if it could assemble the generators with enough American parts to satisfy the requirements. Critics questioned, with just short of half of a machine made from imported pieces, how well it would show off American know-how. The construction delays postponed the generator bidding while the "Buy America" issue faded and the government moderated the requirements so much that the revised guidelines emasculated the intent of the original act. The Westinghouse Company built the first three units, numbers 19, 20, and 21, with each rated at 600,000 Kw. But on August 13, 1973 the Bureau of Reclamation awarded a contract to Canadian General Electric Company, Limited, to furnish and install the larger, 700,000 Kw generators for units 22, 23, and 24.[78]

By conventional standards, the new generators were huge. In their day everyone saw the 108,000 Kw machines in the first two powerhouses as enormous, but the 600,000 Kw machines dwarfed them. The old machines had a diameter of thirty feet, ten inches and weighed 556 tons with penstocks eighteen feet across. The new machines had a diameter of sixty-eight feet, weighed 1,900 tons, and had penstocks forty feet across.[79] A Bureau report pointed out, however, that "despite the high kilowatt rating and enormous physical size, the generators, in most respects, will not have unusual electrical characteristics."[80] In other words, like the dam before them, the generators were not unique; they were just big.

On August 26, 1975 Bureau engineers and the contractors tested unit 19.[81] On October 11, as about 1,500 watched, Senator Henry Jackson and Reclamation Commissioner Gilbert Stamm pushed a button that officially put the new generator into operation.[82] Generator 20 followed in April 1976, and 21 began rotating that December.[83] The Bureau went right on with the second three units already under contract. The 700,000 Kw units required more water, and this exacerbated an old problem. The ebb and flow of water during the day, increased during periods when all the machines operated, again tore away sections of the river bank and caused damage downstream. A temporary solution involved restricting use of the machines during high water months. This reduced the output of Grand Coulee and downstream generators, and the Bureau announced that alterations, and relocating up to 300 families living near the river downstream, would eventually remedy the problem.[84] That work, a four-year, $30 million project to stabilize the river bank, began in the spring of 1982.[85] Even so, the erosion headache continues.

Despite the erosion problem, powerhouse work proceeded. Without fanfare, the first 700,000 Kw generator, unit 22, went into service on May 23, 1978, with 23 following on November 14. The sixth unit went into operation on April 1, 1980, making Grand Coulee Dam again the world's largest power generating station.[86] In May 1980 ash fallout from the Mt. St. Helens volcano slowed the break-in and fine-tuning process.[87] The final unit required adjustments, and all six of the big machines experienced operational difficulties necessitating repair work on their windings and other modifications. During much of 1980 only two of the six generators produced power.[88] To further increase power output workers rewound the stators in the original eighteen generators, raising their rated capacity from 108,000 Kw to 125,000 Kw.[89]

In 1975, at the start-up ceremonies for the first generator in the third powerhouse, Senator Jackson commented that the government could not then build Grand Coulee Dam or anything like it, as people would see it as inflationary.[90] He might have added that environmental concerns would prohibit anything as big and destructive as Grand Coulee Dam. Congress approved the third powerhouse not long before it passed the National Environmental Policy Act which required extensive environmental impact statements. One can only speculate on how that legislation might have— or might yet—influence third powerhouse construction, since the powerhouse remains only half finished. Plans still call for six more generators.

As the nation, and the West, became more sensitive to conservation in the 1960s and 1970s, various federal, state, and local agencies looked at Grand Coulee to see if they could compensate for the impact the dam already made. In June 1968 the Grand Coulee Dam Environmental Advisory Council held its first meeting. Representatives of towns, counties, irrigation districts, the State of Washington, and the federal government drew a plan for the best use of the surrounding area. The organization paid architect Kenneth W.

Brooks $75,000 for a comprehensive design. Presented in 1969, it called for a National Park combined with adjacent tourist facilities, an open air concert arena, an aerial cable to a large exhibition center, electric cars to transport tourists to the powerhouses, a large golf course at Electric City, a model city built by the state and the Indians on Colville Indian Reservation land, campgrounds, swimming pools, and a huge "Omni Bridge" over the river below the dam that would hold a convention center suspended above the churning Columbia.[91] Elaborate and overwhelming, hardly in harmony with most people's notion of environmental protection or preservation, the plan briefly raised eyebrows and then quickly disappeared from view. Copies still languish on the shelves of libraries and government offices, but no one advocates its implementation.

By the late 1960s, however, the government realized that Grand Coulee's existing tourist facilities—the old wooden grandstands used since the 1930s—needed remodeling or replacing. Tourism was the one industry that held promise for the future and it had been among the best ways for the Bureau of Reclamation to demonstrate its accomplishments. The Bureau decided on entirely new accommodations. In 1972 the government purchased the Green Hut Cafe, originally opened in 1938, which sat on a vista point overlooking the dam. In 1976 it destroyed the building and replaced it with a new visitor center that cost over $1.5 million, and which the *Wenatchee World* described as a large "bottle cap perched on the bluff."[92] Across the river, the incline elevator at the third powerhouse began operation in 1977. Like the generators in the powerhouse below, the temperamental mechanism required modifications and adjustments. These changes eventually became components of a self-guided project tour.[93]

The power produced at Grand Coulee was prodigious. It and the other federal dams accounted for one-half of the power from all dams throughout the region and those dams filled 80 percent of the total Pacific Northwest base power load. A shift long anticipated, however, finally came in the late 1970s.[94] Grand Coulee and the other federal dams moved from being constant power producers providing the guaranteed load to the role of peaking plants, coming on and going off during each twenty-four-hour cycle depending on when the demand for power was highest. Although this diminished flood control capacity, the upstream dams in Canada and elsewhere in the United States more than compensated.[95] The switch shifted the base load to thermal plants that are difficult to start, and run more economically if operated constantly. As peaking plants, the generators in the dams can quickly pick up load when required, providing it for local or distant points, generally doing it at times when power is most expensive. In 1975 alone the Southwest bought enough power from the Northwest to save $167 million in fuel costs. The intertie, which allowed for "wheeling" or moving of power from areas of surplus to areas of need, and the program of increased hydroelectric production, integrated the West and helped, at least temporarily, resolve its energy problems.

In 1977 the Jimmy Carter administration saw no need for more irrigation and his environmentalist White House stressed conservation rather than increasing energy production capacity. Carter's men changed the name of the Bureau of Reclamation to the Water and Power Resources Service and directed it to stress preservation rather than reclamation. In line with those priorities, the BPA conducted studies and announced that energy efficiency was six times cheaper than new thermal generating capacity. This change of direction, among other things, undermined nuclear plant construction by the Washington Public Power Supply System (WPPSS) and left the region without a long-term comprehensive program for filling its power needs. It also scuttled any immediate plans to finish the third Grand Coulee powerhouse.

The turmoil had little direct effect at Grand Coulee Dam which, tucked away in eastern Washington, continued producing power, supplying it to consumers from the Canadian border to southern California. On Saturday, January 30, 1982, a few stalwarts gathered at the dam near Franklin Roosevelt's statue to celebrate the 100th anniversary of his birth. Eighteen months later, from July 15-17, 1983, hundreds of old-timers, many who had built the project decades before, assembled near the dam to mark the fiftieth anniversary of the start of construction. They helped dedicate the new powerhouse and reminisced about the old days. That summer the American Society of Civil Engineers cited Grand Coulee's third powerhouse as the outstanding engineering achievement of 1983.

In a curious aside, President Ronald Reagan's Secretary of the Interior, James Watt, attended the 1983 celebrations. At the powerhouse dedication he noticed the plaque commemorating the event.

Watt objected that it did not include his name and he ordered, at a cost of $4,000 to taxpayers, a new plaque with his name and that of Ronald Reagan, but now omitting the names of Presidents Johnson, Nixon, and Carter, whose administrations authorized the building and oversaw its construction.[96] It was a sophomoric act of political pique and self aggrandizement that occasionally typified Watt during his tenure as Interior Secretary.

On a more significant level, the 1983 dedication marked the end of construction—so far—at Grand Coulee Dam. Of course it is not now and never will be finished. Updating equipment, routine maintenance, and potential additions, including at least the second half of the third powerhouse, will change and modernize the plant. The third powerhouse and its six generators, coming after the Columbia River Treaty and the Pacific Northwest/Pacific Southwest intertie, not only made Grand Coulee again the world's largest power plant, but together that intertie and the treaty dams also allowed Grand Coulee's service area to broaden and include almost the entire trans-Rocky Mountain West. Because of its heavy reliance on and abundance of hydroelectric power, at least one historian has dubbed this the "Kilowatt Kingdom."[97] If there is such a kingdom, Grand Coulee Dam is the finest jewel in its crown. The third powerhouse put another facet in that already impressive gem.

Aerial view of Grand Coulee in 1979 shows third powerhouse. *Courtesy Bureau of Reclamation.*

# Afterword

So here it stands, a monument to the idea and the power of an idea; a monument to organization; a monument to cooperation; a monument to opposition; a monument to the United States Army Engineers; a monument to the United States Bureau of Reclamation; a monument to the magic spirit of willing men which accomplishes more than the might of money or the marvels of machinery; a monument to the brains, the intellect of great engineers—and you, class of 1942, could you come back here a thousand years hence, or could your spirit hover around this place ten thousand years hence, you would hear the sojourners talking as they behold this "slab of concrete," and you would hear them say, "Here in 1942, indeed there once lived a great people."

*Rufus Woods to the graduating class of 1942,*
*Grand Coulee High School[1]*

. . . in this region major projects were started which all believed could achieve the planned promised land. . . . The dams along the Columbia were designed to achieve the promised land for the people of the Pacific Northwest.

*Richard Lowitt[2]*

Remember, we play the ball from where it lies.
Attributed to Robert Tyre "Bobby" Jones, Jr.

*by Alistair Cooke[3]*

Rufus Woods titled the eighty-page book that his *Wenatchee Daily World* printed in 1944 *The Twenty-Three Years' Battle for Grand Coulee Dam*. He dated the struggle from 1918, when William Clapp first suggested damming the Columbia River at the Grand Coulee, through 1941, the conclusion of construction on the dam itself.

Woods underestimated the length of the battle he documented. Promoters and visionaries proposed Big Bend reclamation in the 1890s. In its entirety, the effort to build the great dam and the irrigation spanned not twenty-three, but over 100 years, and it is still incomplete.

Through that century many participated in the effort. In 1963 a modest William Clapp objected when someone called him the father of Grand Coulee Dam. "Too many people," he said, "had experienced the vision—construction engineers, farmers, lawyers, senators, local business people, Franklin Roosevelt—for anyone to pass out credits."[4] Clapp was wise. There are many who had a part in parenting the dam. Laughlin MacLean proposed his irrigation scheme in 1892 and David R. McGinnis pushed for irrigation of the Quincy area in the early 1900s. Clapp and Elbert F. Blaine suggested alternatives to water dry Big Bend land. Despite Clapp's reluctance, he and Blaine do deserve credit. Their visions led directly and indirectly to the dam in its present form.

In the 1920s Roy Gill, O. L. Waller, Peter McGregor, Marvin Chase, James A. Ford, J. E. McGovern, Fred Adams, George Washington Goethals, and Harvey Lindley—all connected with the Spokane Chamber of Commerce in one way or another—worked for the gravity plan. They kept the idea alive and deserve recognition for its eventual success, albeit in a form they did not embrace. Rufus Woods, James O'Sullivan, Ed Southard, Nat Washington, Gale Matthews, Albert Goss, Hugh Cooper, and hundreds of unnamed supporters pushed for the pumping plan and the dam.

David Henny, Willis Batcheller, Major John S. Butler, Dr. Elwood Mead, John Page, Arthur Davis, Harry Bashore, Sam Hill, Wesley Jones, Miles Poindexter, Charles Leavy, Clarence Martin, Warren Magnuson, Henry Jackson, Clarence Dill, and Lewis Schwellenbach number among the prominent figures who promoted and backed Grand Coulee. At the national level were Secretary of the Interior Harold Ickes and President Franklin Roosevelt.

The long recitation of names makes the point. Grand Coulee Dam had a legion of "fathers," including the Great Depression of the 1930s. Had Roosevelt's emphasis on planning, controlled land use, and government competition with private power not prevailed when they did, there would be no huge dam at the Grand Coulee today. The Depression provided the conditions that allowed

Roosevelt to build the dam. Hugh Gallagher, one of Roosevelt's biographers, wrote that there are monuments commemorating many of the nation's Presidents, some great and some not so great, but that there is still no memorial to Franklin Roosevelt.[5] Gallagher may be incorrect, for there could be no more fitting monument erected to Roosevelt than Grand Coulee Dam. It embodies the giant vision that Roosevelt had for the nation, his concepts of public works and of public power, of planning, and of service by a government for its people.

And as with so many things in this world, timing was everything. Kirby Billingsley, cousin of Sam B. Hill, wrote in 1962,

> Had Grand Coulee construction not been started when it was, the dam never would have been built for two reasons: one, because the fish people were just coming to life with the realization that hundreds of miles of fish spawning streams would be cut off from the great annual Columbia River salmon migrations and, two, because the private power companies had started construction of a low dam at Kettle Falls and would have proceeded with this. Either one would have stopped Grand Coulee.[6]

Billingsley was correct. In the 1930s the government and project backers attempted to preserve the salmon, but when forced to make a choice they selected the dam and reclamation. They would not have that choice today. Over time, values change, giving each generation a different perspective on the past. Decisions once clear and simple are today questioned. Then the benefits of reclamation outweighed environmental preservation, as it is defined today. Those benefits were viewed as positive conservation measures in their day. Despite any new vision, what was done then cannot be easily undone now.

In 1909 the federal government enacted legislation allowing the states to protect the watersheds of navigable streams. This and creation of the Federal Power Commission in 1920 enabled construction of federal dams on the Colorado and Columbia rivers,[7] leading directly to public/private competition for production and sale of power. The effort to build Grand Coulee Dam played out against the background of that bitter fight of the 1920s and 1930s.[8] Congressional opposition that annually confronted funding of Grand Coulee was only one skirmish in a much larger conflict.[9]

The debate over Grand Coulee encompassed a number of issues. Besides public versus private power, it included reclamation and the wisdom of opening more agricultural land in a time of food surpluses. Grand Coulee construction gave New Deal opponents a chance to denounce big government spending, and it allowed Westerners a forum where they voiced resentment over the intrusion of Eastern politicians in their affairs.[10] This surfaced in the Columbia Valley Authority controversy. Rufus Woods and his minions eagerly sought government money to build their agricultural/industrial empire, but resented any controls suggested by that same government.[11]

Robert Athearn wrote that in the twentieth century Westerners began seeing themselves as poor relatives spiritually and financially dominated by the East.[12] This feeling drove reclamation promoters in the Big Bend. They wanted to build a new, autonomous agricultural/industrial empire that mimicked the humid East, but without its problems. Grand Coulee Dam, they believed, would guarantee ample water and cheap power to pay the bills while serving agriculture and industry. It would make North Central Washington residents independent and prosperous. Yet they wanted the federal taxpayer to cover the cost of that independence.

Wherever people go, they carry with them their cultural baggage.[13] They want any new home to resemble the place they left. They want to replicate the familiar. But they must also accommodate themselves to different conditions. The interaction between the two forces has led to innovation and change. Reclamation is one of these accommodations—the way to re-create the wet East and mimic its agricultural/industrial setting. The price tag for creating such an artificiality has been very high both in terms of money and in environmental damage.

Men like Rufus Woods and James O'Sullivan argued that the national taxpayer should cover the costs. They believed that eventually the project would pay for itself. And it has, due to the power ratepayers throughout the region. That subsidy has hidden the astronomical price for the irrigation water. Without it there would be no Columbia Basin Project and probably no Grand Coulee Dam.

The promoters used any argument to support their cause. They talked about population growth and future needs for increased agricultural land. They emphasized power when that seemed more advantageous. They glibly added easing unemployment, flood control, recreation, navigation, conservation, and, during the Second World War, national security. When it suited their purpose, they embraced the multiple use concept, wrapping everything into one package. Their arguments have continued until the present. In 1973 William Warne wrote, "Meanwhile regional, national, and world populations are increasing. Failure to develop resources to support added numbers of people will certainly cause progressive deterioration of man's environment."[14] A great deal depends on how one defines "environmental protection." It hardly means the same thing in the 1990s as it did in the 1930s and even now there is little agreement among those who advocate it.

The construction of Rock Island, Bonneville, and Grand Coulee dams began what historian Stewart H. Holbrook called the "Dam-Building-Era" in the Northwest.[15] That era is over. It left the Columbia River a series of lakes, and a series of problems that may require dramatic solutions. Those dams, and the irrigation, power, and navigation they provide, have over-burdened the river, devastated fish, polluted waterways, and brought jurisdictional disputes over the finite supply of water. They represent a formidable, technological achievement, yet they have brought formidable unanticipated problems.

More than its technical accomplishment, Grand Coulee Dam represents one of the great political achievements of the century. In a troubled time when other nations chose military conquest to solve their financial and social dilemmas, the United States government turned to large public works projects. Legislators embraced a plan to irrigate one million acres of fertile land and transplant there, from devastated areas elsewhere, thousands of would-be farmers and their families. They planned new cities with industries powered by cheap electricity provided from publicly owned plants. That overdrawn and unrealistic vision did not emerge as hoped, but this should not entirely negate the fact that the government, in a difficult period, moved to fulfill positive goals. Grand Coulee, the jobs it created, and the promise it held, convinced citizens that their political system still worked. Considering the problems of the 1930s, the fact that Grand Coulee Dam stands in the Columbia River today is the most remarkable aspect of its being.

On Saturday, July 18, 1983, 5,000 people stood in the new third powerhouse. The event was the highlight of three days, from July 17 through the 19, when the old-timers celebrated the dam's first half-century and christened its newest addition. They went to picnics and dances, and they marveled at what they had started fifty years before. They took pride in their accomplishment and few doubted that, despite the controversies and environmental debates, the dam was a great success.[16] If limited to evaluating just creation of cheap abundant electricity, they were right.

Because Grand Coulee Dam came at a time of unusual economic circumstances and with a priority on creating jobs, few stopped to consider adverse effects. Native American interests suffered, property owners received short shrift, and the salmon did not command enough attention. Despite appreciable planning, unanticipated problems, such as downstream erosion, have plagued the dam. It is, however, unfair to judge actions of five decades ago by contemporary standards. In the 1930s Americans had less sensitivity toward the integrity of the land and limited understanding of ecosystems or their preservation. They knew that something as large as Grand Coulee Dam would significantly change the region. But few questioned that the benefits far outweighed the drawbacks. The opposition to Grand Coulee then was political and economic— not environmental. And it must be repeated that then most people saw the dams and irrigation as conservation measures that enhanced and protected the environment, a view that has been altered over the ensuing decades.

From that time, when the push to build the dam overwhelmed the opposition, the nation has moved to a place where environmentalists today no longer debate benefits versus losses. They oppose any further dam building and even consider removing some already constructed.[17] Those sentiments, increased demands on the Columbia, and regional population growth are now rapidly ending the days of low-cost hydroelectric power in the Pacific Northwest. The abundance of energy that led to the development initiated during World War II precipitated today's potential shortages. The multiuse nature of Grand Coulee and other Columbia River dams, seen as progressive then and pursued so vigorously and successfully, means that today there is not enough water in the river to generate all the power, irrigate all the land, float all the ships and barges, support all the fish, carry all the sewage, cool the nuclear reactors, and provide all the recreation.[18] What will happen to Grand Coulee Dam in the future because of this remains problematical. The government will not tear it down, but there may never be further additions to its powerhouses. The third powerhouse at Grand Coulee ended the dam building era. Today the benefits no longer clearly outweigh the costs. The present reality dims the past triumph.

Still, Grand Coulee is an unquestioned economic success. The electricity generated there has returned over $4 billion.[19] Since 1942 there has never been any doubt about the ability of the dam to pay for itself and more. The success of the Columbia River dams supported confidence, cockiness, and ultimately over-expansion. It led to the disaster of the Washington Public Power Supply System which projected construction far beyond need. The proponents of power finally did what conservatives feared in the 1920s and 1930—they overwhelmed, or nearly overwhelmed, the market with unneeded power.

The reclamation aspect of the dam is another matter. If the explorer, Colonel Thomas William Symons, had revisited the Columbia Basin in 1982, 100 years after his report to Congress, the changes might have surprised him. Traveling east on Interstate Highway 90, he would have dropped down to the bridge at Vantage. The view from the road is spectacular, with dramatic high, brown cliffs on the east side of the wide and scenic river. Once across, Symons would continue on the Interstate, rising up a winding road until at last he emerged on the plateau several hundred feet above. There he would see a large brown sign welcoming him to the Columbia Basin Project. More impressive, he would find rolling green fields irrigated by giant central pivot sprinklers gently making circles on a landscape that before was dry and sagebrush-covered.

To the south, Symons would recognize the familiar and unchanged Frenchman Hills, but in front of them, the rich land now produces wheat and vegetables. Farther east is Moses Lake, a thriving town where Symons could find recreational facilities along the lake and the nearby Potholes Reservoir. Had Symons left the Interstate and moved south, past Othello and toward Pasco, he would pass more green fields, orchards, and vineyards. Here and there he would cross irrigation canals which carry the water that vitalizes the arid land he once described.

Symons could hardly have foreseen the roads, the canals, the fields, the cities, and the lakes. Rufus Woods, who did foresee them, and much more, might look with mixed emotions. Only half of his agricultural/industrial empire exists. World War II industrialized the West. The resulting dramatic hunger for electricity proved the need for the Columbia River dams. But it also drew that power away from North Central Washington and delivered it to established urban areas. While it solidified the public perception that Grand Coulee Dam came just in time, it killed the industrial aspect of the agricultural/industrial empire locally, although it succeeded on a wider regional level.[20] The transmission system carried the electricity away and left the dam sitting in the middle of nowhere. Although the government built half of the irrigation project, the industry that followed ended up a long way away. Farmers came, but not nearly enough to provide the market that project promoters expected.[21] In 1973 only 2,290 farms operated on the Columbia Basin Project. They supported far fewer than the 80,000 families or 10,000 farms predicted by early visionaries.[22]

That new agricultural domain is itself artificial and it exists because of heavy subsidies.[23] Power ratepayers absorb the cost of irrigation and the federal government adds more benefits because it charges no interest on the money expended to build the project that has not yet been repaid.[24] Yet the farmers repeatedly insist that they cannot survive on small plots and farm size has steadily increased. The more land each farmer works, the larger the individual subsidy. This also means fewer people living in the area and increased seasonal labor. The smaller the number of resident farmers, the less chance for an industrial base. Rufus Woods could hardly have predicted the outcome. His vision was flawed, but no more than others who then looked into the future.

Today Columbia Basin Project farmers see themselves as deserving the water promised long ago, and they demand that others pay the bill. They demonstrated that belief during the contract renegotiations of the 1950s and 1960s and more recently when the government built the Second Bacon Siphon and Tunnel. In 1976, Columbia Basin Development League President Roger Thieme pointed out that it was natural for the farmers to resist paying for the facilities that brought them water. "It is human nature not to spend any more than you have to," he explained.[25] In a sense, the farmers exhibit the combined Physiocrat and century-old Social-ist-Populist notions that the government should support their enterprise if they cannot support it themselves.

Behind the farmers, and behind the project from its inception, is a cadre of professionals, business people, and promoters. Exemplary are the Columbia Basin Irrigation League that the Spokane Chamber of Commerce sponsored in the 1920s, and the Columbia Basin Development League that now lobbies for the project. They still see development of the land as the way to lure industry and stimulate economic growth.[26] In 1933, Major John S. Butler (later Colonel) expressed their hopes, desires, and underlying motives in a section of his report to the House of Representatives:

> The farmer as a primary producer is not necessarily the main beneficiary of irrigation development. Local retailers of every kind, banks, public utilities (both power and railroads), labor, wholesalers, jobbers, manufacturers, and the general public are to a surprising degree dependent upon agricultural production, not only because of the food produced for direct consumption and the raw materials supplied to manufacturers, but because of the general business activity which is created.[27]

Those business people repeatedly demonstrated their weight in Washington state politics from the 1950s through the 1970s. Without difficulty they brought Senators Warren Magnuson and Henry Jackson, both nationally prominent, influential, and powerful, to their aid. With that support, they thwarted the Bureau of Reclamation and coaxed the State of Washington into helping pay recent construction costs. They succeeded in maintaining a successful farming community that, with its formidable subsidies, contributes significantly to the state's economy.[28] The question is whether that contribution exceeds its hidden and apparent economic and environmental costs.

No reclamation project in American history received as much advance study and planning as the Columbia Basin Project. The Joint Investigations alone represented a Herculean undertaking. In 1973, Assistant Reclamation Commissioner William Warne, who worked with them, wrote:

> The farms on the 516,320 acres of land that have been irrigated on the project have made generally excellent progress. As a result of pre-planning, the project has made greater strides than

has any other irrigated area in a similar period in the history of the West. The new farms are wired for electricity and telephone. Schools are conveniently available to all farm families. By and large, settlers in the Columbia Basin Project lands have escaped the austerity and drudgery of pioneers on many other reclamation projects.[29]

The planning, however, was not as successful as Warne indicated. Three factors contributed to its shortcomings. First, World War II drew personnel away from the Bureau of Reclamation and from the Joint Investigations. Because of this, the government did not complete the intensive geologic studies that might have foreseen later drainage difficulties. Despite the unprecedented planning, the project needed more, although it is possible that there could never have been enough.

Second, the war also postponed project development. The result was critical:

World War II delayed settlement for a decade and changed the entire technological structure of United States agriculture. Large scale, low labor requirement, highly specialized, mechanized, efficient farm operating units made the family farm of tradition an outmoded, unrealistic concept. The policies governing the project (and other Bureau of Reclamation projects), however, were unresponsive to the changed post-war conditions. The inflexibility has led to several deplorable aspects as the project's development responded more to outside forces and less and less to control by planners.[30]

Finally, the farmers who settled the area wanted, and continue to want, subsidy without regulation, plans, or limits. From the start, by renting additional units to augment the size of their farms, they thwarted the original project goals. Through less-than-vigorous enforcement, the Bureau of Reclamation allowed the divergence.

Among the goals for the Columbia Basin Project, envisioned in the 1930s, were that it would promote the family farm and prevent the concentration of benefits in the hands of a few individuals. The maximum number of people possible would settle project lands and share the irrigation water subsidized by the power ratepayers. This was the heart of the "Planned Promised Land" concept. And it failed.

A 1974 study found that the top 5 percent of the farmers received 20 percent of the benefits. The upper 25 percent of the tenant-operators, those who both owned and rented land, received 75 percent of the benefits while the bottom 10 percent received no positive net benefits.[31] Some farmers on the project prosper but the benefits are unevenly distributed.

Perhaps increased planning efforts in the post-war years might have anticipated the problems of a changing national economy. This would have allowed the government and the Bureau of Reclamation to reform the Columbia Basin Project Act of 1943. But the New Deal's commitment to planning itself was a scattered and piecemeal effort conducted on the federal and local levels. And the government moved steadily away from such efforts after 1935.[32] By the end of World War II there was little government enthusiasm for planning. The shift left the Columbia Basin Project with antiquated and impractical goals even before the first water flowed in its canals. Consequently, all the participants—the Bureau of Reclamation (covertly and intermittently), the farmers, and the business people and professionals in the area—worked from the start to alter the guidelines and limits laid down so carefully by the planners. The changes that have affected, and liberalized, project restrictions came through the efforts of farmers who challenged the laws. They cajoled Congress into expanding those limits. The adjustments since 1952 resulted from farmers responding to outside forces, not continued planning.

It is fair to ask what the project might be like today had the original planners achieved their goals. It would be a collection of family farms ranging from forty to eighty acres, none of them capable of supplying their owners with a satisfactory living. The area would be a rural slum. It is for the best that this aspect of the project failed.

The lesson of the Columbia Basin Project has wider implications. Not only rapid economic transitions, but also new administrations with new priorities and different visions, affected the direction of the project. From the start, the Bureau of Reclamation and others involved recognized that irrigation of the million-plus acres would take a number of decades. Yet, in the planning they did not allow for the alterations that political, social, and economic change might require. In retrospect, long-term planning, under this country's political system, with its frequent shifts in direction and the alterations in emphasis that come with each new administration, is impossible or, at least, impractical. It is also true that any study of Western development must include Eastern politics. The two are inextricable.

The Columbia Basin does differ markedly from large irrigation projects in California. The anti-speculation legislation laid a foundation for limited ownership that, despite subsequent changes, has remained. There are no huge agribusiness enterprises. In this, the project has so far succeeded.

As viewed in the 1930s, the Columbia River seemingly held an abundance of water rather than the shortage that plagues the Southwest. The only difficulty for North Central Washington was getting that water up onto the land. The river provided the answers. From the start, backers of the Columbia Basin Project linked it directly to Grand Coulee Dam. The dam generated the power to lift the water. It also supplied the electricity that, when sold, paid for the irrigation. In the Columbia they found an abundance that solved their problem.

But this linkage, based on the myth of plenty—that the river would always have a surplus of water—led to a misunderstanding. Repeatedly, promoters claimed that the Columbia Basin Project would amortize itself and not cost the federal taxpayer anything. For thirty years, from the early 1920s through the 1950s, the Bureau of Reclamation, and project backers, like Rufus Woods and James O'Sullivan, insisted that the water would be virtually free. This gave farmers a sense of independence and caused resentment when costs rose. They feel that the power subsidy, built into the project from its inception, allows them the freedom to challenge the Bureau of Reclamation and demand that all costs be covered by the government no matter how high they might climb. They have come, with some justification, to expect something for nothing. The increasing realization that there is no excess water in the Columbia now exposes the myth of its abundance, and the financial reality, brought by shortage, haunts the project's future.

Critics may wonder whether the project was a mistake. From 1952 through 1974 the government's investment of nearly $500 million had directly benefited about 20,000 people.[33] Depending on the method of calculation, those people are repaying anywhere between 5 percent and 15 percent of the total. Power ratepayers, the federal taxpayer, and taxpayers in Washington state cover the rest. The estimated cost for completing the project's irrigation aspect is now over $2 billion. Despite this, however, the answer to whether the government should have built the project at all—and now whether to expand it—depends not on the cost alone but more on the desired outcome.

If farmers, taxpayers, and power ratepayers together agree to cover all costs, no matter how high, the project is a success and it should continue. Although the price of irrigation, added to power rates, is minimal for each household, the willingness of those people to pay the bill is debatable.[34] Because the government charges no interest, the subsidy is considerable, and will never be returned. The project is not and never was self-liquidating. It pays only if the government links irrigation, power, and taxes. If the project is to pay its own way and compete with other farms nationally, then it is a failure and was a mistake. None of this includes the problem of crop surpluses and the actual need for what grows there.

If everyone realized what the final price of reclamation projects might be, it is possible that the public would object and the government would not fund them. Anthony Netboy pointed out that, consequently, Congress builds large things, like the Columbia Basin Project, piecemeal.[35] Once started, despite accelerating estimates, Congress seldom, if ever, stops a project. It is significant, then, that this is exactly the approach that the Bureau is suggesting today for building the project's second half. It is recommending adding facilities in slow steps over a period of years rather than all at once.[36] This would provide incremental costs that might appear more palatable to legislators than the $2 billion projected for full development. Neither this nor the full completion approach considers the possibility of high cost overruns which have consistently appeared in the recent past.

The Bureau of Reclamation—renamed the Federal Water Power Resources Service during the Carter administration and in 1981 changed back to the Bureau of Reclamation—has learned and it has evolved. It is sensitive to criticism that the private sector might do its job more efficiently.[37] On the Columbia Basin Project, it realized that allowing farmers to pay construction costs directly makes them more amenable to higher charges, especially when they compare government charges with the cost of privately obtained water on adjacent lands. When the farmers build more of the delivery facilities themselves, and control operation and maintenance, they understand better the problems involved, and expect less from the government. Since the 1970s the Bureau has moved increasingly in this direction on the Columbia Basin Project. Nationally in the 1990s the agency has begun redefining its role. Under Reclamation Commissioner Dan Beard and the William Clinton administration, it hopes to become a smaller, efficient "water management bureau with a more environmental mission."[38]

In considering project completion, state and federal government officials must today ask difficult questions. How much are power ratepayers willing to subsidize irrigation? Is there sufficient water in the Columbia River to accommodate irrigation, power needs, fish, navigation, and recreation? What further environmental damage might extended irrigation bring? What are Native American needs and rights? Does the nation require additional farm land? What are the real long-term costs of increased irrigation, and who will pay? How much should government subsidize a few farmers producing what will perhaps become surplus crops? The key is to ask the right questions, including the tough environmental ones that are possibly unpopular.

As the Columbia Basin Project stands now, it is the result of visionaries and promoters, like Rufus Woods in Wenatchee and Spokane Chamber of Commerce members who worked for decades to irrigate the Big Bend. That it exists at all is a tribute to their persistence. It is the story of attempted planning, at least partial planning, for land use by the largest number of people. It is the tale of a successful dam that brought cheap power and industry to the greater region if not the immediate vicinity. The unexpected, combined with social, economic, and political changes, has channeled its development, sometimes in directions unintended at the outset. Despite complaints and problems, the government is not likely to tear it down. The pyramids, much smaller than the dam and so often compared with it, have lasted for millennia and Grand Coulee will undoubtedly be around for a long time too. The reality today is a project that continues to hold unrealized potential and at the same time poses difficult and unanswered questions. In answering those questions, planners must "play the ball from where it lies." At Grand Coulee and on the Columbia Basin Project, there is much yet to be done. This is a story not yet completed.

# Appendix

# Physical Structures of the Columbia Basin Project

This appendix will provide technical and factual data about the physical structures on the Columbia Basin Project. Particular items are presented and discussed in the order that one might encounter them if one traveled south from Grand Coulee Dam through the project. All information in this appendix comes from the Bureau of Reclamation's *Project Data Book* as revised in April, 1983.*

### Grand Coulee Dam

Grand Coulee Dam is a concrete gravity dam built from 1933 until 1942. After the addition of the third powerhouse and forebay dam from 1967 through 1974, it contained 11,975,520 cubic yards of concrete. The dam is 550 feet high from bedrock, 500 feet thick at its base, 5,673 feet long. The top of the dam is 1,311.1 feet above sea level.

The dam conjoins three powerhouses. The two original structures each hold nine generators which originally produced 108,000 kilowatts but which were modified in the 1970s so that today they each produce 125,000 kilowatts. The third and newest powerhouse has six generators. Three produce 600,000 kilowatts and three produce 700,000 kilowatts.

### Franklin D. Roosevelt Lake (Reservoir)

The reservoir behind Grand Coulee Dam, named Franklin D. Roosevelt Lake, when full contains 9,386,000 acre-feet of water of

*Bureau of Reclamation, *Project Data Book* (Washington, D.C.: U. S. Government Printing Office, April, 1983), chapter dealing with the Columbia Basin Project, pages 1-24.

which 5,185,400 acre-feet are active or can be used to generate electricity. (One acre-foot is the amount of water that it would take to cover one acre of land to a depth of one foot.) Due to the design of the turbines and the inlet works that allow water into the turbines, not all of the water in the lake can be used to generate power. Once the water falls below a certain level, no power can be produced. The amount of water that can produce power is called the active capacity. The lake, when full, is 151 miles long and stretches to the United States/Canada border.

## The Pump-Generator Plant

The pumps used to raise water out of Franklin D. Roosevelt Lake are located on the left or west side of the lake just behind Grand Coulee Dam. Six of the units are pumps only. Each can lift 1,600 cubic feet of water per second and each raises the water anywhere from 290 to 310 feet, depending on the level of Franklin D. Roosevelt Lake. The pumps produce the equivalent of 65,625 horsepower. One 125,000 kilowatt unit in the left or right powerhouse at Grand Coulee Dam can operate two of the pumps.

The newest six units are a combination pump-generator. When used as pumps they function exactly like the first six units. Each can produce 67,500 horsepower and they can lift 1,700 cubic feet of water per second. When reversed, they act as generators and each can produce 50,000 kilowatts of power.

## Banks Lake, Dams, and Headworks

Banks Lake, south and west of Grand Coulee Dam in the bed of the ancient Grand Coulee (whose scenery, even with the artificially introduced waters of the lake, has an awesome beauty), is an equalizing reservoir into which water is pumped from Franklin D. Roosevelt Lake. At the top of the pump station, the water flows through a short feeder canal into Banks Lake. Banks Lake is formed by two rock-faced, earthfill dams, one at its north end and one at its south end. The north dam is 145 feet high and contains 1,473,000 cubic yards of rock and earth. The south dam, named Dry Falls Dam, contains 1,658,000 cubic yards of rock and earth and is 123 feet high. It is located just a few yards away from the geologic feature from which it gets its name. Banks Lake is twenty-seven miles long and contains 1,275,000 acre-feet of water with an active storage capacity of 715,000 acre feet. Located at the south end of Banks Lake, at Dry Falls Dam, is the headworks unit which controls the flow of water going south onto the Columbia Basin Project. Those headworks are 1,580 feet above sea level and contain six radial gates which are twelve-by-eighteen feet

### The Main Canal and the Bacon Siphons and Tunnels

The Main Canal begins at the headworks at Dry Falls Dam. It is 18.4 miles long including the siphons and tunnels and Billy Clapp Lake, into which it flows. About two miles south of Dry Falls Dam, the water enters the Bacon Siphons and Tunnels. The two parallel siphons are 1,038 and 1,041 feet long respectively. The two parallel tunnels are 10,037 feet and 9,950 feet long respectively. Together they can carry 19,300 cubic feet of water per second.

### Pinto Dam and Billy Clapp Lake

Pinto Dam is an earthfill dam. It contains 1,462,000 cubic yards of earth and rock, and is 130 feet high. The dam is thirty feet wide at the base and 1,900 feet long. The reservoir behind it is Billy Clapp Lake which is six miles long and contains 64,200 acre-feet of water, of which 21,200 acre-feet are active capacity.

### West Canal and East Low Canal

Below Pinto Dam, a bifurcation works divides the flow of water into two systems. The West Canal has a length of 82.2 miles and it skirts the northwest periphery of the project, including a large inverted two-and-one-half mile long siphon around the north end of Soap Lake. From there the canal runs across the Quincy Basin and then south to the northern base of the Frenchman Hills. At that point, the canal enters a 9,280-foot-long tunnel which goes through the Frenchman Hills and then splits into the canals along Royal Slope, including the eight-and-one-half mile long Royal Branch Canal. The West Canal serves 274,000 acres. Its capacity diminishes along its route.

The East Low Canal has an initial capacity of 4,500 cubic feet of water per second. It runs south to Moses Lake and ends near Warden, east of Moses Lake. It is at present 82.4 miles long. When complete, it will run farther south to a point near the Snake River eight miles northeast of Pasco.

## O'Sullivan Dam and Potholes Reservoir

O'Sullivan Dam is a large earthfill dam located on Crab Creek fifteen miles south of Moses Lake. It contains 8,753,000 cubic yards of earth and rock and is 200 feet high. The dam is 19,000 feet, or almost four miles long, and its crest, thirty feet wide, serves as a roadway. Behind it is the Potholes Reservoir which acts as a storage and equalizing reservoir for the southern half of the project. The reservoir, when full, contains 511,700 acre-feet of water, of which 332,200 is usable or active storage. It has a top elevation of 1,028 feet above sea level. A system of wasteways from the areas of the West and East Low Canals provides a method for water to flow into the Potholes Reservoir and this supplements natural and return flows into that lake.

## Potholes Canal

The Potholes Canal begins at the headworks at O'Sullivan Dam and extends 62.4 miles to the south. It has a capacity of 3,900 cubic feet of water per second. Branching off from this canal are the Eltopia Branch Canal, 25.3 miles long, and the Wahluke Branch Canal, 41.2 miles long with its siphon, which is three miles long.

## Other Features

In all, the Columbia Basin Project contains 333 miles of main canals, 1,993 miles of laterals, and 3,498 miles of drains and wasteways. As some of the project lands are higher than the elevations of the canals and laterals, there are pumping plants to further raise water where needed. As of 1982 there were 241 such plants with nearly 500 individual pumps ranging from 3 horsepower to 2,500 horsepower.

# Endnotes

## Endnote Abbreviations

Banks Papers—Papers of Frank L. Banks, Washington State University Library, Pullman

Batcheller Mss.—Papers of Willis T. Batcheller, University of Washington Library, Seattle

BOR Denver Mss.—Files of the Bureau of Reclamation, National Archives, Denver Branch

BOR Ephrata Mss.—Bureau of Reclamation, Columbia Basin Project papers, Project Office, Ephrata, Washington

BOR-FRS Mss.—Records of the Bureau of Reclamation in Federal Records Center, Denver

BOR Office Mss.—Records of the Bureau of Reclamation held in BOR headquarters, Denver

CBD League Mss.—Columbia Basin Development League Papers, Washington State University Library, Pullman

City Light Mss.—Papers of Seattle City Light (formerly called the J. D. Ross Papers), University of Washington Library, Seattle

Columbia Basin Commissions Mss.—Papers of the Columbia Basin Commissions, Washington State Archives, Olympia

Dubuar Scrapbooks—Scrapbooks located in the University of Washington Library microfilm collection

Gill Mss.—Papers of Roy Gill, Washington State University Library, Pullman

Jackson Mss.—Papers of Henry M. Jackson, University of Washington Library, Seattle

Jones Mss.—Wesley L. Jones Collection, University of Washington Library, Seattle

Magnuson Mss.—Papers of Warren G. Magnuson, University of Washington Library, Seattle

O'Sullivan Mss.—Papers of James O'Sullivan, Gonzaga University, Spokane

Papers of the Governors—Papers of various Washington governors at Washington State Archives, Olympia

Project History—Project histories are done yearly by the Bureau of Reclamation for each of its projects. Copies are kept in the local project office, in the regional office (in the case of Grand Coulee, in Boise, Idaho), and in the main office in Denver. An additional set of Grand Coulee project histories exist in the Frank Banks Papers, Washington State University Library, Pullman

Terou Mss.—Scrapbooks of Mrs. Roy (Elma) Terou, at the Coulee Dam City Hall. Most clippings came from the *Spokane Chronicle*.

Waller Mss.—Papers of Osmar Waller, Washington State University Library, Pullman
Woods Mss.—Files of Rufus Woods, held by Wilfred Woods, *Wenatchee World*, Wenatchee, Washington
Both the *Wenatchee Daily World* and the *Wenatchee World* are cited frequently in the following notes. The *Wenatchee Daily World* became the *Wenatchee World* on May 28, 1971.

## Notes for Introduction

1. Richard L. Neuberger, "Man's Greatest Structure," *New York Times Magazine* (August 9, 1942): 14.
2. "The Columbia Basin Project is the largest single irrigation development in the history of reclamation in the United States, and probably also in the world." The irrigation network in California's Central Valley is larger, but it is not a single project. Murray A. Straus and Bernard D. Parrish, *The Columbia Basin Settler: A Study of Social and Economic Resources in New Land Settlement*, Bulletin 566 (Pullman: Washington Agricultural Experiment Stations Institute of Agricultural Sciences, State of Washington, May 1956), p. 1.
3. Department of the Interior, Bureau of Reclamation, *The Story of the Columbia Basin Project* (Washington, D.C.: U.S. Government Printing Office, 1964), p. 5.
4. For a more detailed description of the Columbia Basin Project's physical features see the Appendix.
5. The Bureau of Reclamation accounting system has charged $497 million to irrigation, $888 million to the electric plant, and $302 million to multipurpose aspects of the project. Some features of Grand Coulee Dam itself are credited to irrigation, and some to power generation. Some costs of the pump-generator plant are charged to power generation, and some to irrigation. This system makes it difficult to determine exactly how much the irrigation features of the project actually cost because some of those charges have been attributed to multipurpose aspects such as flood control, recreation, and navigation. The changing value of the dollar further complicates the problem. (Neither here nor anywhere throughout this work have corrections been made to compensate for the changing value of the dollar.)
6. Charles Howard & Associates, Ltd., Arthur Peterson and William Beyers, *Preliminary Socioeconomic Analysis: Second Half of the Columbia Basin Project* (Olympia: Washington State Department of Ecology and State Printing Plant, March 1985), pp. I-9, I-14, I-16, I-18, I-19, I-21, I-23, I-28.
7. Speaking of the same phenomenon in Texas, Donald Worster wrote, "There was nothing uniquely western in [Walter Prescott] Webb's dream of the future. Essentially it amounted to a vision of replication of the East, where Texas would earnestly make the fullest use of their limited water in the pursuit of money and industrial giantism." See Donald Worster, *Rivers of Empire: Water, Aridity and Growth of the American West* (New York: Pantheon Books, 1985), p. 265. Worster might well have said the same about eastern Washington and the Columbia Basin.
8. Richard Lowitt, *The New Deal and the West* (Bloomington: Indiana University Press, 1984), p. 152.
9. *Ibid.*, pp. 138-152, 157.

10. See Wesley Arden Dick, "Visions of Abundance: The Public Power Crusade in the Pacific Northwest in the Era of J. D. Ross and the New Deal" (unpublished Ph.D. dissertation, University of Washington, 1973), esp. pp. 5-37.
11. Tearing down some or all of the dams has been suggested. See Portland *Oregonian* (October 13, 1990), p. A1.

## Notes for Chapter One

1. Ralph W. Johnson, "The Canada-United States Controversy over the Columbia River," *University of Washington Law Review* 41 (August, 1966): 683.
2. Arthur W. Baum, "The Coulee Colossus," *Country Gentleman* 108 (February, 1938): 7.
3. The Columbia is the fourth largest river in North America, surpassed by the Mississippi, Mackenzie, and St. Lawrence, each of which travels farther and carries more water.
4. Project History, 1946, p. 28. Also see George W. Fuller, *A History of the Pacific Northwest: With Special Emphasis on the Inland Empire* (New York: Alfred A. Knopf, 1952 edition), p. 11.
5. Harold Ellsworth Tennant, "The Columbia Basin Project" (unpublished masters thesis, University of Washington, 1937), pp. 34-35.
6. Joseph G. McMacken, *Geology of the Grand Coulee: Grand Coulee Dam and Quincy Basin Irrigation Project* (address before the Northwestern Scientific Association, Spokane meeting, December 27, 1928, reprinted in 1936 at Spokane, Wash.: C. W. Hill Printing Company, 1948), p. 3.
7. Otis W. Freeman, *Grand Coulee and Neighboring Geological Wonders* (Cheney, Wash.: Self-published, 1937), p. 8.
8. Cecil M. Ouellette, "Grand Coulee: Monument to an Ancient River," *National Parks Magazine* 38 (September, 1964): 13.
9. McMacken, *Geology of the Grand Coulee*, p. 3; Tennant, "The Columbia Basin Project," p. 2; Project History, 1946, p. 27.
10. Paul L. Weis and William L. Newman, *The Channeled Scablands of Eastern Washington: The Geologic Story of the Spokane Flood* (Washington, D.C.: U.S. Government Printing Office, 1982), p. 7.
11. Michael James Schulthesis, S.J., "The Struggle for Grand Coulee Dam—Beginnings," (unpublished masters thesis, Gonzaga University, 1961), p. 6.
12. Donald W. Meinig, *The Great Columbia Plain: A Historical Geography, 1805-1910* (Seattle: University of Washington Press, 1968), pp. 423-424. Fuller, *A History of the Pacific Northwest*, pp. 6-7; and John Elliot Allen, Marjorie Burns, and Sam C. Sargent, *Cataclysms on the Columbia: A Layman's Guide to the Features Produced by the Catastrophic Bretz Floods in the Pacific Northwest* (Portland, Ore.: Timber Press, 1986), pp. 107-116.
13. For example, see McMacken, *Geology of the Grand Coulee*, p. 5. Also see J. Harlan Bretz, *The Grand Coulee* (New York: American Geographical Society, 1932), p. 39; Freeman, *Grand Coulee and Neighboring Geological Wonders*, pp. 8-10.
14. Fuller, *A History of the Pacific Northwest*, p. 19.
15. *Spokane Spokesman-Review* (June 1, 1927), O'Sullivan Mss.; *Wenatchee Daily World* (July 13, 1928), p. 1; (August 13, 1933), p. 1.
16. J. Harlen Bretz, "The Lake Missoula Floods and the Channeled Scabland," *Journal of Geology* 77 (September, 1969): 517-527.

17. Ray J. Schrick, "'Scablands' in the Columbia Basin," *Reclamation Era* 39 (May 1953): 89-90.
18. Weis and Newman, *The Channeled Scablands*, pp. 11-16.
19. Bretz, "The Lake Missoula Floods and the Channeled Scabland," p. 505.
20. Allen, *Cataclysms on the Columbia*, pp. 80-82.
21. *Ibid.*, pp. 84, 97.
22. *Ibid.*, p. 114.
23. *Ibid.*, pp. 69-71.
24. For the best telling of the Bretz story, see *Ibid.*
25. George Macinko, "Types and Problems of Land Use in the Columbia Basin Project Area Washington" (unpublished Ph.D. dissertation, University of Michigan, 1961), p. 7; and Elbert Ernest Miller, "Geography of Grant County, Washington (unpublished masters thesis, University of Washington, 1947), p. 27.
26. George Sundborg, *Hail Columbia: The Thirty-Year Struggle for Grand Coulee Dam* (New York: The Macmillan Company, 1954), p. xv.
27. Stewart H. Holbrook, *The Columbia* (New York: Rinehart and Co., 1956), p. 17; and Schulthesis, "The Struggle for Grand Coulee Dam," pp. 9-10.
28. Oscar Osburn Winther, *The Great Northwest: A History* (New York: Alfred A. Knopf, 1952), p. 222.
29. Meinig, *The Great Columbia Plain*, pp, 156, 160; and Craig E. Holstine, *Forgotten Corner: A History of the Colville National Forest, Washington* (Colville, Wash.: *Colville Statesman-Examiner*, Inc., 1987), pp. 11-15.
30. The Nez Perce did not surrender or allow the government to confine them to reservations until 1877 when Chief Joseph admitted defeat in Montana. In 1878 Chief Moses agreed to live with his people on the Colville Reservation. They signed a treaty in 1884. See Grand Coulee Dam Bicentennial Association (Vesta Seiler, Chairman), *From Pioneers to Power: Historical Sketches of the Grand Coulee Dam Area* (Nespelem, Wash.: RIMA Printing and Graphics, 1976), p. 9.
31. Macinko, "Types and Problems of Land Use in the Columbia Basin Project Area Washington," p. 11.
32. John Fahey, *The Inland Empire: Unfolding Years, 1879-1929* (Seattle: University of Washington Press, 1986), p. 12, and Alexander Campbell McGregor, *Counting Sheep: From Open Range to Agribusiness on the Columbia Plateau* (Seattle: University of Washington Press, 1982), p. 13. Also see Albert N. Williams, *The Water and the Power: Development of the Five Great Rivers of the West* (New York: Duell, Sloan and Pearce, 1951), p. 339 and Henry Nash Smith, "Rain Follows the Plow: The Notion of Increased Rainfall for the Great Plains, 1844-1880," *Huntington Library Quarterly* 10 (Spring 1947): 171.
33. Robert C. Nesbit and Charles M. Gates, "Agriculture in Eastern Washington, 1890-1910," *Pacific Northwest Quarterly* 37 (October 1946): 281; and Charles M. Gates, "A Historical Sketch of the Economic Development of Washington Since Statehood," *Pacific Northwest Quarterly* 39 (July 1948): 219.
34. Tennant, "The Columbia Basin Project," p. 51.
35. For a full description of the geology of the Grand Coulee site see Richard Foster Flint and William H. Irwin, "Glacial Geology of Grand Coulee Dam, Washington," *Bulletin of the Geological Society of America* 50 (May, 1939): 661-680.

## Notes for Chapter Two

1. This chapter only mentions a few of the schemes that proposed irrigating or actually irrigated parts of eastern Washington.
2. Thomas William Symons, *The Symons Report on the Upper Columbia River & Great Plain of the Columbia* (Fairfield, Wash.: Ye Galleon Press, 1967), pp. 98-133 (esp. p. 121). This is a reprint of the original document published in 1882 as 47th Congress, 1st Session, Executive Document #186. When Symons made the survey he held the rank of First Lieutenant but the Army elevated him to Colonel before publication of the report.
3. Carlos A. Schwantes, *The Pacific Northwest: An Interpretive History* (Lincoln: University of Nebraska Press, 1989), p. 204.
4. Oscar Osburn Winther, *The Great Northwest: A History* (New York: Alfred A. Knopf, 1947), p. 305.
5. William D. Miner, "A History of the Columbia Basin Projects" (unpublished Ph.D. dissertation, University of Indiana, 1950), p. 29. Also see Israel C. Russell, "A Geological Reconnaissance in Central Washington," *U. S. Geological Bulletin* 18 (1893): 99.
6. Daniel Miller Ogden, Jr., "The Development of Federal Power Policy in the Pacific Northwest" (unpublished Ph.D. dissertation, University of Chicago, 1949), p. 120. Also see *Spokesman-Review* (April 17, 1939), p. 12; and Ralph E. Dyar, *News for an Empire: The Story of the Spokesman-Review of Spokane, Washington, and the Field it Serves* (Caldwell, Ida.: The Caxton Printers, 1952), pp. 400-401.
7. Bruce Mitchell, *The Story of Rufus Woods and the Development of Central Washington* (Wenatchee: *Wenatchee Daily World*, 1965), p. 10; and Bruce Mitchell, *Flowing Wealth: The Story of Water Resource Development in North Central Washington, 1870-1950* (Wenatchee: *Wenatchee Daily World*, March 6, 1967; reprinted January 1980), pp. 8, 23. Also see *Wenatchee Daily World* (July 8, 1983), p. A 5.
8. Ogden, "The Development of Federal Power Policy in the Pacific Northwest," p. 537.
9. Click Relander, "The Battleground of National Irrigation," *Pacific Northwest Quarterly* 52 (October, 1961): 146. Also see Roy M. Robbins, *Our Landed Heritage: The Public Domain, 1776-1936* (New York: Peter Smith, 1942), p. 332.
10. Michael C. Robinson, *Water For the West: The Bureau of Reclamation, 1902-1977* (Chicago: Public Works Historical Society, 1979), p. 16. Theodore Roosevelt strongly backed the "Homes for the Homeless" ideal. He stressed reclamation for its social significance: ". . . our aim should be not simply to reclaim the largest area of land and provide homes for the largest number of people, but to create for this new industry the best possible social and industrial conditions; and this requires that we not only understand the existing situation, but avail ourselves of the best experience of the time in solution of its problems." Roosevelt as quoted in Yahaya Doka, "Policy Objectives, Land Tenure, and Settlement Performance: Implications for Equity and Economic Efficiency in the Columbia Basin Irrigation Project" (unpublished Ph.D. dissertation, Washington State University, 1979), p. 29. Also see Samuel P. Hayes, *Conservation and the Gospel of Efficiency: The Progressive Conservation Movement, 1890-1920* (Cambridge, Mass.: Harvard University Press, 1959), passim.

11. The Army Corps of Engineers also opposed passage of the act and formation of the Reclamation Service, which it saw as competition.
12. George Sundborg, *Hail Columbia: The Thirty-Year Struggle for Grand Coulee Dam* (New York: The Macmillan Company, 1954), p. 16.
13. Dyar, *News For An Empire*, p. 339.
14. *Spokane Spokesman-Review* (October 28, 1903), p. 1, O'Sullivan Mss.
15. Department of the Interior, Geological Survey, *First Annual Report of the Reclamation Service, June 17 to December 1, 1902* (Washington, D.C.: U.S. Government Printing Office, 1903), p. 304.
16. T. A. Noble to F. H. Newell, May 6, 1903 and September 23, 1903, BOR Denver Mss. Also see "Preliminary Report on the Feasibility of Taking Water from the Spokane River or Lake Coeur D'Alene for Irrigating Land Under the Big Bend Project," report dated January 12, 1904, BOR Office Mss.
17. Noble estimated that each acre irrigated would exceed $1,625. John Fahey, *The Inland Empire: The Unfolding Years, 1897-1929* (Seattle: University of Washington Press, 1986), p. 92. Also see Mitchell, *Flowing Wealth*, p. 23; Michael James Schulthesis, S. J., "The Struggle for Grand Coulee Dam— Beginnings" (unpublished masters thesis, Gonzaga University, 1961), p. 92.
18. *Spokane Spokesman-Review* (June 28, 1905), O'Sullivan Mss.
19. Fahey, *The Inland Empire*, p. 104.
20. Mitchell, *The Story of Rufus Woods*, p. 10; and Sundborg, *Hail Columbia*, p. 17.
21. *Spokesman-Review* (April 2, 1906), O'Sullivan Mss.
22. Engineers, promoters, and Reclamation Service personnel increasingly recognized the need to link power generation and irrigation, with the former paying the costs of the latter. On April 16, 1906, Congress passed legislation that addressed this idea. The Reclamation Project Townsite Act gave the Reclamation Service the authority to build hydroelectric power plants and to sell the energy, preferably to public agencies, with the proceeds offsetting irrigation repayment charges. The Salt River Project near Phoenix, Arizona, became one of the first to benefit from the law when, in 1909, electricity generated at Roosevelt Dam pumped its water and at the same time supplied power to the city. This established the legal foundation on which Grand Coulee Dam rested almost thirty years later. Robinson, *Water For the West*, p. 27.
23. Project History—Grand Coulee Dam, 1933, p. 68.
24. Bruce Charles Harding, "The Columbia Basin Irrigation League: Its Origins and Operations" (unpublished masters thesis, State College of Washington, 1951), pp 5-6.
25. Miner, "A History of the Columbia Basin Projects," pp. 43-58; Winther, *The Great Northwest*, pp. 314-315; and *Spokesman-Review* (September 18, 1908), O'Sullivan Mss.
26. *Spokesman-Review* (May 20, 1909), O'Sullivan Mss.
27. The measure failed by 189,065 to 102,315. Ogden, "The Development of Federal Power Policy in the Pacific Northwest," p. 123; and Harding, "The Columbia Basin Irrigation League," p. 6. Also see *Spokesman-Review* (August 29, 1914), O'Sullivan Mss.
28. Stewart H. Holbrook, *The Columbia* (New York: Rinehart and Company, 1956), p. 304.
29. Theodore Saloutos, *The American Farmer and the New Deal* (Ames: Iowa State University Press, 1982), p. 3; and Russell Blankenship, *And There Were Men* (New York: Alfred A. Knopf, 1942), p. 168.

30. *Wenatchee Daily World* (June 1, 1918), p. 4.
31. The telling of the Billy Clapp legend here is drawn largely from a typed manuscript done by W. Gale Matthews dated 1961, titled "Beginnings of the Columbia Basin Reclamation Project," and contained in the O'Sullivan Mss. A second copy exists in the files of the Bureau of Reclamation's regional office in Boise, Idaho. There are other versions of the story. For a biography of Matthews see Rita Seedorf, "Building the Big Bend Country: The Saga of the Matthews Family 1890-1960" (unpublished masters thesis, Eastern Washington University, 1980).
32. *Wenatchee Daily World* (March 19, 1918), p. 7.
33. The *Wenatchee Daily World* had actually run many other stories covering proposed irrigation ideas and some of them received as much or more coverage than the Clapp idea. For examples, see *Wenatchee Daily World* (July 1, 1918), p. 2 for a 13,000-acre idea to irrigate areas in Okanogan County or *Wenatchee Daily World* (August 9, 1918), p. 6 for a story on the Marvin Chase Project of 12,000 acres that was to have been located near Tonasket.
34. Rufus Woods to Marlen Pew, editor, *New York Times*, January 4, 1934, Woods Mss. Also Rufus Woods to Wellington Pegg, December 4, 1930, Woods Mss. The usual quote attributed to Steiner is, "Dam the Columbia? Verily, Baron Munchausen, Thou art a piker!" (The Baron Munchausen reference was to a character known for his exaggerated stories.) See Rufus Woods, *The 23-Years' Battle for Grand Coulee Dam,* (Wenatchee: *Wenatchee Daily World,* 1944), p. 62.
35. Rufus Woods to Wellington Pegg, December 4, 1930, Woods Mss.
36. For a complete study of the subject, see Victor Howard Bagnall, "The *Wenatchee World* and Its Influence on the Columbia Basin Project" (unpublished masters thesis, University of Washington, 1961), passim, and esp. pages 7-9. Also see Harding, "The Columbia Basin Irrigation League," p. 10.
37. Ogden, "The Development of Federal Power Policy in the Pacific Northwest," p. 127.
38. James O'Sullivan's biographer, George Sundborg, writes that the *Wenatchee Daily World* did not mention Grand Coulee Dam again until nine months after the original story. This is accurate except for the one mention referred to here. See Sundborg, *Hail Columbia,* p. 21.
39. *Wenatchee Daily World* (December 3, 1918), p. 8.
40. George W. Dilling to Columbia Basin Commission, January 22, 1949, letter in Columbia Basin Commission Mss.
41. Perhaps the best telling of the E. F. Blaine story is by Wilbur W. Hindley, "The Story of Grand Coulee Dam—1890-1842," *Spokane Spokesman-Review,* Progress Edition, Part I, January 25, 1942, pp. 3-23.
42. John Fahey, "Spokane's Grand Plan to Pre-empt Coulee Dam," *Spokane Magazine* 5 (July, 1981): 17.
43. George W. Dilling to Columbia Basin Commission, January 22, 1949, Columbia Basin Commission Mss.
44. James A. Ford of the Spokane Chamber of Commerce said that Blaine spoke before 150-200 people. See William Fred Bohrnsen, "A History of the Grand Coulee Dam and the Columbia Basin Reclamation Project" (unpublished masters thesis, University of Washington, 1942), p. 10. Some doubt that such a meeting occurred as no record of it remains. However, others argue that during the epidemic of Spanish influenza immediately after World

War I, the Spokane Chamber of Commerce held only small, informal gatherings, and this explains the lack of evidence backing Blaine's claim. Historian Daniel Ogden, Jr. doubts that this actually happened. Ogden, "The Development of Federal Power Policy in the Pacific Northwest," p. 129. The *Wenatchee Daily World* had noted the rise in influenza cases that fall. See: *Wenatchee Daily World* (September 27, 1918), p. 2.
45. Harding, "The Columbia Basin Irrigation League," pp. 10-12.

## Notes for Chapter Three

1. Rodman Paul, *The Far West and the Great Plains in Transition, 1859-1900* (New York: Harper & Row, Publishers, 1988), p. 299.
2. Daniel Miller Ogden, "The Development of Federal Power Policy in the Pacific Northwest" (unpublished Ph.D. dissertation, University of Chicago, 1949), p. 126.
3. John Fahey, *The Inland Empire: Unfolding Years, 1879-1929* (Seattle: University of Washington Press, 1986), p. 105.
4. John R. Donald, S. J., "The Early Years: Grand Coulee Dam vs. Water from the Pend Oreille 1919-1922" (unpublished masters thesis, Gonzaga University, 1967), p. 15. The Washington Water Power Company was controlled by Electric Bond and Share, a John Pierpont Morgan holding company. See Gene Tollefson, *BPA and the Struggle for Power at Cost* (Portland: Bonneville Power Administration, 1987), p. 82.
5. Historian Michael K. Green suggested that the gravity plan actually originated with the Washington Water Power Company as a direct response to the July 1918 story about a dam at the Grand Coulee. Michael Knight Green, "Politics and Kilowatts: The Washington Water Power Company and Public Power, 1918-1941" (unpublished masters thesis, University of Idaho, 1962), p. 20. Evidence in the files of the Bureau of Reclamation backs Green's suggestion. By October 1918, W. P. Romans of the Spokane Chamber of Commerce had already made contacts with the Reclamation Service and initiated discussions about what they then called the "Big Bend Project." Blaine did not deliver his idea to the Chamber of Commerce until November (see previous chapter). It is clear, then, that members of the Chamber of Commerce, many of whom also worked for the Washington Water Power Company, had been looking over the Big Bend Project before Blaine "caught them by surprise" with the idea. See F. E. Weymouth to W. P. Romans, October 30, 1918, BOR Denver Mss.
6. The gravity plan ended Colonel Hugh Lincoln Cooper's proposed hydroelectric project in Z Canyon farther downstream on the Pend Oreille River, something that would also compete with the Washington Water Power Company. On November 18, 1920, Cooper visited Grand Coulee and later declared his support for it and the Z Canyon projects. The *Spokesman-Review* accused Cooper of selfishly backing the pumping plan to lure the government away from the Pend Oreille River. In return, Cooper threatened to "go down to Washington and fight the state's Pend Oreille diversion project for twenty-four hours and ten minutes a day." "With reference to the attack on my motives in approaching the Columbia Basin [Survey] Commission as printed in the Spokesman's Review [*sic*] I have nothing to say at this time," Cooper wrote to Nat Washington. "The final history . . . when it is written

will be all of the reply to the Spokesman's Review's [*sic*] attitude I will care for." See George Sundborg, *Hail Columbia: The Thirty-Year Struggle for Grand Coulee Dam* (New York: The Macmillan Company, 1954), p. 29; *Wenatchee Daily World* (November 20, 1920), p. 1; William Fred Bohrnsen, "A History of the Grand Coulee Dam and the Columbia Basin Reclamation Project" (unpublished masters thesis, University of Washington, 1942), p. 18; Hugh L. Cooper to Nat Washington, December 20, 1920, Woods Mss.

7. Alfred R. Golzé, *Reclamation in the United States* (New York: McGraw-Hill Book Co., 1952), p. 13.

8. Sundborg, *Hail Columbia*, p. 23. "At least 50,000 families could be accommodated on the lands mentioned in the project." said Lister. See *Spokesman-Review* (December 1, 1918), O'Sullivan Mss.

9. Ralph E. Dyar, *News For an Empire: The Story of the Spokesman-Review of Spokane, Washington and the Field it Serves* (Caldwell, Ida.: Caxton Printers, Ltd., 1952), p. 404.

10. Bruce C. Harding, "Water From the Pend Oreille: The Gravity Plan for Irrigating the Columbia Basin," *Pacific Northwest Quarterly* 45 (April 1954): 53-54.

11. Fahey, *The Inland Empire*, p. 106. Also see *Wenatchee Daily World* (December 23, 1918), p. 1.

12. Fahey, *The Inland Empire*, p. 106.

13. *Wenatchee Daily World* (February 12, 1919), pp. 3, 7.

14. *Ibid.* (February 20, 1919), p. 5.

15. Later in March the legislature passed other legislation which included the right of the state to take private property for reclamation and resettlement purposes. This passed a vote of the people in November 1920. The state also levied a tax to raise money to fund reclamation. See *Wenatchee Daily World* (March 4, 1919), p. 8 for comment on the State Reclamation Act. Historian Bruce Mitchell writes that the Washington Reclamation Act was generally a failure, since much of the money, which ultimately amounted to $3,471,204.25, was lost in projects that became financial disasters such as the Whitestone Project and the White Bluffs-Hanford Projects. See Bruce Mitchell, *Flowing Wealth: The Story of Water Resource Development in North Central Washington, 1870-1950* (Wenatchee: *Wenatchee Daily World*, March 6, 1967, reprinted January 1980), p. 8.

16. Governor Lister died on June 4, 1919 of heart and kidney disease. His death proved a loss for the Columbia Basin Project as he had been a staunch supporter, while Louis Hart did not feel so strongly about the irrigation venture. In appointing the commission, Hart requested and received considerable advice. For example, William Hutchinson Cowles, editor of the Spokane *Spokesman-Review* and head of the prominent Spokane family which owned that newspaper, urged that the governor appoint, among others, Arthur D. Jones and N. W. Durham of Spokane, Elbert F. Blaine, Osmar Lysander Waller of the State College in Pullman, or Professor Henry Landes. W. H. Cowles moved west from Chicago in 1891 and became owner of the *Spokesman-Review* in 1894. He made it a progressive reform-minded newspaper until the 1920s when he steered it increasingly in a conservative direction. See *Wenatchee Daily World* (June 4, 1919), p. 1; Carlos A. Schwantes, *The Pacific Northwest: An Interpretive History* (Lincoln: University of Nebraska Press, 1989), p. 229; W. H. Cowles to Louis F. Hart, February 24, 1919, Columbia Basin Commissions Mss.

17. Louis F. Hart to Peter McGregor, March 7, 1919, letter found in the papers of Louis F. Hart, Papers of the Governors. Also see Hart to Marvin Chase, March 4, 1919.
18. On March 20, Franklin E. Weymouth, Chief Engineer of the Reclamation Service, met and spoke to the members of the Columbia Basin Survey Commission in Spokane. At a public session someone asked Weymouth how long he thought it would take to complete the project. "Well, gentlemen, this project is a little bigger than the Panama Canal, and that took one hundred years." In making that prediction, Weymouth came closer to the mark than anyone then could have guessed. See *Wenatchee Daily World* (March 14, 1919), p. 2; (March 22, 1919), p. 2; Sundborg, *Hail Columbia*, p. 26.
19. For a complete biography of James O'Sullivan, see Sundborg, *Hail Columbia* and James O'Sullivan, "The Struggle For Grand Coulee Dam," typed manuscript in the O'Sullivan Mss., p. 14.
20. Frederick Haynes Newell and Daniel William Murphy, *Principles of Irrigation Engineering: Arid Lands, Water Supply, Storage Works, Dams, Canals, Water Rights, and Products* (New York: McGraw-Hill Book Company, Inc., 1913), pp. 127-134.
21. An editorial in the *Wenatchee Daily World* gave O'Sullivan credit for linking power sales and irrigation. See *Wenatchee Daily World* (February 28, 1930), p. 4 editorial.
22. Ronald Albert Weinkauf, "The Columbia Basin Project, Washington: Concept and Reality, Lessons for Public Policy" (unpublished Ph.D. dissertation, Oregon State University, 1973), p. 7.
23. *Wenatchee Daily World* (May 14, 1919), p. 7.
24. For reference, see *Ibid.* (July 8, 1983), p. A12.
25. Sundborg, *Hail Columbia*, p. 42.
26. O. L. Waller to F. E. Weymouth, April 26, 1919, Columbia Basin Commissions Mss.; *Wenatchee Daily World* (May 13, 1919), p. 3.
27. Green, "Politics and Kilowatts," p. 24.
28. In the background, the 1920 presidential election approached. Democrat Nathaniel Willis Washington of Ephrata, a direct descendant of John Augustine Washington, the brother of George Washington, and Frank Bell, also of Ephrata and aid to Representative Clarence Cleveland Dill, went as delegates to the Democrat's convention in San Francisco. There Washington advocated irrigation in general and the Columbia Basin Project in particular. Most sources argue that he met and convinced Franklin Delano Roosevelt, soon selected as his party's vice presidential candidate, to support the pro-reclamation resolution that the convention ultimately adopted. During the subsequent fall campaign, Roosevelt toured parts of Washington state and commented on its irrigation potential. The meeting with Nat Washington, if it happened, and certainly the Roosevelt trip through Washington, had ramifications that became apparent over a decade later. When Senator Warren Gamaliel Harding, the Republican convention's eventual choice, announced his support for reclamation projects, Rufus Woods felt relieved. Both parties took the stand he favored. C. C. Dill served as a representative from Washington in the Sixty-Fourth and Sixty Fifth Congresses (1915-1919) but he lost a re-election bid in 1918 due to his stand against American participation in World War I. Dill was elected as a senator in 1922 and served until 1935, and would become important in the Grand Coulee story. See Russell Blankenship, *And There Were Men* (New York: Alfred A.

Knopf, 1942), p. 171; Murray Morgan, *The Dam* (New York: The Viking Press, 1954), p. 22; Click Relander, "The Battleground of National Irrigation," *Pacific Northwest Quarterly* 52 (October 1961): 149; Frank Freidel, "Franklin Roosevelt in the Northwest: Informal Glimpses," *Pacific Northwest Quarterly* 76 (October 1985): 122-131; *Spokane Daily Chronicle* (June 25, 1920), O'Sullivan Mss.

29. Minutes of the Columbia Basin Survey Commission, July 1, 1920, page 137, Columbia Basin Commissions Mss.

30. *Wenatchee Daily World* (July 6, 1920), p. 7.

31. Actually, as proponents of the dam pointed out, the treaty only said that British subjects would receive the same treatment as American citizens respecting navigation of the river. The commission's report did not include this part of the treaty. For a discussion of this in detail see Green, "Politics and Kilowatts," pp. 26-27. For the actual treaty in question see William M. Malloy, *Treaties, Conventions, International Acts, Protocols, and Agreements Between the United States of America and Other Powers* (Washington, D. C.: U. S. Government Printing Office, 1910-1938), Vol. 1, p. 657. Professor Waller later dismissed the problem of the treaty as insignificant. See Waller to J. N. Clover of the *Seattle Post-Intelligencer*, January 27, 1922, Waller Mss.

32. Columbia Basin Survey Commission, *The Columbia Basin Project* (Olympia, Wash.: Frank M. Lamborn, Public Printer, 1920), passim.

33. D. C. Henny to F. Weymouth, June 22, 1920, BOR Denver Mss.

34. William D. Miner, "A History of the Columbia Basin Projects" (unpublished Ph.D. dissertation, University of Indiana, 1950), p. 119.

35. F. E. Weymouth to Chief Engineer A. P. Davis, July 13, 1920, BOR Denver Mss.

36. Among the legends that surround Grand Coulee Dam is the story that Nat Washington's father told him in 1910 (when he was six years old) that some day a dam would rise at the Grand Coulee and irrigate the land. See *Wenatchee Daily World* (July 8, 1983), p. A-2; (December 2, 1920), p. 8.

37. Bruce Charles Harding, "The Columbia Basin Irrigation League: Its Origins and Operations" (unpublished masters thesis, State College of Washington, 1951), p. 34.

38. The gathering was also called the Pacific Northwest Irrigation Congress and the Reclamation Congress. See *Wenatchee Daily World* (July 7, 1920), p. 1; (September 16, 1920), p. 1.

39. *Ibid.* (October 2, 1920), p. 1.

40. Quincy Valley Irrigation District, H. L. Johnson, Secretary to A. P. Davis, October 9, 1920; and A. P. Davis to H. L. Johnson, October 19, 1920, Columbia Basin Commissions Mss. Copies of these letters also appear in Jones Mss.

41. Harding, "The Columbia Basin Irrigation League," p. 34. Also see Fred A. Adams to Arthur Powell Davis, October 20, 1920, BOR Denver Mss.

42. A. P. Davis to Columbia Basin Survey Commission, October 30, 1920, BOR Denver Mss. Copies also in the Waller Mss.

43. Bruce Mitchell, *Flowing Wealth*, p. 26.

44. O'Sullivan to Davis, December 22, 1920, O'Sullivan Mss. In an earlier letter to Rufus Woods, O'Sullivan admitted, "we had gone too far or too unreservedly with reference to what A. P. Davis said." See O'Sullivan to Woods, November 29, 1920, Woods Mss.

45. D. C. Henny to Munn, handwritten note dated December 15, 1920, BOR Denver Mss.

46. Board of Engineers to Chief Engineer, Denver, December 13, 1920, p. 14, Waller Mss.
47. Sundborg, *Hail Columbia*, p. 52.
48. *Wenatchee Daily World* (March 14, 1921), pp. 10-11; and Harding, "The Columbia Basin Irrigation League," p. 125.
49. "I am reliably informed that the Columbia Basin Committee of the Spokane Chamber of Commerce will be consulted on the appointment of the Supervisor for the Columbia Basin Survey," Arthur J. Turner said in a confidential handwritten note to Osmar Waller six weeks before the legislation passed. Turner argued that he should be that supervisor. "However, certain members of the Columbia Basin Committee of the Chamber of Commerce may feel that the Supervisor should be a Publicity man, not an engineer." The Turner note indicates that throughout, the Spokane people orchestrated the progress of the bill in the legislature, as well as the composition of the people subsequently put in charge of the Columbia Basin Project. See A. J. Turner to O. L. Waller, January 31, 1921, Waller Mss.
50. *Wenatchee Daily World* (July 7, 1921), p. 1.
51. Sundborg, *Hail Columbia*, p. 57. Also see Green, "Politics and Kilowatts," p. 32.
52. Ogden, "The Development of Federal Power Policy in the Pacific Northwest," pp. 142-143.
53. *Spokesman-Review* (June 27, 1921), BOR Denver Mss. Also see *Spokesman-Review* (July 8, 1921), p. 4.
54. Reports actually came from both Fred Adams and Ivan Goodner, the state geologist. Green, "Politics and Kilowatts," p. 34.
55. *Spokane Daily Chronicle* (July 20, 1921), O'Sullivan Mss.
56. D. C. Henny to A. P. Davis, August 5, 1921, BOR Denver Mss.
57. Green, "Politics and Kilowatts," p. 36.
58. Ivan E. Goodner to Willis T. Batcheller, July 29, 1921, Columbia Basin Commissions Mss. The Chief Joseph Dam has since been built at the Foster Creek site.
59. For more on Batcheller, see Richard C. Berner, *Seattle 1921-1940: From Boom to Bust* (Seattle: Charles Press, 1992), pp. 307, 328, 420-421, 426-427.
60. Willis T. Batcheller, "Report of Willis T. Batcheller," February 10, 1922, p. 2, Columbia Basin Commissions Mss. Also see *Spokesman-Review* (August 12, 1921), O'Sullivan Mss.
61. Born June 29, 1858, in Brooklyn, New York, Goethals entered the army in 1882, and, as a first lieutenant, served at Vancouver (Washington) Barracks with the army's Department of the Columbia. For two years Goethals conducted surveys and explorations in Washington and Idaho. He examined the Mountain or Petite Trail from Kettle Falls to Lake Osoyoos and he guided General William Tecumseh Sherman when the general visited Fort Colville. In speculating on where the suggestion to hire Goethals had originated, D.C. Henny cited a clipping from the November 1, 1921 issue of the *Journal of Electricity* that hinted at the Goethals contact. Historian Daniel M. Ogden argues that the suggestion to hire Goethals came out of a meeting held by the Spokane Chamber of Commerce in the Davenport Hotel in Spokane. Senators Wesley Jones and Miles Poindexter attended that meeting. James A. Ford is credited with the idea to hire Goethals, whom he saw as having a name significant enough to counter Colonel Hugh L. Cooper. See D. C. Henny to Weymouth, November 7, 1921, BOR Denver Mss.; Ogden, "The Development of Federal Power Policy in the Pacific Northwest," p. 147;

Mitchell, *Flowing Wealth*, p. 26; Walter R. Griffin, "George W. Goethals, Explorer of the Pacific Northwest, 1882-1884," *Pacific Northwest Quarterly* 62 (October 1971): 129; Craig E. Holstine, *Forgotten Corner: A History of the Colville National Forest, Washington* (Colville, Wash.: *Colville Statesman-Examiner*, 1987), p. 13.

62. Goethals's biographer points out that the George W. Goethals & Co., Inc. never succeeded in getting all the work it wanted and yet in ten years as a consulting engineer the general completed an "appalling" amount of work. "His typewritten reports on various projects cram four drawers of a large steel cabinet." Just before he died in 1928 the General said, "my intentions are good, but my deeds are nil." See Joseph Bucklin Bishop and Farnham Bishop, *Goethals: Genius of the Panama Canal* (New York: Harper & Brothers, 1930), pp. 419, 424-425.

63. The Spokane people remained bitter for years that the Seattle Chamber of Commerce refused to help pay for the Goethals study and in fact tried to obstruct the work. "That attitude is, of course, untenable and was prompted by ignorance, selfishness or petty malice." See *Spokesman-Review* (June 25, 1922), BOR Denver Mss.

64. Miner, "A History of the Columbia Basin Projects," p. 142.

65. *Spokesman-Review* (January 5, 1922) (February 1, 1922), O'Sullivan Mss.; Harding, "The Columbia Basin Irrigation League," pp. 54-55; *Wenatchee Daily World* (January 14, 1922), p. 1. Even in Portland and in New York, some questioned the wisdom of hiring Goethals for a study that seemed beyond his ability. See Henny to Willis T. Batcheller, January 22, 1922, Batcheller Mss.

66. Harding, "The Columbia Basin Irrigation League," p. 57; Sundborg, *Hail Columbia*, p. 65.

67. Green, "Politics and Kilowatts," p. 38.

68. *Wenatchee Daily World* (April 7, 1922), p. 1. Also see George W. Goethals, "The Columbia Basin Project," March 30, 1922, Columbia Basin Commissions Mss.

69. *Spokesman-Review* (April 9, 1922), p. 1. Also see *Portland Oregonian* (April 9, 1922), p. 1; and *Wenatchee Daily World* (April 10, 1922), pp. 1, 8. Goethals also estimated that project maintenance would run about forty-eight cents an acre.

70. Goethals, "The Columbia Basin Project," p. 23. Among the legends surrounding Grand Coulee Dam is the story that in private General Goethals said, "It's a perfect dam site. If I were twenty years younger I'd like nothing better than to tackle the job of building a dam here." As the story goes, Goethals actually knew what he was doing, but he had agreed privately to find in favor of the gravity plan, was upset about what happened, yet nevertheless continued to endorse the gravity idea. See Wilbur W. Hindley, "The Story of Grand Coulee Dam—1890-1942," *Spokesman-Review*, Progress Edition, (January 25, 1942), p. 7; and Harding, "The Columbia Basin Irrigation League," pp. 58-59 (footnotes). This story does not entirely hold up. On April 6, 1922, Goethals wrote to Marvin Chase, "The more I have studied the project the more I believe in it and the more I am amazed at the amount of data that you were able to collect." Goethals to Chase, April 6, 1922, Columbia Basin Commissions Mss.

71. Green, "Politics and Kilowatts," p. 40; and Ogden, "The Development of Federal Power Policy in the Pacific Northwest," p. 150. Also see Harding, "The Columbia Basin Irrigation League," pp. 58-59.

72. Ogden, "The Development of Federal Power Policy in the Pacific Northwest," p. 150.
73. Sundborg, *Hail Columbia*, p. 67. Historian Robert E. Ficken points out that the *Wenatchee Daily World* had been aided by a loan from Spokane's W. H. Cowles, who held stock in the paper until at least 1918. This made it difficult for Woods to argue with Spokane and accounts for some of the favorable early mentions of the gravity scheme. Also, Ficken continues, Woods knew that open conflict between Spokane and Wenatchee would make acquiring federal funding more difficult. Also, despite the activity between the Goethals report and 1929, many felt that the pumping plan was a dead issue. Ficken to Pitzer, April 5, 1993.
74. Henny to Weymouth, April 10, 1922, BOR Denver Mss.
75. Concerning Batcheller's notes and other work, Scott specifically directed Ivan Goodner to "see that such remain in the office of the Columbia Basin Survey and that no copies are removed from your office without my permission." Goodner had reason to comply since he also nudged Batcheller toward the Grand Coulee site, suggesting that Batcheller "fudge" his figures to prove the pumping plan unworkable. Batcheller, instead, produced numbers that strongly supported the pumping plan. David Henny wrote to A. P. Davis, "I will say that Mr. Batcheller told me that about two months after he started with his work Mr. Goodner had requested him, using his own terms, to fudge his report so as to make a pumping project seem impracticable, which of course, Mr. Batcheller had refused to do." Scott to Goodner, January 16, 1922, Columbia Basin Commissions Mss.; . Miner, "A History of the Columbia Basin Projects," p. 130; Henny to A. P. Davis, February 15, 1922, BOR Denver Mss.
76. *Seattle Daily Times* (February 12, 1922), p. 10. Batcheller Mss.
77. Mitchell, *Flowing Wealth*, p. 26. Also see Willis T. Batcheller, "Report on Columbia River Pumping and Power Project," February 10, 1922, typed manuscript in the Columbia Basin Commissions Mss., pp. 99-100.
78. In January 1921 the Federal Power Commission, which came into being June 10, 1920, ordered a survey of the upper Columbia River. Colonel James B. Cavanaugh of the Army Corps of Engineers headed the project working with, among others, Washington state Supervisor of Hydraulics Marvin Chase. Once formed, the Cavanaugh survey proceeded without fanfare and remained very much in the background. See Project History—Grand Coulee Dam, 1933, p. 31.
79. Paul Curtis Pitzer, "Visions, Plans, and Realities: A History of the Columbia Basin Project" (unpublished Ph.D. dissertation, University of Oregon, 1990), pp. 27-28.
80. Department of the Interior and Washington State University, *The Economic Significance of the Columbia Project Development* (Washington, D. C.: U. S. Government Printing Office, 1965), pp. 21-22. The number of farms in 1909 had been 2,350. Also see Theodore Saloutos, *The American Farmer and the New Deal* (Ames: Iowa State University Press, 1982), p. 8. In 1922 the *Wenatchee Daily World* noted that in 1912 seventeen farm families lived along the road from Othello to Cunningham. In 1922 only five lived there. See *Wenatchee Daily World* (February 7, 1922), p. 1.

## Notes for Chapter Four

1. William D. Miner, "A History of the Columbia Basin Projects" (unpublished Ph.D. dissertation, University of Indiana, 1950), p. 167.
2. Clarence C. Dill, *Where Water Falls* (Spokane, Wash.: C. W. Hill/Printers, 1970), p. 141.
3. The Washington Water Power Company used its employees to solicit funds for the Columbia Basin Irrigation League. They justified this by saying that every good citizen desired to see a greater population and prosperity. This would increase the power load carried by the company and mean cheaper electricity for everyone. See Daniel Miller Ogden, "The Development of Federal Power Policy in the Pacific Northwest" (unpublished Ph.D. dissertation, University of Chicago, 1949), p. 153. Also see *Wenatchee Daily World* (April 21, 1922), p. 1; (April 22, 1922), p. 1.
4. Late in November 1922 a group of pump plan supporters, headed by C. R. Berry of the Fruitland Publication Company, formed an organization they called the Columbia Basin Reclamation Association. It was one of three or four such organizations that people, mostly on the west side of the project area, formed during the early 1920s. They lasted only a short time and existed more in name than in reality. See Miner, "A History of the Columbia Basin Projects," p. 138.
5. *Wenatchee Daily World* (November 8, 1922), p. 1; (November 9, 1922), p. 1. At thirty-eight, this made Dill the youngest member of the Senate.
6. *Spokesman-Review* (October 30, 1922), O'Sullivan Mss.; and Bruce C. Harding, "Water from the Pend Oreille: The Gravity Plan for Irrigating the Columbia Basin," *Pacific Northwest Quarterly* 45 (April 1954): 58.
7. The congressional hearings into the Columbia Basin Project made little use of the Federal Power Commission report done by Colonel James B. Cavanaugh. Cavanaugh worked on his study through late 1922. He held hearings in Spokane that deteriorated into a debate over the merits of the pumping and gravity plans. On June 30, 1922, Cavanaugh completed his report and sent it to the Federal Power Commission. That body submitted it to the Secretary of War on February 17, 1923. In his report, Cavanaugh urged that no power permits be issued on the Columbia River from the Canadian border down to Foster Creek (well below the Grand Coulee) until completion of further investigations, and certainly not before officials determined if they might need the Grand Coulee site to irrigate the Columbia Basin. Washington state officials, notably Marvin Chase, the Supervisor of Hydraulics, objected, and despite Cavanaugh, on June 26, 1922 the Washington Water Power Company received its preliminary permit for the Kettle Falls site. The bill hit some snags in the House when Republican Representative William Henry Stafford of Wisconsin, among others, argued that it needed considerable study. Stafford was typical of many in the Midwest and East who opposed Western reclamation. This bill had the effect of officially making the Columbia Basin Project a Federal Reclamation Project under the Reclamation Service. *Spokane Daily Chronicle* (December 13, 1922), O'Sullivan Mss.; *Wenatchee Daily World* (February 7, 1923), p. 14. Also see Miner, "A History of the Columbia Basin Projects," pp. 149-151;

*Spokesman-Review* (February 21, 1923), O'Sullivan Mss.; Project History—Grand Coulee Dam, 1933, p. 32; Bruce Charles Harding, "The Columbia Basin Irrigation League" (unpublished masters thesis, State College of Washington, 1951), pp. 125-129.

8. Pumping plan backers in Ephrata protested Goodwin's place as they felt, correctly, that he represented Spokane interests. Nat Washington wrote to Thomas B. Hill of the Seattle Chamber of Commerce, "We were advised today that the Spokane people are making a strong effort to secure the appointment of General George W. Goethals, Mr. Marvin Chase, and Mr. Goodwin . . . on the commission for which the $100,000 appropriation was recently made." Washington asked Seattle people to use their influence to stop such appointments. See Washington to Hill, March 7, 1923, Washington papers contained in O'Sullivan Mss.

9. Harding, "The Columbia Basin Irrigation League," pp. 130-141. Harding's telling of the details of Senate Bill 3745, reintroduced as S 3808, is particularly detailed. Also see Project History—Grand Coulee Dam, 1933, p. 35. Davis gave considerable thought to selection of Homer Gault. Weymouth had suggested consulting engineer Joseph Jacobs of Seattle and commented, "That project is so large and the problems are so huge and complicated that we should have a very strong man in charge of it." Weymouth to Davis, March 7, 1923, BOR Denver Mss.

10. During his tenure as president of the CBIL, Lindley made numerous trips to Washington, D.C., spent three winters there at his own expense, and hours working with congressmen promoting the project.

11. Miner, "A History of the Columbia Basin Projects," p. 157; *Wenatchee Daily World* (May 2, 1923), p. 10.

12. Rumors circulated that the Spokane people, irked at his activity exposing the truth about the pumping plan and forcing the drilling operation, had exerted their influence, resulting in the Interior Department firing A. P. Davis. These were probably exaggerations as Davis's fate rested more with the whim of the new, more conservative, and less reclamation-prone Coolidge administration.

13. Michael C. Robinson, *Water For the West: The Bureau of Reclamation, 1902-1977* (Chicago: Public Works Historical Society, 1979), p. 49.

14. For background on Hoover Dam see Joseph E. Stevens, *Hoover Dam: An American Adventure* (Norman: University of Oklahoma Press, 1988), especially pages 18-19.

15. For example, see Weymouth to Gault, December 14, 1923, BOR Denver Mss.

16. In this Gault was wrong. See Chapter Eighteen.

17. First estimates put the irrigable acreage at around 2,500,000. The state study and Goethals set it at 1,753,000 acres. Gault reduced this to 1,133,000 acres. Draft of Gault Report and Advisors Report, April 6, 1924, typed draft found in BOR Denver Mss. The advisors did argue that drainage would be a problem on parts of the project, and they did not think that the proposed storage reservoir in the Grand Coulee would be water-tight.

18. Goodwin to Weymouth, April 14, 1924, BOR Denver Mss. In the letter Goodwin expresses dismay and surprise at the Gault figures and the concurrence by the review board.

19. Bruce Mitchell, *Flowing Wealth: The Story of Water Resource Development in North Central Washington, 1870-1950* (Wenatchee: *Wenatchee Daily World*,

1967 and reprinted, 1980), p. 28. Also see Senate Committee Print, 69th Congress, 2nd Session, *Columbia Basin Project: Report of Special Commission, August 25, 1925, Board of Engineers' Report, February, 1925, Board of Engineers' Report, April 6, 1924, Gault Report, March, 1924* (Washington, D.C.: Government Printing Office, 1927), passim.

20. Elwood Mead became Commissioner of Reclamation on April 3, 1924. William E. Warne, *The Bureau of Reclamation* (New York: Praeger Publishers, 1973), p. 239; Raymond Moley, *What Price Federal Reclamation?* (New York: American Enterprise Association, Inc., 1955), p. 10. Compilation of Reports on the Columbia Basin Project, Board of Review, February 25, 1925, typed reports found in BOR Denver Mss.; Project History—Grand Coulee Dam, 1933, pp. 41-44; Miner, "A History of the Columbia Basin Projects," pp. 168-175; Department of the Interior Memorandum to the Press, August 26, 1925, BOR Denver Mss.; "Columbia Project to Cost Two Hundred Millions," *Engineering News Record* 94 (March 5, 1925): 400.

21. Miner, "A History of the Columbia Basin Projects," p. 174-177; Murray Morgan, *The Dam* (New York: The Viking Press, 1954), p. 23; *Wenatchee Daily World* (October 23, 1925), p. 1; (November 10, 1925), p. 1; (November 19, 1925), p. 1.

22. Michael Knight Green, "Politics and Kilowatts: The Washington Water power Company and Public Power, 1918-1941" (unpublished masters thesis, University of Idaho, 1962), p. 30. There are unsubstantiated stories about Billy Clapp and other pumpers traveling quietly around Idaho talking to residents about the damage that "theft" of their water would cause should the gravity plan be adopted.

23. Mitchell, *Flowing Wealth*, p. 29; *San Poil Eagle* (February 3, 1927), O'Sullivan Mss.; *Wenatchee Daily World* (January 24, 1927), p. 1

24. O. L. Waller wrote to Fred Adams, "I cannot see why storing water by gravity to the Columbia Basin for irrigation should not be an asset to the State of Idaho in the matter of power development rather than a loss, and I do not see why Idaho should be interested in a pumping plant at the head of the Grand Coulee." Waller to Fred Adams, February 20, 1926, Waller Mss. On May 18, 1926, Glenn L. Parker put out a report for the State of Idaho entitled "Storage Regulation in Flathead Basin for Power and Its Effect on the Columbia Basin Project," in which he reported favorably on the prospect of using water from Idaho for the Columbia Basin Project. Compilation of Reports, 1941, BOR Denver Mss. And Eugene Logan and Glenn L. Parker completed a report in which they wrote, "The results of this investigation of power possibilities on Priest River disclose the fact that a greater amount of storage is required for power than would be required to fulfill the total requirements of the Columbia Basin Project regardless of the amount of storage used in Pend Oreille Lake for that purpose." Eugene Logan, "Preliminary Report Columbia Basin Project Water Power Analysis Power Possibilities of Priest River, Idaho," typed report, p. 4, Columbia Basin Commissions Mss. William Clapp complained to State Supervisor of Hydraulics Ross K. Tiffany that Logan was a former employee of the Washington Water Power Company, and his reports on power possibilities might not be objective. Clapp to Tiffany, December 13, 1926, Columbia Basin Commissions Mss. Also see Ogden, "The Development of Federal Power Policy in the Pacific Northwest," p. 158.

25. Project History—Grand Coulee Dam, 1933, p. 46; Ogden, "The Development of Federal Power Policy in the Pacific Northwest," p. 159; *Wenatchee Daily World* (November 12, 1929), p. 5.

26. Emmett Kaiser VandeVere, "History of Irrigation in Washington" (unpublished Ph.D. dissertation, University of Washington, 1948), p. 215. VandeVere gives Jones considerable credit for handling the details of these bills in Washington, D.C., while at the same time quieting the arguing factions back in Washington state. Subsequent reports on the Columbia River over the years have all come to be known as "308 Reports" after this first 1926 document.

27. Gus Norwood, *Columbia River Power For the People: A History of Policies of the Bonneville Power Administration* (Washington, D.C.: U.S. Government Printing Office, 1981), p. 44.

28. *Wenatchee Daily World* (July 12, 1926), p. 1.

29. Rufus Woods, aware of the interest the government increasingly showed in power resources, selected this time to alter the masthead of his newspaper. For years the *Wenatchee Daily World* had billed itself as, "The Greatest Daily in the World for Cities Under 10,000 Population." Then, on August 26, 1925, the *World*'s new masthead declared, "Published in the Apple Capital of the World and in the Buckle of the Power Belt of the Great Northwest." Woods had written, "We are located in the very heart of the water power belt—the greatest in the world," in an editorial that appeared in the *World* on November 22, 1920. The comment on power referred here more to the expectations at the Priest Rapids site than anything yet imagined from the Columbia Basin Project. *Wenatchee Daily World* (August 26, 1925), p. 1. In 1926, Woods ran a front-page story indicating that power projects would eventually bring giant-scale industry to Washington and the Northwest. The agricultural/ industrial empire became a theme that he used to promote the Columbia Basin Project, Grand Coulee Dam, and the region in general. *Wenatchee Daily World* (April 10, 1926), p. 1.

30. George Sundborg, *Hail Columbia: The Thirty Year Struggle for Grand Coulee Dam* (New York: The Macmillan Company, 1954), p. 79. CBIL expenses for the tours amounted to $5,407.63, including a ten-dollar daily spending allowance for each legislator.

31. *Wenatchee Daily World* (August 30, 1927), p. 1; *Spokane Spokesman-Review* (September 2, 1927), O'Sullivan Mss.; Miner, "A History of the Columbia Basin Projects," p. 212.

32. *Spokane Daily Chronicle* (November 18, 1927), O'Sullivan Mss.

33. *Spokesman-Review* (December 6, 1927), O'Sullivan Mss.; Ogden, "The Development of Federal Power Policy in the Pacific Northwest," p. 162.

34. The files of Wesley Jones contain numerous letters from businessmen, all expressing support for the files. For example, see Hall to Jones, March 6, 1928, and other letters in that file, Jones Mss.

35. J. W.. Summers to Jones, July 5, 1928, *ibid.*

36. *Wenatchee Daily World* (December 19, 1928), p. 1.

37. *Ibid.* (January 4, 1929), p. 1; Harding, "The Columbia Basin Irrigation League," p. 163; *Spokesman-Review* (February 22, 1929), O'Sullivan Mss.

38. Martin F. Seedorf and Rita G. Seedorf, "James O'Sullivan, the 'Pumpers' & the Fight for Grand Coulee Dam, 1918-1933," paper delivered at the Pacific Northwest History Conference, April 13-15, 1989, Moscow, Idaho. The Seedorf paper points out correctly that this was just one of many organizations that came and went over the years. This one, however, set the stage

for a more serious development to come. Also see Sundborg, *Hail Columbia*, p. 88.

39. *Wenatchee Daily World* (March 5, 1929), p. 1.
40. Sundborg, *Hail Columbia*, p. 96.
41. Bruce Mitchell, *The Story of Rufus Woods and the Development of Central Washington* (Wenatchee: *Wenatchee Daily World*, 1965), p. 20.
42. *Wenatchee Daily World* (June 4, 1929), p. 1.
43. From an article reprinted in the *Grant County Journal* (February 20, 1929), O'Sullivan Mss.
44. Southard to Woods, June 12, 1929, Woods Mss.
45. Green, "Politics and Kilowatts," p. 50; Harding, "The Columbia Basin Irrigation League," p. 95.
46. Major Butler and the Army Corps of Engineers actually officially completed their preliminary study on the Columbia River on February 11, 1929 and began their formal detailed investigation on June 4, 1929. Sundborg, *Hail Columbia*, p. 91; Ogden, "The Development of Federal Power Policy in the Pacific Northwest," p. 161; Mitchell, *Flowing Wealth*, p. 29; Miner, "A History of the Columbia Basin Projects," p. 239; *Wenatchee Daily World* (January 23, 1929), p. 1; (April 24, 1929), p. 1; (July 20, 1929), p. 1.
47. *Wenatchee Daily World* (November 9, 1929), p. 1; (November 12, 1929), p. 5.
48. The national Grange, however, remained as opposed to the project as ever. Albert S. Goss, editorial, *The Grange News* (August 5, 1929), O'Sullivan Mss.; *Wenatchee Daily World* (August 10, 1929), p. 1; (November 23, 1929), p. 1.
49. Green, "Politics and Kilowatts," p. 53. Also see Norwood, *Columbia River Power for the People*, p. 33, and John Kenneth Galbraith, *The Great Crash* (Boston: Houghton Mifflin Co., 1961), pp. 138 and 196. For a discussion of the contest between public and private power see: Myron Kendall Jordan, "The Kilowatt Wars: James D. Ross, Public Power and the Public Relations Contest for the Hearts and Minds of Pacific Northwesterners" (unpublished Ph.D. dissertation, University of Washington, 1991).
50. Green, "Politics and Kilowatts," p. 53; Ogden, "The Development of Federal Power Policy in the Pacific Northwest," p. 167; *Wenatchee Daily World* (June 6, 1930), p. 1.
51. *Wenatchee Daily World* (October 11, 1929), p. 1.
52. *Ibid.* (July 1, 1929), p. 1.
53. Columbia Basin Executive Committee Meeting Minutes, July 11, 1929, p. 2, Waller Mss.
54. *Ibid.*
55. Sometime during 1930 the CBIL changed its formal stationery. For years every letter sent out by any official in the organization carried on its reverse side a map showing the proposed Columbia Basin irrigation project. The canals and waterways all originated with a diversion out of Lake Pend Oreille in Idaho. They ended in the southern half of the Big Bend country where the figure "1,883,000 ACRES" indicated the ultimate size of the project. Nowhere did the map show any proposal for a dam at the Grand Coulee. Then, suddenly in the summer of 1930, the map on the stationery included a small rectangle across the Columbia north of Wilbur and above it, in a faint print, the words "PROPOSED DAM GRAND COULEE." Samples of the stationary are available in the BOR Denver Mss., in the Columbia Basin Commissions Mss., and in the Gill Mss. Note especially Woodruff to R.F.

Walter, August 7, 1930, BOR Denver Mss. for a first example of the stationary with the dam included.

56. This competition between the two agencies will not be detailed here. It is worthy of a study in itself. See Grant McConnell, *Private Power and American Democracy* (New York: Alfred A. Knopf, 1966), pp. 222-223, 226. Also see Charles McKinley, *Uncle Sam in the Pacific Northwest: Federal Management of Natural Resources in the Columbia River Valley* (Berkeley: University of California Press), esp. pp. 65-156.

57. *Wenatchee Daily World* (August 2, 1930), p. 1; (August 4, 1930), p. 1.

58. *Spokesman-Review* (August 21, 1930), O'Sullivan Mss.; *Wenatchee Daily World* (August 21, 1930), p. 1.

59. "O'Sullivan," Woods wrote in June 1930, "I have found the nigger in the woodpile. The President of our Chamber of Commerce [Wenatchee] is a heavy holder of [Washington] Water power [Co.] securities. He bought $10,000 worth at one time." For Rufus Woods that conflict of interest and others explain his own city's reluctance to help in the effort toward the pump plan, as much as he would have liked. In another letter Rufus Woods wrote, "When the people began to talk about this big power project seriously there was organized the Columbia Basin League in Spokane with its 134-mile gravity canal. We all accepted that proposal in good faith for years but we have begun to seriously doubt the good faith of that enterprise so far as certain interests are concerned. Recently we asked the Washington Water Power Company to keep its fingers out of the Grand Coulee dam. This brought Frank G. Post, president, to my office. I told him frankly what we suspected as evidenced by the fact that Engineer [Arthur J.] Turner of his company was on a board which turned in an adverse report on the dam. The consulting engineer of the Columbia Basin League it developed was a heavy stockholder in the Washington Water Power Company. . . . The tactics have been to make fun of James O'Sullivan and of Batcheller and to kill off any engineer who favored the dam." Woods to O'Sullivan, June 30, 1930, O'Sullivan Mss.; Woods to Wellington Pegg, December 4, 1930, Woods Mss.

60. *Wenatchee Daily World* (October 30, 1930), p. 1; (November 26, 1930), p. 5; (December 8, 1930), p. 1; (December 10, 1930), p. 4.

61. Paul C. Pitzer, "The Atmosphere Tasted like Turnips: The Pacific Northwest Dust Storm of 1931," *Pacific Northwest Quarterly* 79 (April, 1988): 50-55.

62. Sundborg, *Hail Columbia*, p. 146. Ed Southard wrote to Elwood Mead, "We have just gone thru the worst wind storm in history, and from 30 to 50% of the sown grain is said to be either blown out or badly damaged, and that damage extends thru the entire Big Bend Country. Soil erosion has reached the point where we can see the entire ruin of the country unless we get water. It also seems to me that the farmers on the border lands (I mean semi-arid lands) are doomed, and some way must be found to put them on irrigated lands. I think modern methods adopted by Russia will ruin our day land wheat farmers." Southard to Mead, April 27, 1931, BOR Denver Mss.

63. For comment on the nature of dust storms and the debate over their causes see Donald Worster, *Under Western Skies: Nature and History in the American West* (New York: Oxford University Press, 1992), pp. 93-105.

64. The *Spokesman-Review* ran pictures of the event and reported that about 3,000 attended, but the accompanying article omitted mention of a dam at the Grand Coulee. The *Wenatchee Daily World* placed the number at 6,000,

the *Grant County Journal* estimated closer to 8,000, and they both said a great deal about the dam. *Wenatchee Daily World* (June 29, 1931), p. 1; *Grant County Journal* (July 3, 1931), O'Sullivan Mss.; *Spokesman-Review* (July 5, 1931), O'Sullivan Mss.

65. In its story about Mead and his request to the Federal Power Commission, the *Wenatchee Daily World* interpreted the action as evidence that Mead favored the pump plan. *Wenatchee Daily World* (May 18, 1931), p. 1.

66. Sundborg, *Hail Columbia*, p. 167; Miner, "A History of the Columbia Basin Projects," p. 275; *Wenatchee Daily World* (July 21, 1931), p. 1.

67. *Wenatchee Daily World* (July 23, 1931), p. 1.

68. See *Columbia River and Minor Tributaries: Letter from the Secretary of War Transmitting Pursuant to Section 1 of the River and Harbor Act Approved January 21, 1927, . . . Containing a General Plan for the Improvement of the Columbia River and Minor Tributaries for the Purposes of Navigation and Efficient Development of Water-Power, the Control of Floods, and the Needs of Irrigation*, 2 Vols. (Washington, D.C.: U.S. Government Printing Office, 1933), 73rd Congress, 1st Session, House Document No. 103.

69. *Wenatchee Daily World* (September 21, 1931), p. 1.

70. This was not exactly correct. The Butler report said that both the gravity and pump plans were technically feasible but that the gravity plan was not economically feasible. The pump plan was economically feasible only if the irrigation aspect was paid for with money from the sale of power.

71. The *Seattle Daily Times* also wrote the story as if the Bureau of Reclamation had produced its report first and then the Army study had merely agreed with it. The *Times* wrote, "These decisions have been reached on the basis of the report of the Army engineers, which supplements, and in every essential respect sustains, the preceding studies of the Bureau's own representatives." *Seattle Daily Times* (October 23, 1931), BOR Denver Mss. On that clipping, someone unidentified wrote, "The underlined words [which were 'which supplements,' and 'sustains'] made Maj. Butler real peevish. He said they made it look like the tail was wagging the dog. Ha Ha."

72. The report became House Document No. 103-7311. The gravity plan included some small power generating stations along its route, but nothing that compared with the size of Grand Coulee Dam.

73. *Columbia River and Minor Tributaries*, pp. 481-483, 500-509, 517-521. Also see VandeVere, "History of Irrigation in Washington," p. 217; Project History—Grand Coulee Dam, 1933, pp. 48-49.

74. See endnote 78 for this chapter.

75. Project History—Grand Coulee Dam, 1933, pp. 58-63.

76. Harding, "The Columbia Basin Irrigation League," p. 197; Miner, "A History of the Columbia Basin Projects," p. 276. A confidential report within the Bureau of Reclamation estimated that power generation could produce enough revenue to return the cost of the project in fifty years. Bureau of Reclamation confidential report, September 30, 1931, BOR Denver Mss.

77. Despite his frequent irritation with James O'Sullivan and his professional posture of objectivity, Butler saw not only the merits of the pumping plan but also the tactics that the gravity forces had used to stifle it. In a handwritten note, sent to James O'Sullivan a few days before news of his report became public, he said, "I think you and your Associates have put up a fine fight against great odds and I . . . believe that everything will come out

O.K.. some day." Norwood, *Columbia River Power for the People*, p. 45; Butler to O'Sullivan, September 10, 1931, O'Sullivan Mss.

78. Historian Michael Knight Green argues that the Columbia Basin Irrigation League did not entirely give up the gravity plan despite the Butler report. Green wrote, "Evidently the results of the Butler report reached the Washington Water Power Company's president, Frank Post, in a most unusual manner. Sometime in September, 1931, the Columbia Basin Irrigation League held a small confidential meeting in Spokane to discuss plans for entertaining the House Irrigation and Reclamation Committee, which the League had invited out again on another expense paid junket. Present at the meeting were Post, Senators Wesley Jones and Clarence C. Dill and a few officials of the CBIL. While discussing the plans for entertaining the congressional committee, Senator Jones accidentally mentioned that Butler's report, which he had seen, endorsed the building of Grand Coulee Dam, whereupon President Post is reported to have jumped to his feet with the following outburst—'He found for the dam? Impossible! Where is Roy Gill? Have him get in touch with Mead.' Several years later when the files of the CBIL accidentally fell into O'Sullivan's hands the reason why Post in a moment of passion would have cried aloud the names of Roy Gill of the CBIL and Dr. Elwood Mead, head of the Bureau of Reclamation, became more apparent. The evidence indicates that Dr. Mead, under pressure from the gravity boosters, particularly Gill, had agreed to go ahead with the gravity plan despite the recommendations of the Army." See Green, "Politics and Kilowatts," pp. 54-55. Green cites Sundborg, *Hail Columbia*, as a partial source. Green also argues that Elwood Mead may not have been as interested in the gravity plan as he let on, since this period marked the start of the feud between the Bureau of Reclamation and the Army Corps of Engineers over control of and jurisdiction in federal dam construction. Pumpers had developed a distrust of Mead and felt unclear about where he stood. Roy Gill did go to Washington, D.C., in October to see Mead. See *Seattle Daily Times* (October 23, 1931), BOR Denver Mss. Historian William Miner implies that after these discussions, Gill realized that the gravity cause was lost, and that it was Gill who put a stop to plans by the Spokane people to attempt to discredit the Butler report and to face realistically the defeat of the gravity plan. See Miner, "A History of the Columbia Basin Projects," p. 279. For a discussion of the influence of organized interest groups on political power in the United States see McConnell, *Private Power and American Democracy* esp. pp. 3, 22, 28.

79. *Seattle Post Intelligencer* (October 24, 1931), O'Sullivan Mss.

80. For examples of this position see Rufus Woods, *The 23-Year's Battle for Grand Coulee Dam* (Wenatchee: *Wenatchee Daily World*, 1944); George Sundborg, "Grand Coulee Dam Where the Good Guys Won," *Public Power* 41 (July-August 1983): 28-39.

81. Historian Daniel Miller Ogden wrote, "From the evidence which the author has gone over, . . . it seems apparent that the Columbia Basin Irrigation League and its supporters did not seek an objective determination of the relative merits of all reasonable plans. Indeed, there is extensive evidence to support the conclusion that the promoters of the gravity plan went to great lengths to insure that no proposals, other than the gravity plan, were seriously considered or promoted. It further appears that their actions were taken deliberately, in part at least, to protect the Washington Water Power

Company by preventing any federal dam in the area." See Ogden, "The Development of Federal Power Policy in the Pacific Northwest," p. 182.

82. *Spokesman-Review* (July 22, 1933), Golden Anniversary Edition, p. 3.

83. Among those included were University of Washington President Dr. Henry Suzzalo, State Senator Frederick Steiwer of Oregon, *Oregon Journal* editor Marshall N. Dana, and prominent Seattleites Charles D. Stimson, Asahel Curtis, L. C. Gillman, and Elbert F. Benson.

84. Ogden, "The Development of Federal Power Policy in the Pacific Northwest," p. 181. Also see Columbia Basin Executive Committee Meeting, minutes, May 23, 1929, Waller Mss.

85. Sam Hill, among others, told the CBIL that without their ongoing support, the Columbia Basin Project would never have come about. See Harding, "The Columbia Basin Irrigation League," p. 184.

86. Sundborg, *Hail Columbia*, p. 141.

87. *Wenatchee Daily World* (December 6, 1932), p. 3.

## Notes for Chapter Five

1. As quoted in, Earl Clark, "Rufus Woods: Grand Coulee Promoter," *Montana: The Magazine of Western History* 24 (October 1979): 39.

2. The Bureau of Reclamation claimed that its projects stabilized families on the land in that it made acres unsuitable for dry farming useful and dependable. This allowed a farmer to depend on his land, and hence it kept the population from moving. In light of the coming dust bowl, the argument fell on receptive ears during the 1930s. See Richard Lowitt, *The New Deal and the West* (Bloomington: Indiana University Press, 1984), pp. 91-93.

3. H. W. Lincoln, War Department Engineer to District Office, Seattle, November 25, 1931, BOR Denver Mss.

4. Mead to Bashore, October 7, 1931, *ibid.*

5. Woods to Judge L. C. Gillman, September 23, 1931, Woods Mss.

6. *Wenatchee Daily World* (October 9, 1931), p. 1; Bruce Mitchell, *Flowing Wealth: The Story of Water Resource Development in North Central Washington, 1870-1950* (Wenatchee: *Wenatchee Daily World*, March 6, 1967, reprinted January 1980), p. 33.

7. Mead to Harper, October 29, 1931, BOR Denver Mss.

8. Daniel Miller Ogden, "The Development of Federal Power Policy in the Pacific Northwest" (unpublished Ph.D. dissertation, University of Chicago, 1949), p. 190. Also see Wilbur W. Hindley, "The Story of Grand Coulee Dam—1890-1942," *Spokane Spokesman-Review* progress edition (January 25, 1942), p. 11.

9. *Wenatchee Daily World* (November 17, 1931), p. 1; *Seattle Post-Intelligencer* (November 12, 1931), O'Sullivan Mss.; William D. Miner, "A History of the Columbia Basin Projects" (unpublished Ph.D. dissertation, University of Indiana, 1950), p. 285.

10. Woods to O'Sullivan, December 10, 1931, Woods Mss.

11. With good reason, Woods and O'Sullivan feared that Gill would work for a dam at the Grand Coulee, but for a small, low dam unable to generate a large surplus of power. This would both preserve the Kettle Falls site, still under consideration, and keep government power from competing with the Washington Water Power Company.

12. Introduced on January 7 and 11, 1932, the measure became HR 7446, backed by Representative Samuel Billingsley Hill of Waterville, and S 2860 sponsored jointly by Senators Jones and Dill. Project History—Grand Coulee Dam, 1933, p. 69. Also see *Wenatchee Daily World* (January 5, 1932), p. 1.
13. Ogden, "The Development of Federal Power Policy in the Pacific Northwest," p. 179.
14. *Wenatchee Daily World* (January 16, 1932), p. 1.
15. As quoted in Ernest R. Abrams, *Power in Transition* (New York: Charles Scribner's Sons, 1940), p. 50. Also see "Statement Submitted by Arthur M. Hyde, Secretary of Agriculture, to the Board of Engineers for Rivers and Harbors, War Department, January 30, 1932," p. 4, copy available at Washington State University Library, Pullman, Washington.
16. *Wenatchee Daily World* (March 5, 1932), p. 4; and O'Sullivan to Robinson, March 8, 1932, O'Sullivan Mss.
17. Mead to General Lytle Brown, War Department, March 19, 1932, BOR Denver Mss.
18. *Wenatchee Daily World* (June 21, 1932), p. 1.
19. Woods to Southard, Clapp, Matthews, Patrick, February 11, 1932, Woods Mss.
20. *Wenatchee Daily World* (July 27, 1932), p. 1.
21. Gus Norwood, *Columbia River Power for the People: A History of Policies of the Bonneville Power Administration* (Washington, D. C.: U. S. Government Printing Office, 1981), p. 26; Hindley, "The Story of Grand Coulee Dam," p. 11; Mitchell, *Flowing Wealth*, p. 34; Abrams, *Power in Transition*, p. 22; Miner, "A History of the Columbia Basin Projects," p. 313; Ogden, "The Development of Federal Power Policy in the Pacific Northwest," p. 192.
22. Carlos A. Schwantes, *The Pacific Northwest: An Interpretive History* (Lincoln: University of Nebraska Press, 1989), p. 304.
23. *Wenatchee Daily World* (November 8, 1932), p. 1.
24. *Ibid.* (November 19, 1932), p. 1.
25. *Ibid.* (November 12, 1932), p. 10.
26. Miner, "A History of the Columbia Basin Projects," p. 309.
27. *Wenatchee Daily World* (November 19, 1932), p. 7.
28. Albert S. Goss to Rufus Woods, December 15, 1932, O'Sullivan Mss.
29. *Wenatchee Daily World* (January 30, 1933), p. 1.
30. *Ibid.* (January 7, 1933), p. 1; (January 11, 1933), p. 1.
31. *Spokesman-Review* (March 3, 1933), O'Sullivan Mss.
32. George Sundborg, *Hail Columbia: The Thirty-Year Struggle for Grand Coulee Dam* (New York: The Macmillan Company, 1954), pp. 234-236. Sundborg writes that O'Sullivan found damaging evidence in the files that showed how unscrupulous the CBIL had been in its fight for the gravity plan, but that at a later date he claimed that someone robbed the office and removed the files.
33. Roy Gill to Dr. Elwood Mead, March 31, 1933, Gill Mss.
34. Roy Gill, Chairman of the Spokane Chamber of Commerce Executive Committee to the Trustees (comment on his testimony), February 8, 1933, typed copy in the BOR Denver Mss. Also see Hindley, "The Story of Grand Coulee Dam" p. 11.
35. Mead to Gill, March 10, 1933, Gill Mss.
36. *Spokane Chronicle* (March 10, 1933), O'Sullivan Mss.

37. Goss to Woods, October 28, 1942, Woods Mss. A telegram dated April 13, 1933 from Goss to James O'Sullivan confirms some of what Goss remembered in the 1942 letter to Woods. Goss to O'Sullivan, April 13, 1933, O'Sullivan Mss.

38. How much Roosevelt knew about the Bureau of Reclamation/Army Engineers conflict over the river is unknown. Roosevelt sending Dill to the Bureau of Reclamation may be the origin of the opinion that Roosevelt decided to share the projects, giving Grand Coulee to Reclamation and Bonneville to the Army.

39. Dill reported later that Roosevelt suggested a $50 to $60 million figure. He mentioned those numbers in a wire he sent to Governor Martin on April 17, 1933. The way Dill told the story in his book, he saw Roosevelt alone on April 17, which is clearly in error. See Clarence C. Dill, *Where Water Falls* (Spokane, Wash.: C. W. Hill, Printers, 1970), p. 170. Also see Sundborg, *Hail Columbia*, p. 238. On November 25, 1935, Rufus Woods typed a memo to himself which he left in his files. In part it read, "I have just had another visit with A. S. Goss, former master of the Washington Stae [*sic*] Grange. For hisotira [*sic*] reference I am writing this for my confidential files. Goss says, 'On the night of election in November, 1932, [Louis McHenry] Howe, private secretary of Governor Franklin D Roosevelt called me from Albany, N. Y. He said that they had it on good authority that Taber, master of the national grange, was about to go on the national radio hook-up for Hoover for the reason that Hoover was a dry.'" Then Woods added, "My own opinion is that the above shows the trump cards played by A. S. Goss seconded by that of C. C. Dill [that] brought the Grand Coulee into being." Memo dated November 25, 1935, Woods Mss. In this, Woods gives Goss the major credit for convincing Roosevelt to back the dam. It is debatable whether Goss was more active in the Roosevelt meeting than Dill and which had the greater influence.

40. Goss to Martin, April 17, 1933; Goss to O'Sullivan, April 17, 1933, both in O'Sullivan Mss.

41. Goss to O'Sullivan, April 17, 1933, *ibid.*

42. Dill indicated in the letter that Roosevelt was looking at the project for the first time, which does not square with Dill's later assertion that he had long talks with Roosevelt about the Columbia Basin years before. In his later reporting of the April 17 meeting, Dill has Roosevelt telling *him* about the project. See Dill, *Where Water Falls*, pp. 149, 168. It is clear that events as reported by Dill, especially when recollected years later, should be examined very closely. For a view of C. C. Dill very different from this author's, which credits Dill with a very positive attitude toward Grand Coulee and much more influence with Roosevelt in getting it built, see Kerry E. Irish, "Clarence Dill: The Life of a Western Politician" (unpublished Ph.D. dissertation, University of Washington, 1994).

43. Mead to Walter, April 17, 1933, BOR Denver Mss.

44. Dill to Mead, April 21, 1933, *ibid.*

45. *Spokane Chronicle* (May 2, 1933), O'Sullivan Mss.; *Wenatchee Daily World* (May 4, 1933), p. 7. At this point Roy Gill wrote to Joseph A. Swalwell that Senator Dill was using the Columbia Basin to further his campaign for re-election and that it was unfortunate that they had not yet been able to prove that Dill was not sincere about the project. Gill to Joseph A. Swalwell, May

4, 1933, Gill Mss. And Gill wrote to Mae Schnurr, Mead's assistant, "Mr. Goss was here on the 9th, he reported that Dr. Mead was enthused over the new idea of a low dam. He thinks Senator Dill is really in earnest about Columbia Basin. My guess is that Dill will maneuver it around so he can use it as his chief Campaign issue for re-election. 'If you want Columbia Basin, re-elect Dill.' His chief issue six years ago was, 'I have 34 Democratic Senators pledged to vote for Columbia Basin.' I tried every possible way to get Senator Jones to get the Bill into the Senate to call a 'showdown' on Dill, but was unable to make him move." Gill to Schnurr, May 11, 1933, Gill Mss.

46. *Wenatchee World* (July 8, 1983), p. A36.

47. Joseph E. Stevens, *Hoover Dam: An American Adventure* (Norman: University of Oklahoma Press, 1988), pp. 38-39; *Wenatchee Daily World* (October 20, 1931), p. 1.

48. Emmett Kaiser VandeVere, "History of Irrigation in Washington" (unpublished Ph.D. dissertation, University of Washington, 1948), p. 221.

49. Gill to Schnurr, May 6, 1933, Gill Mss.

50. *Wenatchee Daily World* (May 16, 1933), p. 6; *Spokane Chronicle* (May 18, 1933), O'Sullivan Mss.; Ogden, "The Development of Federal Power Policy in the Pacific Northwest," p. 198; Sundborg, *Hail Columbia*, 239; Michael Knight Green, "Politics and Kilowatts: The Washington Water Power Company and Public Power, 1918-1941" (unpublished masters thesis, University of Idaho, 1962), p. 59.

51. Woods to O'Sullivan, May 24, 1933, telegram, O'Sullivan Mss.

52. Goss to Woods, June 15, 1933, Woods Mss.

53. See Paul C. Pitzer, *Building the Skagit: A Century of Upper Skagit Valley History 1870-1970* (Lake Oswego, Ore.: Galley Press, 1978), pp. 57, 79.

54. Mark Mendenhall to James O'Sullivan, May 2, 1933, Columbia Basin Commissions Mss.

55. Dill to Ross, March 14, 1933, letter, and July 6, 1933, telegram, City Light Mss.

56. From Roosevelt's papers and the records kept during the first months of the New Deal, it is impossible to say for certain who actually saw the President. See Roy Talbert Jr., *FDR's Utopian: Arthur Morgan of the TVA* (Jackson: University Press of Mississippi, 1987), p. 80.

57. Ross to Frank Fitts, et. al., July 8, 1933, City Light Mss. Also see Richard C. Berner, *Seattle 1921-1940: From Boom to Bust* (Seattle: Charles Press, 1992), pp. 420-422.

58. When Dill found out about claims that he owned land in the Columbia Basin he wrote to Seattle City Commissioner Ralph Nichols, who had repeated the rumor, "Whether you said it, I do not know, and if you did say it, I have no knowledge of where you secured your information. I am sure you do not want to do me an injustice or tell anything that isn't true about me. For that reason I am writing you to say I do not own a foot of land in the Columbia Basin Project, that I haven't any real property in the world with the exception of my home in Spokane, some lots in Spokane and an interest in some property just south of Seattle." Nichols wrote to Dill and apologized. Dill to Nichols, June 12, 1933, Columbia Basin Commissions Mss.

59. *Bellingham Herald* (July 11, 1933), Woods Mss.

60. Sundborg, *Hail Columbia*, 225.

61. Unidentified clipping (May 5, 1933), O'Sullivan Mss.; Miner, "A History of the Columbia Basin Projects," p. 318.

62. J. H. Secrest to Clarence D. Martin, June 30, 1933, memo, BOR Denver Mss.; *Wenatchee Daily World* (June 30, 1933), p. 1.
63. *Spokesman-Review* (June 30, 1933), O'Sullivan Mss.
64. *Wenatchee Daily World* (July 3, 1933) p. 1.
65. *Ibid.* (July 6, 1933), p. 1; (July 11, 1933), p. 1; Mitchell, *Flowing Wealth*, p. 37; Project History—Grand Coulee Dam, 1933, p. 72; *Spokane Daily Chronicle* (July 11, 1933), O'Sullivan Mss. A copy of the contract is available in the files of the Columbia Basin Commissions Mss.
66. *Wenatchee Daily World* (July 17, 1933), p. 1; (July 15, 1933), p. 1; *Spokane Press* (July 15, 1933), O'Sullivan Mss.; Project History—Grand Coulee Dam, 1933, p. 72; *Spokane Press* (July 17, 1933), O'Sullivan Mss.; Stuart Chase, "Great Dam," *Atlantic Monthly* 162 (November 1938): 198.
67. *Spokesman-Review* (July 17, 1933), O'Sullivan Mss.
68. Norwood, *Columbia River Power for the People*, p. 36; Ogden, "The Development of Federal Power Policy in the Pacific Northwest," p. 201; *Spokane Press* (July 25, 1933), O'Sullivan Mss.
69. Dill to O'Sullivan, July 25, 1933, telegram, O'Sullivan Mss.
70. *Wenatchee Daily World* (July 26, 1933), p. 1. Dill later reported that he had had a much more lengthy talk with Roosevelt during which Roosevelt called Ickes into the office and told the Secretary of the Interior to proceed with the Grand Coulee Dam. "If that is what you want, it will be done," Dill reported as Ickes's reply. See Sundborg, *Hail Columbia*, p. 248; Hindley, "The Story of Grand Coulee Dam," p. 14. In 1949 historian Daniel Miller Ogden and writer Ralph E. Dyar of the *Spokesman-Review* attempted to verify the Dill version of the meeting. Ogden contacted Ickes, who denied knowing anything about it. See Ogden, "The Development of Federal Power Policy in the Pacific Northwest," p. 203. Also see Ralph E. Dyar, *News for An Empire: The Story of the Spokesman-Review of Spokane, Washington, and the Field it Serves* (Caldwell, Ida.: Caxton Printers, 1952), p. 412. Ickes insisted that he did talk privately with Roosevelt about the project, but never in a meeting with Dill. Ickes's diary gives evidence that this is true. His entry for July 27 reads, "I told them [a meeting of the Special Board of Public Works] at the outset that the President wanted us to pass on two or three of the big Federal projects. The first one was the Coulee Dam on the Columbia River. . . . With these facts and the added one that the President wanted it done, I put an end to the discussion. . . . In the end we passed it." See Harold L. Ickes, *The Secret Diary of Harold L. Ickes: The First Thousand Days 1933-1936*, (New York: Simon and Schuster, 1953), pp. 70-71. Ickes made no mention of seeing Dill on the previous day. If Roosevelt had called Ickes into the meeting, Dill would certainly have mentioned it to the press and in his telegrams back to Washington. There is no discussion of any such encounter; only Dill's brief comment that he was encouraged. All of the evidence indicates that the event, as remembered by Dill, did not happen. Yet Dill remembered the incident with even more detail when he wrote his book in 1970, and in a footnote he commented on the Dyar inquiry. Dill speculated that Ickes had really forgotten the incident. See Dill, *Where Water Falls*, p. 178. See endnote 93 below.
71. *Wenatchee Daily World* (July 28, 1933), p. 1; *Spokesman-Review* (July 29, 1933), O'Sullivan Mss. The Public Works Administration formally awarded the $63,000,000 to the State of Washington on September 12, 1933. It stated that the project was not to compete with, but was to supplement existing

power-generating facilities. It awarded $20,000,000 for construction of Bonneville Dam on September 29, 1933. "The actual harnessing of the Columbia River at Bonneville resulted more from political expediency than from economic demand." Roosevelt bowed to the demand from Portland, and also from the Corps of Engineers, that they too have a project. The government kept Bonneville power rates low by writing off over 20 percent of the cost of the dam to navigation, thus creating the cheap "yardstick" rate that Roosevelt wanted. See Abrams, *Power in Transition*, pp. 55-57. On August 5, 1933 the Public Works Administration appointed Marshall N. Dana as one of its ten regional advisors. Dana also led the effort for planning in the Pacific Northwest and as such worked closely with Frederick A. Delano, Chairman of the National Planning Board and uncle of President Roosevelt.

72. Dill to O'Sullivan, July 27, 1933, Columbia Basin Commissions Mss. Also see *Spokane Daily Chronicle* (July 27, 1933), O'Sullivan Mss. Ickes's biographer points out that Roosevelt had the final say on each PWA project. See T.H. Watkins, *Righteous Pilgrim: The Life and Times of Harold L. Ickes, 1874-1952* (New York: Henry Holt and Co., 1990), p. 369.

73. Ickes, *The Secret Diary of Harold L. Ickes: The First Thousand Days 1933-1936*, p. 72.

74. *Spokane Daily Chronicle* (July 28, 1933), O'Sullivan Mss.

75. *Ibid.* (August 1, 1933), O'Sullivan Mss.; *Spokesman-Review* (August 2, 1933), O'Sullivan Mss.; *Wenatchee Daily World* (August 2, 1933), p. 5.

76. *Wenatchee Daily World* (September 2, 1933), p. 1. It may or may not be significant to note that Representative Samuel Billingsley Hill's cousin, Kirby Billingsley, worked for the *Wenatchee Daily World*.

77. The depression with its unemployment, and news that work might begin along the Columbia River, had brought a small population to the area where the dam would be built. In addition, a small gold rush of sorts had started along the Columbia. In the summer of 1932 the *Wenatchee Daily World* reported that between 300 and 500 people worked looking for gold from Wenatchee and upriver. This activity actually increased in the summer of 1933.

78. The Columbia Basin Commission decided that it needed its own engineer to oversee the project and guarantee that the Bureau of Reclamation accommodated state concerns. James O'Sullivan contacted Major John S. Butler and asked if he would take the position. Butler showed interest and asked the salary, but before O'Sullivan could answer, declined the position. He found he could not simultaneously work for the state and retain his position with the Army Corps of Engineers. O'Sullivan looked elsewhere for an engineer. On October 2, 1933, Governor Martin did appoint a board of engineers to work with the Bureau of Reclamation and report to the Columbia Basin Commission. O'Sullivan to Butler, July 3, 1933; Butler to O'Sullivan, July 10, 1933; Butler to O'Sullivan, July 17, 1933, all in Columbia Basin Commissions Mss.

79. Ogden, "The Development of Federal Power Policy in the Pacific Northwest," p. 206.

80. Actually some work did proceed at the dam site with the landowners making an unofficial agreement to allow use of their property while at the same time posting notice of "No Trespassing" and making claims of trespass to protect their legal rights.

81. There was a second, more complicated problem. An opinion from the state's legal counsel all but voided the contract between the state and the Bureau

of Reclamation. The main issue centered on whether the state, the federal government, or some combination of the two, would build the project, or the federal government build it with the Columbia Basin Commission leasing it after completion. This had resulted from a previous opinion of the state solicitor suggesting that the Washington state legislature would have to pass a law authorizing the indebtedness of 70 percent of the $63,00,000, a provision required by the PWA, and that it would have to be referred to a vote of the people. He deemed this infeasible. State Solicitor Margold letter, October, 1933, BOR Denver Mss. Also see Raymond F. Walter to Elwood Mead, October 6, 1933, *ibid.*; *Wenatchee Daily World* (October 9, 1933), p. 9.

82. Stoutmeyer to Mead, October 17, 1933, BOR Denver Mss.
83. *Spokesman-Review* (October 23, 1933) editorial. The editorial did not share Dill's optimism, stating that an alliance of state taxpayers, especially those in Seattle backing the Skagit development, might not approve of the state doing the job, that they might disrupt plans for state funding, and that if the federal government built the dam it would create a huge surplus of electricity.
84. Mead to Walter, October 17, 1933, BOR Denver Mss.
85. The press announced the release of $1,000,000 on September 14, but the action did not become official until Ickes signed "Release No. 122 of the Federal Emergency Administration of Public Works" October 9, 1933. A copy found in BOR Denver Mss.
86. *Wenatchee Daily World* (October 25, 1933), p. 1.
87. The *Wenatchee Daily World* announced Ickes's recommendation and complained that the Public Works Board itself insisted that the state adopt the 70 percent plan that caused the trouble which it now used as the lever to gain control of the dam.
88. Harold L. Ickes to Federal Emergency Administrator of Public Works, November 1, 1933, copy in Gill Mss.
89. *Wenatchee Daily World* (November 1, 1933), p. 8; *Grant County Journal* (November 3, 1933), p. 1; *Spokesman-Review* (November 1, 1933), O'Sullivan Mss.
90. *Wenatchee Daily World* (November 27, 1933), p. 1.
91. *Spokesman-Review* (November 26, 1933), O'Sullivan Mss.
92. *Ibid.* (November 10, 1933), O'Sullivan Mss.
93. *Ibid.* (November 17, 1933), O'Sullivan Mss. It is very likely this confrontation with Ickes that Dill remembered when he spoke about the meeting with Roosevelt in May.
94. Richard L. Neuberger, *Our Promised Land* (New York: The Macmillan Company, 1938), pp. 82, 90.
95. Grant McConnell, *Private Power and American Democracy* (New York: Alfred A. Knopf, 1966), p. 69; Lowitt, *The New Deal and the West*, pp. 77, 89.
96. Robert S. McElvaine, *The Great Depression: America, 1929-1941* (New York: Times Books, division of Random House, Inc., 1984), p. 152.
97. Norris Hundley, Jr., *Water and the West: The Colorado River Compact and the Politics of Water in the American West* (Berkeley: University of California Press, 1975), p. xvi.
98. Hundley, *Water and the West*, p. xiv. Also see Martin T. Farris and Roy J. Sampson, *Public Utilities: Regulation, Management and Ownership* (Boston: Houghton Mifflin Co., 1973), pp. 265-268.
99. Raymond Moley, *What Price Federal Reclamation?* (New York: American Enterprise Association, Inc., 1955), p. 16.
100. Abrams, *Power in Transition*, p. 24.

## Notes for Chapter Six

1. Wanda Wilson Whitman, ed. *Songs that Changed the World*, (New York: Crown Publishers, 1969), pp. 85-86. Lyrics of "Brother, Can you Spare a Dime?"© 1932 by Harms, Inc.
2. *Wenatchee Daily World* (July 13, 1933), p. 1; George Sundborg, *Hail Columbia: The Thirty Year Struggle for Grand Coulee Dam* (New York: The Macmillan Company, 1954), p. 245.
3. Project History—Grand Coulee Dam, 1933, pp. 8, 73.
4. "Grand Coulee First Big NRA Project for Pacific Northwest," *Pacific Builder and Engineer* 38 (August 5, 1933): 19.
5. *Wenatchee Daily World* (December 1, 1933), p. 1.
6. *Columbia Basin Project News* 8:9 (September 1950): 2.
7. "All-Star Staff Directs Engineering on Grand Coulee Project," *Pacific Builder and Engineer* 39 (November 4, 1933): 20; L. Vaughn Downs, *The Mightiest of Them All: Memories of Grand Coulee Dam* (Fairfield, Wash.: Ye Galleon Press, 1986), p. 24
8. Much of the biographical material on Banks and the quote appeared as part of an insert in an article written by Banks. See Frank A. Banks, "The Grand Coulee High Dam . . . And the Irrigation of the Columbia Basin," *Pacific Builder and Engineer* 40 (September 1, 1934): 17.
9. *Spokesman-Review* (April 19, 1934), special edition, Dubuar Scrapbooks, #9, p. 9; *Wenatchee Daily World* (April 18, 1934), p. 5; "Last Minute News Dispatch from Grand Coulee Dam," *Pacific Builder and Engineer* 39 (April 14, 1934): 31.
10. *Wenatchee Daily World* (August 22, 1933), p. 1; (August 29, 1933), p. 8; "Grand Coulee," *Pacific Builder and Engineer* 38 (August 5, 1933): 17, editorial.
11. *Spokesman-Review* (September 3, 1933), O'Sullivan Mss.; Bruce Mitchell, *Flowing Wealth: The Story of Water Resource Development in North Central Washington, 1870-1950* (Wenatchee: *Wenatchee Daily World*, March 6, 1968, reprint January 1980), pp. 37-38.
12. Project History—Grand Coulee Dam, 1933, p. 78.
13. *Spokesman-Review* (November 20, 1931), O'Sullivan Mss.; *Wenatchee Daily World* (August 3, 1933), p. 1; (September 8, 1933), p. 1; Sundborg, *Hail Columbia*, p. 244; L. C. Gilman to R. F. Walter, August 9, 1933, BOR Denver Mss.
14. *Wenatchee Daily World* (October 2, 1933), p. 1; (October 10, 1933), p. 1; (November 2, 1933), p. 1.
15. *Wenatchee Daily World* (November 14, 1933), p. 6; (February 2, 1934), p. 6; *Grant County Journal* (November 17, 1933), p. 1; Project History—Grand Coulee Dam, 1934, p. 9.
16. *Wenatchee Daily World* (March 16, 1934), p. 1; (April 10, 1934), p. 1; (April 14, 1934), p. 1; (May 22, 1934), p. 10; (June 12, 1934), p. 1; Project History—Grand Coulee Dam, 1934, pp. 10-11; "U.S. Construction Railroad: Odair to Grand Coulee Dam Site," *Pacific Builder and Engineer* 40 (May 5, 1934): 25.
17. *Wenatchee Daily World* (September 5, 1933), p. 2.
18. "Field Notes on the Grand Coulee Dam," *Pacific Builder and Engineer* 39 (August 19, 1933): 7.
19. "At a recent conference with Senator C. C. Dill and the Columbia Basin Commission the possibility of getting started on some construction work in connection with the Columbia River dam for the purpose of giving prompt

relief to unemployment and speeding up as much as possible the work under the Public Works Administration was discussed." Banks to Mead, September 6, 1933, BOR Denver Mss.

20. Project History—Grand Coulee Dam, 1933, p. 78; Mitchell, *Flowing Wealth*, pp. 37-38; *Wenatchee Daily World* (September 2, 1933), p. 8. Actual copies of the contracts with the two companies, dated September 8, 1933, are available in Columbia Basin Commissions Mss.

21. Clarence Cole, Althe Thomas, Fred Berry, Ceylon G. Rossman, and Harold Sheerer.

22. *Wenatchee Daily World* (September 9, 1933), p. 10; Walter A. Averill, "The Site for Grand Coulee Dam," *Pacific Builder and Engineer* 39 (October 7, 1933): 14; *The Story of the Columbia Basin Project* (Washington, D.C.: U.S. Government Printing Office, 1978), p. 8.

23. *Wenatchee Daily World* (September 25, 1933), p. 1; (September 26, 1933), p. 6.

24. *Spokesman-Review* (September 28, 1933), O'Sullivan Mss.

25. *Wenatchee Daily World* (October 11, 1933), p. 10.

26. *Spokesman-Review* (November 4, 1933), O'Sullivan Mss.

27. Joe Mehan to James O'Sullivan, November 8, 1933, Columbia Basin Commissions Mss.

28. Mehan managed to evict a few of the squatters but one Al Meyers refused to move. Meyers had put up a shanty business that he called the "Dam Site Cafe." Mehan ordered Meyers to get out, but the would-be restaurateur refused. "With Mr. Banks concurrence I boarded up his place and he now threatens me with arrest and also that he will shoot me," wrote Mehan. "But this should not be taken too serious [*sic*] as I probably would just take his gun away if he did show one." The next day Meyers ripped down the boards and opened for business as usual. When Mehan investigated later he found the restaurant operating. "Mehan, according to Meyers, early today entered the cafe and ordered not only Meyers, but some patrons, to vacate the premises. Meyers refused to do so unless Mehan had a court order and when none appeared, he tossed Mehan out." A few days later Frank Banks sent a contractor's bulldozer to push down Meyers's cafe. The dispossessed, but not defeated, businessman moved some distance away and built a "casino" that lasted for a few months. Joe Mehan to James O'Sullivan, November 23, 1933), Columbia Basin Commissions Mss.; *Spokesman-Review* (November 25, 1933), (January 10, 1934), O'Sullivan Mss.

29. Banks to Mead, October 4, 1933, BOR Denver Mss.

30. Walter A. Averill, "Stripping of Overburden at Grand Coulee Dam Six Weeks Ahead of Schedule," *Pacific Builder and Engineer* 39 (April 14, 1934): 24.

31. *Grant County Journal* (October 17, 1933), p. 1; *Wenatchee Daily World* (November 2, 1933), p. 1; *Spokesman-Review* (October 27, 1933), O'Sullivan Mss.

32. Banks to O'Sullivan, April 20, 1934, Columbia Basin Commissions Mss.

33. Ellsworth C. French to C. C. Dill, November 20, 1933, Columbia Basin Commissions Mss.

34. "Winter Contracts for Grand Coulee," *Pacific Builder and Engineer* 39 (November 4, 1933): 20.

35. *Wenatchee Daily World* (November 20, 1933), p. 1; *Grant County Journal* (November 24, 1933), p. 1.

36. Project History—Grand Coulee Dam, 1933, p. 9; *Wenatchee Daily World* (December 5, 1933), p. 1.

37. *Wenatchee Daily World* (January 4, 1934), p. 11.

38. W. C. Morse, Consulting Engineer, to Columbia Basin Commission, February 5, 1934, Columbia Basin Commissions Mss.
39. O'Sullivan to Mehan, October 17, 1933), Columbia Basin Commissions Mss.
40. Project History—Grand Coulee Dam, 1934, p. 39; *Wenatchee Daily World* (February 3, 1934), p. 8; (April 6, 1934), p. 1; *Spokesman-Review* (January 10, 1934), O'Sullivan Mss.
41. *Grant County Journal* (January 5, 1934), p. 1; *Spokesman-Review* (January 18, 1934), O'Sullivan Mss.; Banks to Walter, January 5, 1934 and Walter to Banks, January 16, 1934, both in BOR Denver Mss.; Mehan to O'Sullivan, December 27, 1933, Columbia Basin Commissions Mss.
42. *Spokesman-Review* (November 23, 1933), O'Sullivan Mss.
43. Dill to Banks, August 7, 1933, BOR Denver Mss.; *Wenatchee Daily World* (August 18, 1933), p. 15.
44. *Grant County Journal* (January 5, 1934), p. 1.
45. *Wenatchee Daily World* (November 15, 1933), p. 1; (December 14, 1933), p. 1; (January 6, 1934), p. 6; (February 10, 1934), p. 1; (February 23, 1934), p. 12; (February 28, 1934), p. 1; (April 11, 1934), p. 7; *Spokesman-Review* (November 23, 1933); (January 20, 1934), O'Sullivan Mss.; Hurney, "History of Grand Coulee Dam," p. 36; "What's Doing at Grand Coulee Dam," *Pacific Builder and Engineer* 40 (June 9, 1934): 35.
46. *Wenatchee Daily World* (October 9, 1933), p. 9.
47. Marion Zioncheck to Mead, December 12, 1933, BOR Denver Mss.
48. *Wenatchee Daily World* (February 5, 1934), p. 3.
49. Instability of the earth on the banks of the Columbia River was probably the most significant reason for the change from a suspension to a cantilever bridge. See "Slides in West Abutment Area are Problem at Grand Coulee Dam," *Western Construction News* 10 (September, 1935): 258.
50. *Wenatchee Daily World* (November 17, 1933), p. 16; *Spokesman-Review* (November 30, 1933), O'Sullivan Mss.; Project History—Grand Coulee Dam, 1933, p. 80; A. Gilbert Darwin, "Grand Coulee Dam and Power Plant Specifications," *Western Construction News* 9 (April 1934): 110; "Preliminary Data on Design and Construction Features of Grand Coulee Dam," *Pacific Builder and Engineer* 39 (April 14, 1934): 23.
51. *Wenatchee Daily World* (June 14, 1934), p. 6.
52. *Ibid.* (January 30, 1934), p. 10.
53. "Geology of the Rock Foundation of Grand Coulee Dam, Washington," *Bulletin of the Geological Society of America* 49 (November 1, 1938): 1633-1635; "Problems of Slide Control Handled at Grand Coulee," *Western Construction News* 15 (June 1940): 204.
54. *Wenatchee Daily World* (April 10, 1934), p. 7.
55. Project History—Grand Coulee Dam, 1934, p. 10; *Wenatchee Daily World* (March 27, 1934), p. 1; I. E. Stevenson, "Slide on Schedule No. 1," *Pacific Builder and Engineer* 40 (April 14, 1934): 29; Walter A. Averill, "Moving a Mountain a Mile—at Grand Coulee," *Pacific Builder and Engineer* 41 (October 12, 1935): 37.

## Notes for Chapter Seven

1. Murray Morgan, *The Dam* (New York: The Viking Press, 1954), p. xviii.
2. *Wenatchee Daily World* (June 15, 1934), p. 16; *Seattle Post Intelligencer* (August 4, 1934), special edition, clipping found in Dubuar Scrapbooks, #9, p. 54.

3. *Wenatchee Daily World* (March 3, 1934), p. 1; (April 20, 1934), p. 1; A. Gilbert Darwin, "Grand Coulee Dam and Power Plant Specifications," *Western Construction News* 9 (April 1934): 103.

4. George Sundborg reports that in all, twenty companies acquired copies of the plans and specifications for Grand Coulee Dam. See Sundborg, *Hail Columbia* (New York: The Macmillan Company, 1954), p. 257; *Wenatchee Daily World* (April 30, 1934), p. 3; (May 10, 1934), p. 6.

5. For information on the Six Companies at this time and on Boulder Dam see Joseph E. Stevens, *Hoover Dam: An American Adventure* (Norman: University of Oklahoma Press, 1988), esp. pp. 44-46.

6. McGovern owned an insurance business and he knew, through his contacts, that bonding companies felt uneasy about Grand Coulee because of the tricky river diversion. See James Edward O'Sullivan, "The Struggle for the Grand Coulee Dam," typed manuscript found in O'Sullivan Mss., pp. 54-55. O'Sullivan maintained that James McGovern, in return for giving inside information to representatives of the Six Companies, received assurances that his insurance company would gain all Six Company business.

7. G. E. Bjork, "Bids Opened at Spokane for Grand Coulee Dam and Power Plant," *Western Construction News* 9 (July 1934): 224.

8. *MWAK Columbian* (August 9, 1935), Vol 1, p. 2, copy found in Banks Mss.; *Wenatchee Daily World* (June 25, 1935), p. 12.

9. The Columbia does not usually exhibit sudden flash floods like some other rivers. Its changes in volume usually occur over a period of days or weeks. In the extreme, the Columbia River at the site of Grand Coulee Dam can rise from a low of around 17,000 cubic feet of water per second to a high of up to and even over 500,000 cubic feet per second.

10. An article in the *Wenatchee Daily World* stated that James O'Sullivan had suggested the unique method of river diversion that had lowered the cost of the MWAK bid. The point is debatable. O'Sullivan did write an article that appeared in a trade journal in which he suggested a method for diverting the river similar to that which the MWAK Company eventually used. It is quite possible that based on this article, or on conversations with O'Sullivan, the idea had its origin, or vice versa. See *Wenatchee Daily World* (June 21, 1934), p. 10. Also see James O'Sullivan, "Two Suggested Methods for Diverting the Columbia River at Grand Coulee Dam," *Pacific Builder and Engineer* 40 (June 9, 1934): 29-30.

11. "Brief Digest of Specifications for Grand Coulee Dam," *Pacific Builder and Engineer* 40 (May 5, 1934): 23.

12. *Wenatchee Daily World* (June 18, 1934), p. 1; (June 19, 1934), p. 8; (June 25, 1934), p. 10; Bjork, "Bids Opened," p. 224.

13. The Nez Perce Indian representatives honored Frank Banks, Elwood Mead, and Raymond Walter. The three engineers donned full headdress regalia while the proud Native Americans dubbed Mead Chief Lek Ne Kus or "Plenty Water;" Walter became Chief Tues Tenat or "High Builder;" and Banks received the designation Chief Weatakis or "High Mogul." *Wenatchee Daily World* (July 19, 1934), p. 1.

14. *Spokesman-Review* (June 19, 1934), O'Sullivan Mss.

15. "Construction Starts Soon on Grand Coulee Dam," *Pacific Builder and Engineer* 40 (July 7, 1934): 16.

16. *Seattle Star* (July 30, 1934), Dubuar Scrapbooks, #9, p. 21.

17. There would be an additional 500,000 cubic yards in the powerhouse and other structures. Behind the dam, a lake holding 1,300,000 acre feet of water would stretch over seventy-five miles. Figures on how high this dam was to be vary depending on the source. The problem seems to be that some determine the height from bedrock while others measure from the level of the river. This figure is from bedrock. A later figure of 297 feet, which appears near the end of this chapter, was calculated from the low-river level. The top of the low dam would be at 1,116 feet above sea level. "General Information Concerning the Columbia Basin Project," information sheet dated July 15, 1934, BOR Denver Mss. Also see Project History—Grand Coulee Dam, 1934, p. 20.

18. *Seattle Star* special edition (July 30, 1934), Dubuar Scrapbooks, #9, p. 29.

19. *Grant County Journal* (August 3, 1934) p. 1; (August 10, 1934), p. 1.

20. Harold L. Ickes, *The Secret Diary of Harold L. Ickes: The First Thousand Days* (New York: Simon and Schuster, 1954), p. 183.

21. Richard L. Neuberger, "Power as an Issue," *Current History* 44 (September 1936): 69. Hubert "Hu" Blonk covered the President's visit. An immigrant from Holland who came to the United States as a youngster, Blonk attended the University of Washington's School of Journalism. After graduation he found a job at Grand Coulee where he reported for both the *Wenatchee Daily World* and a competitor, the *Spokane Chronicle*. Roosevelt's visit was Blonk's first big story, and he tried to stand as close to the chief executive's car as Secret Service men allowed. Blonk stayed with Grand Coulee through 1941 and much that is detailed here comes from his stories in those newspapers. *Wenatchee World* (July 8, 1983), p. A41; Hu Blonk interview, July 16, 1988. Also see Hu Blonk, *Behind the By-Line Hu: A Fiesty Newsman's Memoirs* (Wenatchee: n. p., 1992).

22. *Post-Intelligencer* special edition (August 4, 1934,); *Seattle Star* special edition (July 30, 1934), both in Dubuar Scrapbooks, #9, pp. 44, 28.

23. The full volume of the excavation would not be realized until after the government changed plans from the small dam to the high dam. This change is covered in the next chapter.

24. *Wenatchee Daily World*, (August 15, 1934), p. 14.

25. *Ibid.* (October 12, 1934), p. 20; (October 24, 1934), p. 10; (October 29, 1934), p. 7.

26. *Ibid.* (January 22, 1935), p. 10.

27. *Ibid.* (January 23, 1935), p. 12.

28. "Work Starts on Cofferdams on Coulee Project," *Pacific Builder and Engineer* 41 (January 5, 1935): 40.

29. The pier tipped out of line by nine inches.

30. *Wenatchee Daily World* (October 3, 1936), p. 12.

31. "Earth Pressure Tilts Pier of Grand Coulee Bridge," *Engineering News Record* 115 (November 7, 1935): 646; *Wenatchee Daily World* (August 26, 1935), p. 10; (September 16, 1935), p. 7; (October 29, 1935), p. 7.

32. Kugelman to O'Sullivan, January 25, 1936, Columbia Basin Commissions Mss.; Project History—Grand Coulee Dam, 1936, p. 15; *Star Newspaper* (Grand Coulee, 1983), 50th anniversary edition, p. 2.

33. Project History—Grand Coulee Dam, 1934, pp. 13, 15; 1935, p. 16; *Wenatchee Daily World* (September 1, 1934), p. 10; (March 5, 1935), p. 12; (May 15, 1935), p. 7; "MWAK Purchases World's Largest Concrete Plant for Grand Coulee Dam," *Pacific Builder and Engineer* 41 (June 8, 1935): 26.

34. Stevens, *Hoover Dam*, pp. 117-158.
35. Chapter eleven discusses the communities that grew around the dam.
36. *Wenatchee Daily World* (January 28, 1935), p. 10.
37. Walter A. Averill, "U. S. Railroad for Grand Coulee Nears Completion," *Pacific Builder and Engineer* 40 (December 1, 1934): 40.
38. *Wenatchee Daily World* (October 13, 1934), p. 12; Project History—Grand Coulee Dam, 1934, p. 14.
39. These rents were considered high by many. It meant that only the top layer of the work crew would live in Mason City. This gave hope to nearby towns which had looked to the labor force to swell their ranks. This will be discussed in Chapter Eleven. See *Wenatchee Daily World* (December 4, 1934), p. 12.
40. O. G. F. Markhus, "Grand Coulee Contractors Build Mason City," *Reclamation Era* 25 (April, 1935): 70.
41. *Wenatchee Daily World* (November 12, 1934), p. 12; (November 19, 1934), p. 10.
42. *Ibid.* (September 8, 1934), p. 1; (November 13, 1934), p. 12; Averill, "U. S. Railroad Nears Completion," p. 20.
43. Project History—Grand Coulee Dam, 1934, p. 15; *Wenatchee Daily World* (December 5, 1934), p. 12; (December 10, 1934), p. 1.
44. *Grant County Journal* (July 26, 1935), p. 1; *Wenatchee Daily World* (July 27, 1935), p. 10; (July 29, 1935), p. 1; (July 31, 1935), p. 1; "The Rails Reach Out," *Reclamation Era* 25 (September 1935): 183.
45. "World's Largest Construction Conveyor Speeds Excavation at Grand Coulee," *Western Construction News* 10 (March 1935): 80.
46. *Wenatchee Daily World* (December 13, 1934), p. 12; (March 16, 1935), p. 12; "MWAK Rushes West Side Cofferdam: World's Largest Belt Conveyor Successful," *Pacific Builder and Engineer* 41 (March 2, 1935): 36; O. G. F. Markhus, "The Belt Conveyor System at Grand Coulee," *Reclamation Era* 25 (June, 1935), p. 109.
47. "Belt Conveyors for Spoil Transport," *Engineering News Record* 115 (August 1, 1935): 152.
48. *Wenatchee Daily World* (November 26, 1934), p. 1.
49. R. J. Kugelman to James O'Sullivan, January 12, 1935, Columbia Basin Commissions Mss.
50. *Wenatchee Daily World* (January 22, 1935), p. 1.
51. *Ibid.* (March 30, 1935), p. 1.
52. "Grand Coulee Excavating Program Modified to Meet Changed Design," *Western Construction News* 10 (September 1935): 253.
53. "Two Months' Construction Progress," *Engineering News Record* 115 (August 1, 1935): 146.
54. Kugelman to O'Sullivan, November 19, 1934, Columbia Basin Commissions Mss.
55. *Wenatchee Daily World* (November 19, 1934), p. 10. On January 28, 1935, 500,000 cubic yards of material tumbled into the area that would eventually hold the dam's forebay. On February 20 a second slide of 100,000 cubic yards dropped material into the south end of the excavation.
56. See Kugelman to O'Sullivan, February 23, 1935, Columbia Basin Commissions Mss.
57. Kugelman to O'Sullivan, May 4, 1935, Columbia Basin Commissions Mss.
58. "Work Starts on Cofferdams on Coulee Project," p. 39.

410 *Grand Coulee: Harnessing a Dream*

59. John L. Savage, "Dam Stress and Strains Studied by Slice Models," *Engineering News Record* 113 (December 6, 1934): 721.
60. First the contractor erected a rock-filled timber breakwater, or jetty, some 375-feet-long directly upstream from the axis of the dam. It diminished water velocity along the line of the west side where the cofferdam would sit and provided a foundation for derricks. Started in October, this structure was in place on December 26. Getting the steel-sheet pilings to the site proved a problem. Each steel sheet pile weighed 3,100 pounds, or 38.8 pounds per lineal foot and they were eighty feet long, fifteen inches wide, three-eighths of an inch thick, and six together made an unwieldy load weighing ten tons. With the railroad still months away, trucks carried the steel and their drivers needed considerable skill to maneuver the lengthy piles over the last primitive thirty miles. Using considerable caution the vehicles moved at a top speed of only three-and-one-half miles an hour. Project History—Grand Coulee Dam, 1935, p. 115; *Wenatchee Daily World* (November 18, 1934), p. 20; Walter A. Averill, "World's Largest Cellular Cofferdam," *Pacific Builder and Engineer* 41 (May 4, 1935): 25, 27.
61. Project History—Grand Coulee Dam, 1935, p. 99.
62. Kugelman to O'Sullivan, February 2, 1935, Columbia Basin Commissions Mss. For additional comment on the same problem see L. Vaughn Downs, *The Mightiest of Them All: Memories of Grand Coulee Dam* (Fairfield, Wash.: Ye Galleon Press, 1986), p. 79.
63. Project History—Grand Coulee Dam, 1935, p. 114.
64. *Wenatchee Daily World* (February 12, 1935), p. 1.
65. *Grant County Journal* (February 22, 1935), p. 1; *Wenatchee Daily World* (February 19, 1935), p. 1.
66. *Wenatchee Daily World* (March 6, 1935), p. 1; *Spokesman-Review* (March 7, 1935), O'Sullivan Mss.
67. *MWAK Columbian*, March 1, 1935, Banks Mss.
68. In all the 3,000-foot-long structure combined four configurations. The first and last sections consisted of a single line of interlocking pilings backed by a shore-side rear wall of timbers. At the north, or downstream end, the future location of the dam's tailrace, the next section held uniform oval-shaped cells ninety feet long with a thirty-six-foot arc forming each end. Slightly smaller cells, forty-by-fifty feet, made up the main part of the cofferdam which crossed the area that the dam would later occupy. In all the finished structure held 16,000 tons of steel-sheet piles, enough to stretch 131 miles if laid end to end, 23,000 total tons of steel including reinforcing material, and two million board feet of timbers. Dump trucks, and later conveyor belts, poured over 253,000 cubic yards of earth and debris into the cells giving them increased substance. *Wenatchee Daily World* (March 25, 1935), p. 12; "Constructing the First Cofferdam," *Engineering News Record* 115 (August 1, 1935), p. 148; Project History—Grand Coulee Dam, 1935, p. 114.
69. Averill, "World's Largest Cellular Cofferdam," p. 24; "Achievement at Coulee," *Pacific Builder and Engineer* 41 (May 4, 1935): 12, editorial.
70. *Wenatchee Daily World* (April 16, 1935), p. 12.
71. O. G. F. Markhus, "Diversion of the Columbia River—Grand Coulee Dam," *Reclamation Era* 27 (February 1937): 30.
72. "MWAK Purchases World's Largest Concrete Plant for Grand Coulee Dam," p. 26; "Constructing the First Cofferdam," p. 151; *Wenatchee Daily World*

(May 21, 1935), p. 7; (June 4, 1935), p. 10; Project History—Grand Coulee Dam, 1935, p. 120.

73. *Wenatchee Daily World* (May 1, 1935), p. 14; Project History—Grand Coulee Dam, 1935, p. 131.
74. Kugelman to O'Sullivan, August 17, 1935, Columbia Basin Commissions Mss.
75. Kugelman to O'Sullivan, September 28, 1935, *ibid.*
76. *Wenatchee Daily World* (September 3, 1935), p. 10; *Spokesman-Review* (October 24, 1935), O'Sullivan Mss.; "The Grand Coulee Dam," *The Engineer* 165 (February 4 and 11, 1938): 155.
77. Dill to Martin, July 11, 1934, Clarence D. Martin papers, Papers of the Governors.
78. *Spokesman-Review* (November 11, 1934), Woods Mss.
79. Bohannan to Woods, November 15, 1934, *ibid.*
80. *Wenatchee Daily World* (April 26, 1935), p. 1; (June 5, 1935), p. 7.
81. *Ibid.* (August 17, 1934), p. 18.
82. *Ibid.* (October 8, 1934), p. 3.
83. Kugelman to O'Sullivan, October 20, 1934, Columbia Basin Commissions Mss.
84. *Wenatchee Daily World* (January 21, 1935), p. 10.

## Notes for Chapter Eight

1. Herbert Corey, "Kilowatts May Go Begging," *Public Utilities Fortnightly* 13 (March 29, 1934): 384.
2. "Grand Coulee Dam and the Columbia Basin," *Pacific Builder and Engineer* 39 (April 14, 1934): 4, editorial.
3. Corey, "Kilowatts May Go Begging," p. 379.
4. George Sundborg, *Hail Columbia: The Thirty Year Struggle for Grand Coulee Dam* (New York: The Macmillan Company, 1954), p. 225.
5. Dook Stanley, "Grand Coulee—A Great Power Threat," *Barron's* 13 (December 25, 1933): 3. *Business Chronicle* wrote, "Ostensibly a plan 'to provide work and lift the depression,' the raiders of the public treasury don't go far enough. Why not take the proposed $200,000,000 to $250,000,000 which the completed Grand Coulee project is supposed to cost and spend it right away; and, then turn around and squander an equal amount to be provided by taxes or bonds or printing-press money to tear the dam [*sic*] thing down as soon as it is completed? That would put twice the amount of money in circulation, [and] provide twice as much work." See *Business Chronicle* (September 20, 1933), typed copy of article found in Woods Mss.
6. *Chicago Sunday Tribune* (July 29, 1934), Woods Mss.
7. Owen P. White, "Spare That Desert!" *Collier's* 93 (June 16, 1934): 10. Closer to home the *Bremerton News Searchlight* wrote: "Probably nothing more visionary or impractical or ruthless with waste of public funds has come in the mass of schemes to use taxpayers' money than the Coulee Dam project. Originally the plan was to make possible the development of needed lands; practical businessmen thought the expenditure of twenty-three millions of dollars would accomplish this. Then it was discovered land was not required. When things got so that it seemed wise to start in wasting public money

someone revived the Coulee Dam. They not only revived it, but added millions to the cost of a project exclusively for use in generating power, something not before even comprehended, and an additional hundred feet or so and millions to more than one hundred. And there it stands now." *Bremerton News Searchlight* (November 6, 1934), Woods Mss.

8. Report by Alvin F. Darland on conferences held in Denver, December 2 and 4, 1933, Columbia Basin Commissions Mss.

9. John W. Kelly as quoted in Sundborg, *Hail Columbia*, p. 269.

10. In discussing the Tennessee Valley Authority Roosevelt wrote that it was "charged with the broadest duty of planning . . . for the general social and economic welfare of the Nation. . . . It is time to extend planning to a wider field." See Roy Talbert, Jr., *FDR's Utopian: Arthur Morgan of the TVA* (Jackson: University Press of Mississippi, 1987), p. 128; and Franklin D. Roosevelt, *On Our Way* (New York: John Day Co., 1934), pp. 53-56. "Columbia Basin Tracts for Mid-West Farmers is U.S. Plan," reported the *Wenatchee Daily World* (August 13 1934), p. 1. Impoverished farmers from the drought-stricken Midwest were to be moved to the fertile Columbia Basin.

11. Marquis James, "The National Domain and the New Deal: An Interview by Marquis James with Harold L. Ickes, Secretary of the Interior," *Saturday Evening Post* 206 (December 23, 1933): 10; also see "Ickes Lifts His Ban," *Pacific Builder and Engineer* 39 (December 16, 1933): 3, editorial.

12. "Grand Coulee—The Key to Columbia River Power," *Electrical West* 72 (February 1934): 19; *Wenatchee Daily World* (February 2, 1934), p. 14; (January 29, 1934), p. 8; "Sand and Gravel May be Hauled Six Miles to Grand Coulee Dam by Rail or Tram," *Pacific Builder and Engineer* 39 (February 3, 1934): 25.

13. C. C. Dill to O'Sullivan, February 4, 1934, O'Sullivan Mss.

14. Then Dill accused Ickes of being unable to keep up with his work. Ickes was already upset about charges, emanating largely from J. D. Ross of Seattle, that Dill and his ex-secretary Frank Bell, then Commissioner of Fisheries, secretly owned land in the project area. Consequently Ickes had his assistants investigate Dill's private affairs. Suddenly, on July 14, 1934, Dill announced that he would not run for re-election in November. The liberal Lewis Baxter Schwellenbach won Dill's seat and held it for one term. Dill returned to his private law practice in Spokane. James O'Sullivan never forgave Dill for his activity concerning Grand Coulee Dam. In 1940 Dill ran for governor. O'Sullivan opposed him with a vengeance. Representative Charles Leavy wrote to O'Sullivan, "I am deeply concerned about a rumor which has reached me to the effect that you are opposing Senator Dill for the Governorship and supporting Mayor [Arthur] Langlie. The rumor is to the effect that you feel that Dill is not the friend of Grand Coulee. I know your deep devotion to this undertaking and the exceptional part that you have played in its being brought into existence. I know, too, that some years ago you expressed some concern about Senator Dill's attitude on the High Dam. You will recall that I told you then that Senator Dill was an unquestioned friend of the project. During the four years that I have been in Congress and the Senator has been here in the private practice of law, he has consulted with me on many, many occasions and it was due to his great legislative experience and splendid advice given me that I was able to accomplish what I did on behalf of Grand Coulee. I can state unqualifiedly that there is no better friend of this project or no one that would preserve it for the people to a greater degree than Dill if he were Governor of our state. I hope that the report that came to

me was unfounded, and if not, I want to urge you to refrain from questioning the good faith of Senator Dill, because it would be like questioning my own." The flap over Dill became somewhat public. Dill's opponent, Arthur Langlie, charged that Dill had not only not helped in the struggle for Grand Coulee Dam, but that he had also fought against the high dam. Albert S. Goss, former head of the state grange, announced that Dill had indeed been cool toward Grand Coulee and that the real support for the dam came from Sam B. Hill. In the end, Langlie won the election. How much the debate over Grand Coulee played in the victory is debatable. Dill ran unsuccessfully for Congress in 1942. He served as an appointed member of the Columbia Basin Commission from 1945 until 1948 and he died in 1978. See Leavy to O'Sullivan, October 1, 1940 and James Edward O'Sullivan, "The Struggle for the Grand Coulee Dam," p. 58, both in O'Sullivan Mss.; Sundborg, *Hail Columbia,* pp. 271-276; *Wenatchee Daily World* (July 14, 1934), p. 8; (October 7, 1940), p. 6; (October 31, 1940), p. 1. Also see Richard W. Larsen, "How C. C. Dill Interested F. D. R. in Grand Coulee Dam," *The Seattle Times* (January 3, 1971), magazine section, pp 12-13.

15. The composition of the commission had recently changed—E. F. Banker, Rufus Woods, and McGovern remained but new were Judge John C. Bowen and Senator W. G. Ronald.

16. O'Sullivan, "The Struggle for the Grand Coulee Dam," p. 54-56.

17. *Ibid.*

18. "The Secretary [wrote O'Sullivan about himself] of the Commission showed the President the exhibit at Soap Lake and then said, 'Mr. President, my name is James O'Sullivan,' and the President replied, 'You don't need to tell me who you are. I know all about you. You are doing fine work on this Project; keep it up; I am with you.' " O'Sullivan, "The Struggle for Grand Coulee Dam," p. 57. Also see Sundborg, *Hail Columbia,* p. 267. Sundborg reports the conversation with Roosevelt slightly differently and comments that O'Sullivan scattered vegetables in the laps of the people in Roosevelt's car.

19. In 1948 Ickes wrote to Ralph E. Dyar of the *Spokesman-Review,* "It was on this trip that I began to understand the importance and size of this project. All of the route from Ephrata to Grand Coulee was placarded with 'we want the high dam.' Still I had no conviction. But when the President and I, from a car well up the side of Coulee, followed Frank Banks' eloquent forefinger and were able to have some appreciation of the vision that he was pointing out, I began to see with the eye of understanding. When I got back to Ephrata, I was committed in my own mind to the high dam, although I did not say so then, even to the President." See Daniel Miller Ogden, "The Development of Federal Power Policy in the Pacific Northwest" (unpublished Ph.D. dissertation, University of Chicago, June, 1949), pp. 272-273.

20. *Wenatchee Daily World* (August 11, 1934), p. 1.

21. *Spokesman-Review* (August 15, 1934), O'Sullivan Mss.; *Grant County Journal* (August 17, 1934).

22. In a meeting in Seattle Ross declared that Skagit installed horsepower would cost $88 while Grand Coulee installed horsepower would be $495, or five times as much. Harrison W. Mason to James O'Sullivan, November 24, 1934, Columbia Basin Commissions Mss. Also see Sundborg, *Hail Columbia,* p. 264; Paul C. Pitzer, *Building the Skagit: A Century of Upper Skagit Valley History 1870-1970* (Lake Grove, Ore.: Galley Press, 1978), p. 57.

23. Holden to Woods, October 22, 1934, typed copy in Woods Mss. For further background see Martin T. Farris and Roy J. Sampson, *Public Utilities: Regulation, Management and Ownership* (Boston: Houghton Mifflin Co., 1973), p. 266; and William O'Dell Sparks, "J. D. Ross and Seattle City Light 1917-1932" (unpublished masters thesis, University of Washington, 1964), passim.
24. Woods to Ross, November 27, 1934, Woods Mss.
25. O'Sullivan to Woods, February 12, 1935, O'Sullivan Mss.
26. James Rorty, "Grand Coulee," *Nation* 140 (March 20, 1935), p. 329.
27. *Wenatchee Daily World* (January 17, 1935), p. 1.
28. The change meant that although one excavation would accommodate the entire job, MWAK had to increase its size by about two million cubic yards with most of that on the river's east side. The Bureau altered the MWAK contract so that it described a much wider foundation 177 feet high rather than the 297-foot finished structure. Designer John L. Savage and his assistant, Sinclair O. Harper, rushed final drawings for the huge dam while Leslie N. McClellan worked on specifications for the electrical equipment. On January 30, 1936, Ickes revealed a revised MWAK contract that allowed the contractor $7 million more for the new work, mostly the placement of an additional 3,400,000 cubic yards of concrete and the increased overburden removal. "Grand Coulee Excavating Program Modified to Meet Changed Design," *Western Construction News* 10 (September 1935): 253; *Wenatchee Daily World* (June 8, 1935), p. 1; (January 31, 1936), p. 13; (June 6, 1935), p. 1; "Agreement Reached on Grand Coulee Contract," *Reclamation Era* 26 (March 1936): 65; Project History—Grand Coulee Dam, 1935, p. 19; *Spokesman-Review* (June 9, 1935), O'Sullivan Mss; Charles H. Carter, "Change of Plan for Grand Coulee Dam," *Reclamation Era* 25 (July 1935): 135.
29. Project History—Grand Coulee Dam, 1935, pp. 25, 27.
30. With a copy of the report that went to Elwood Mead, engineer Raymond F. Walter included a letter that said, in part, "The change from the low dam to the high dam at Grand Coulee would make it possible to purchase the large turbines, generators, transformers and other power plant machinery as soon as the necessary specifications would be prepared and this would afford a large amount of employment in the factories in the East and the construction of the main canals and laterals of the irrigation development would afford a large amount of local employment because such construction can be carried on simultaneously at a large number of points." Ogden, "The Development of Federal Power Policy in the Pacific Northwest," p. 213; Walter to Mead, December 22, 1934, BOR Denver Mss.
31. Mead to Ickes, December 27, 1934, BOR Denver Mss.
32. Donald C. Swain, "The Bureau of Reclamation and the New Deal, 1933-1930," *Pacific Northwest Quarterly* 61 (July 1970): 139. Also see Donald Worster, *Rivers of Empire: Water, Aridity, and the Growth of the American West* (New York: Pantheon Books, 1985), pp. 184-187.
33. Woods to Martin, June 1, 1935, Woods Mss.
34. Hill to Woods, May 23, 1935, *ibid.*
35. *Spokesman-Review* (May 30, 1935), O'Sullivan Mss.
36. In a letter to Governor Martin, Woods wrote, "There has been plenty of propaganda here to kill off the entire project. In fact, that very thing came up before a committee of the leaders, and a suggestion was then made to drop the whole Grand Coulee Dam, after ten million dollars has been expended. So you see we are up against the same old thing that we have had for years. This matter of

speculation in lands I believe has been over emphasized, although there are quite a number here in Washington who have been buying the land at $20.00 an acre and who expect to get $1,000.00 an acre out of it. Dr. Mead tells me that the boy who brought him down in the taxi the other morning was one of these." Woods to Martin, June 1, 1935, Woods Mss. Also see *Wenatchee Daily World* (June 7, 1935), p. 1; (June 10, 1935), p. 3.

37. O'Sullivan to Chambers of Commerce, March 1, 1935, Columbia Basin Commissions Mss.; Bruce Mitchell, *Flowing Wealth the Story of Water Resource Development in North Central Washington* (Wenatchee: *Wenatchee Daily World*, 1967), p. 40; *Wenatchee Daily World* (March 23, 1935), p. 1.

38. 295 *Statutes at Large* U.S. 174 (1935) *United States v Arizona*. Sam Hill wrote to James O'Sullivan, "As a result of the Parker Dam decision it is thought by the Public Works Administration that the Grand Coulee and other dams in navigable waters may be involved in some legal difficulty. The prevailing opinion as to this appears to be that there is authority for the Grand Coulee dam but in order to be on the safe side it is well to have specific authority from Congress therefore. Inasmuch as it was necessary to have legislation authorizing the Parker Dam the attorneys in the Department of the Interior prepared a bill for that purpose and included it in provisions covering the Grand Coulee and other dams in navigable waters projected by the Public Workd [*sic*] Administration." Hill to O'Sullivan, May 14, 1935, Woods Mss. Also see Mitchell, *Flowing Wealth*, p. 41; Gus Norwood, *Columbia River Power for the People: A History of Policies of the Bonneville Power Administration* (Washington, D. C.: U. S. Government Printing Office, 1981), p. 38; *Wenatchee Daily World* (April 30, 1935), p. 10.

39. Ogden, "The Development of Federal Power Policy in the Pacific Northwest," pp. 215-217.

40. Historian Daniel Miller Ogden writes, "Most of the concern and controversy over the bill was centered on Parker dam, and, indeed, only Dr. Elwood Mead, Commissioner of Reclamation, who spoke for the entire Bureau but centered most of his attention on Parker dam and its problems, could be said to have in any way represented or been concerned with the interests of Grand Coulee dam at either hearing. Senator McNary of Oregon, a member of the Senate Committee on Commerce which heard the testimony, attended, but made no effort to get Grand Coulee into the proceedings." See Ogden, "The Development of Federal Power Policy in the Pacific Northwest," p. 217. For a different view of McNary see Richard Lowitt, *The New Deal and the West* (Bloomington: Indiana University Press, 1984), p. 160, where Lowitt credits McNary with seeing President Roosevelt and for having some responsibility for the decision of the PWA to fund Grand Coulee in the first place.

41. The railroad pension act decision, the National Recovery Act decision, and the Frazier-Lemke farm mortgage act decision all contributed to the problem.

42. *Spokesman-Review* (May 30, 1935), O'Sullivan Mss; *Wenatchee Daily World* (May 30, 1935), p. 10.

43. *Wenatchee Daily World* (June 17, 1935), p. 1; *Spokesman-Review* (June 22, 1925), O'Sullivan Mss.

44. *Wenatchee Daily World* (July 9, 1935), p. 1.

45. Radio Address by Francis D. Culkin, June 15, 1935, typed copy found in Woods Mss. Also see O'Sullivan to Goss, July 2, 1935, Columbia Basin Commissions Mss.; *Wenatchee Daily World* (July 17, 1935), p. 3.

46. Woods to Culkin, July 19, 1935, Woods Mss.
47. Hill to O'Sullivan, August 26, 1935, O'Sullivan Mss.
48. *Wenatchee Daily World* (August 20, 1935), p. 1.
49. Schwellenbach to James O'Sullivan, August 19, 1935, telegram in Columbia Basin Commissions Mss.
50. Hill to Woods, August 19, 1935, Woods Mss.
51. Project History—Grand Coulee Dam, 1935, p. 32; "Authorization, Parker and Grand Coulee Dams," *Reclamation Era* 25 (October 1935): 198.

# Notes for Chapter Nine

1. "Details of Concreting Procedure at Grand Coulee Dam," Western Construction News 11 (July 1936): 209.
2. "Concrete Placing Started on Grand Coulee Dam," *Western Construction News* 11 (January 1936): 22.
3. *Oregonian* (December 22, 1935), Dubuar Scrapbooks, #8, p. 15; *Walla Walla Sunday Union Bulletin* (June 9, 1935), Dubuar Scrapbooks, #10, p. 11; *Wenatchee Daily World* (December 2, 1935), p. 7; (December 17, 1935), p. 7; "Concrete Placing Started on Grand Coulee Dam," p. 22.
4. Project History—Grand Coulee Dam, 1935, p. 87.
5. "Manufacturing 4,500,000 c. y. of Concrete for Coulee Dam," *Pacific Builder and Engineer* 42 (January 4, 1936): 34.
6. *Post-Intelligencer* (October 29, 1935), Dubuar Scrapbooks, #8, p. 8.
7. The first cubic yard of concrete, which eventually became part of Grand Coulee Dam, was poured on October 10, 1935. At first it served as a base for the supports for the west-side mixing plant but in time the material was molded into the downstream face of the dam. See *Wenatchee Daily World* (October 10, 1935), p. 12. Also see Project History—Grand Coulee Dam, 1935, p. 19; Charles Thompson, "Grand Coulee Dam Concreting Plant Is Notable for Design Efficiency," *Western Construction News* 11 (February 1936): 31.
8. *Wenatchee Daily World* (October 5, 1935), p. 7; "Manufacturing 4,500,000 c. y. of Concrete for Coulee Dam," p. 30.
9. Forty-seven percent of the cement came from Concrete, Washington; 19 percent came from Bellingham; 13 percent from Metaline Falls; 11 percent from Grotto; 10 percent from Spokane. *Wenatchee Daily World* (August 21, 1935), p. 8; (August 26, 1935), p. 1; (September 9, 1935), p. 8.
10. *Ibid.* (December 24, 1935), p. 7.
11. The entire west-side foundation consumed 42,000 sacks of concrete. In January 1936 grouting started on the east side of the river. Project History—Grand Coulee Dam, 1935, p. 23; "Progress Notes from the Grand Coulee Dam Project," *Western Construction News* 11 (March 1936): 89; *Seattle Star* (July 30, 1934), Dubuar Scrapbooks, # 9, p. 36; *Spokesman-Review* (December 25, 1935), O'Sullivan Mss.; *Wenatchee Daily World* (January 24, 1936), p. 13.
12. "Concrete Mixing and Placing at Grand Coulee Dam," *Engineering News-Record* 116 (January 23, 1936): 120; *Wenatchee Daily World* (December 19, 1935), p. 1 including picture; "Grand Coulee Dam, U.S.A.," *The Engineer* (British) 171 (March 28, 1941): 208; "Steel Trestles To Be Imbedded in Concrete at Grand Coulee Dam," *Steel* 98 (May 11, 1936): 58.

13. *Wenatchee Daily World* (November 26, 1935), p. 6.
14. Kugelman to O'Sullivan, November 30, 1935, Columbia Basin Commissions Mss. Also see Project History—Grand Coulee Dam, 1935, p. 53.
15. *Spokesman-Review* (November 23, 1935), O'Sullivan Mss.; *Wenatchee Daily World* (November 29, 1935), p. 1; (November 30, 1935), p. 6.
16. Bruce Mitchell, *Flowing Wealth: The Story of Water Resource Development in North Central Washington 1870-1950* (Wenatchee: *Wenatchee Daily World*, March 6, 1967, reprinted January 1980), p. 42.
17. *Wenatchee Daily World* (December 6, 1935), p. 1.
18. *Spokesman-Review* (December 7, 1935), O'Sullivan Mss.; *Oregonian* (December 8, 1935), Dubuar Scrapbooks, #8, p. 13. Seven months earlier, in May, the State of Washington began collecting a sales tax. One dollar bought 500 aluminum tokens that customers used to pay the increment added to each purchase.
19. "Concrete Placing Started on Grand Coulee Dam," p. 22.
20. *Oregonian* (December 8, 1935), Dubuar Scrapbooks, #8, p. 13.
21. *Wenatchee Daily World* (December 9, 1935), p. 9; (December 14, 1935), p. 12.
22. *Ibid.* (December 20, 1935), p. 12; "Concreting Block 40 at Grand Coulee Dam," *Western Construction News* 11 (March 1936): 92.
23. Project History—Grand Coulee Dam, 1936, p. 15.
24. "Manufacturing 4,500,000 c. y. of Concrete for Coulee Dam," p. 30.
25. *Wenatchee Daily World* (February 26, 1936), p. 7.
26. "Progress Notes from the Grand Coulee Dam Project," p. 89.
27. *Wenatchee Daily World* (January 27, 1936), p. 1.
28. For a biography of Mead see James R. Kluger, *Turning on Water with a Shovel: The Career of Elwood Mead* (Albuquerque: University of New Mexico Press, 1992).
29. *MWAK Columbian* 2 (April 17, 1936): 1.
30. Four months later the State of Washington attempted to levy an inheritance tax on Mason's $2 million estate. His widow fought the action claiming that they were residents of Kentucky. The case drug on for some months with the state finally realizing only a portion of its claim. *Wenatchee Daily World* (April 18, 1936), p. 1; *Spokesman-Review* (April 19, 1936), O'Sullivan Mss.
31. Daniel McFarland, "Cleaning Concrete Surfaces to Ensure Bond," *Engineering News-Record* 118 (June 17, 1937): 922.
32. For a similar discussion in connection with Hoover Dam see Joseph E. Stevens, *Hoover Dam: An American Adventure* (Norman: University of Oklahoma Press, 1988), p. 221.
33. *Wenatchee Daily World* (December 11, 1936), p. 18.
34. For fuller details see "Details of Concreting Procedure At Grand Coulee Dam," pp. 209-213; H. B. Hodgins, "Electrically Cured Concrete for Grand Coulee Dam," *Electric Journal* 33 (April 1936): 189-190; "Half-Million Yards of Concrete Poured at Grand Coulee," *Pacific Builder and Engineer* 42 (June 6, 1936): 30; "Concrete Dispatching System Used at Grand Coulee Dam," *Engineering News-Record* 117 (September 10, 1936): 377-379; "Concrete Cooling—Memorandum Report," May 22, 1936, BOR Denver Mss.
35. "Grand Coulee Dam, U.S.A.," p. 208.
36. The *Wenatchee Daily World* identified August 12, 1936 as the day the one millionth cubic yard of concrete was placed. The Bureau of Reclamation project history for Grand Coulee Dam gave the date as August 14. See

*Wenatchee Daily World* (August 12, 1936), p. 7; Project History—Grand Coulee Dam, 1936, p. 18.

37. Kugelman to O'Sullivan, April 25, 1936, Columbia Basin Commissions Mss.
38. Unidentified clipping found in the scrapbooks of Elma Terou, Coulee Dam City Hall Archives, Coulee Dam City Hall, Coulee Dam, Washington.
39. Grant Gordon, "Freezing Arch Across Toe of East Forebay Slide at Grand Coulee Dam," *Reclamation Era* 27 (January 1937): 12.
40. Kugelman to O'Sullivan, August 1, 1936), Columbia Basin Commissions Mss.
41. The process had been pioneered by a Prussian named F. H. Poetsch. Grant Gordon, Associate Engineer, to Frank Banks, "Freezing Arch Across Toe of East Forebay Slide—Grand Coulee Dam," undated report, Banks Mss.
42. *Spokesman-Review* (October 21, 1936), O'Sullivan Mss.; *Wenatchee Daily World* (August 19, 1936), p. 7.
43. Kugelman to O'Sullivan, October 3, 1936, Columbia Basin Commissions Mss. Also see Grant Gordon, "Freezing Arch Across Toe of East Forebay Slide," pp. 12-16; Grant Gordon, "The Use of Refrigeration in Building the Grand Coulee Dam," *Refrigerating Engineering* 33 (January-February, 1937): 13-16; Grant Gordon, "Arch Dam of Ice Stops Slide," *Engineering News-Record* 118 (February 11, 1937): 211-215; "Freezing a Dam," *Power Plant Engineering* 41 (July 1937): 433-434; F. D. McHugh, "Huge 'Ice Pack' Saves Dam," *Scientific American* 156 (February 1937): 403.
44. *Seattle Daily Journal of Commerce* (January 29, 1937), Dubuar Scrapbooks, #8, p. 50.
45. Grant Gordon, Associate Engineer, to Frank Banks, "Freezing Arch Across Toe of East Forebay Slide—Grand Coulee Dam," undated report, Banks Mss.
46. *Wenatchee Daily World* (April 20, 1937), p. 7.
47. "Progress Notes from the Grand Coulee Dam Project," p. 89; *Wenatchee Daily World* (February 14, 1936), p. 7.
48. *Wenatchee Daily World* (August 5, 1936), p. 7.
49. "Method of Sounding for the Grand Coulee Cofferdams," *Western Construction News* 11 (October 1936): 332-333.
50. Donald O. Nelson, "Cross River Cofferdams at Coulee," *Pacific Builder and Engineer* 42 (March 6, 1937): 34.
51. *Wenatchee Daily World* (November 28, 1936), p. 12.
52. Unlike the west side of the river, where the pilings had to touch bedrock wherever possible, drilling indicated that the bed of the Columbia contained dense, impervious, stable material which would support core-wall cofferdams without allowing too much seepage. See "MWAK Starts Diversion of the Columbia," *Pacific Builder and Engineer* 41 (December 5, 1936): 40; "Diverting the Columbia At Grand Coulee with Timber Cribs and Gravel Fills," *Western Construction News* 11 (December 1936): 387.
53. "MWAK Starts Diversion of the Columbia," p. 40.
54. *Wenatchee Daily World* (September 24, 1936), p. 1.
55. *Ibid.* (August 12, 1936), p. 7.
56. "Channel Excavation at Grand Coulee," *Western Construction News* 11 (February 1937): 70.
57. *Spokesman-Review* (November 8, 1936), O'Sullivan Mss.
58. *MWAK Columbian* (October 30, 1936); (November 6, 1936).
59. The event had a symmetry to it as the first pouring of concrete had been on Thanksgiving Day one year before. Kugelman to O'Sullivan, November 28, 1936, Columbia Basin Commissions Mss.

60. *Wenatchee Daily World* (December 15, 1936), p. 1.
61. *Grant County Journal* (December 18, 1936), pp. 1, 3. Also see *Post-Intelligencer* (December 18, 1936), Dubuar Scrapbooks, #8, p. 15.
62. As Quoted in *Spokesman-Review* (November 21, 1936), O'Sullivan Mss. Also see *MWAK Columbian* (November 20, 1936): 4.
63. Project History—Grand Coulee Dam, 1937, pp. 11, 43; *Wenatchee Daily World* (December 30, 1936), p. 7; (January 4, 1937), p. 1; (January 9, 1937), p. 12; (January 26, 1937), p. 7; (February 12, 1937), p. 1; (February 20, 1937), p. 7.
64. *Wenatchee Daily World* (April 3, 1936), p. 11; (April 7, 1936), p. 7.
65. R. F. Walter to Robert C. Weaver, March 5, 1936, BOR Denver Mss.
66. Banks to Construction Engineer, January 17, 1936; Walter to Banks, January 22, 1936, both in *ibid.*
67. Walter to Banks, February 14, 1936, *ibid.*
68. Banks to Walter, February 21, 1936, *ibid.*
69. Page to Banks, March 21, 1936, *ibid.*
70. Banks to Walter, March 27, 1936, *ibid.*
71. Walter to Banks, March 30, 1936, *ibid.*
72. Walter to Banks, April 7, 1936, *ibid.* Secretary of the Interior Harold L. Ickes had previously been involved with the Six Companies at Boulder Dam. He criticized their failure to hire black workers. Very few blacks ever secured jobs at the Boulder Project. See Stevens, *Hoover Dam*, pp. 176-177. Also see T. H. Watkins, *Righteous Pilgrim: The Life and Times of Harold L. Ickes 1874-1952* (New York: Henry Holt and Company, 1990), pp. 289-290, 641-653, and Kluger, *Turning on Water with a Shovel*, p. 145.
73. Project History—Grand Coulee Dam, 1937, p. 72; L. Vaughn Downs, *The Mightiest of Them All: Memories of Grand Coulee Dam* (Fairfield, Wash.: Ye Galleon Press, 1986), p. 50.
74. Interview with Ida Bartles, June 15, 1988, Grand Coulee, Washington.
75. Banks to Page, July 8, 1937, BOR Denver Mss.
76. In defense of Beery, Frank Banks wrote a long, confidential letter to Reclamation Commissioner John C. Page. In it he commented, "He [H. D. Huxley who represented the Department of Labor] also called me and told me that he had had a man here for two weeks investigating Captain Beery and that Beery was going to be let out. I asked him what the charges were and he stated that he drank intoxicating liquor, didn't pay his debts, and on one occasion sent non-union men to a contractor, who had asked for union men. Beery does have some difficulty in getting along on the $150 per month that he gets and I have heard of a couple of times that he got pretty well lit up . . . ; but in the two years that we have been located together on this job, I have never seen him take a drink, nor seen him when he had the appearance of having been drinking.

"The unions all over the country apparently feel that this is a psychological time to improve their positions and are making the most of it, and we can probably expect considerable activity. I understand that they are particularly anxious to obtain control of the Spokane, Coulee Dam, Yakima, and Seattle office of the National Reemployment Service, because of the volume of placements being handled through these offices.

"I am not writing you this because of having any prejudices one way or another for that is not the case. I am in favor of the organization of labor, in fact I once belonged to the Iron Molders' Union; but I am not in sympathy with some of their racketeering tactics. I have carefully studied the rights of the contractor and labor, both union and non-union, and will endeavor to see that they are properly administered; but I wanted you to know what was going on since some phases of the matter seem to be reaching you."

Banks was more surprised when he learned that Commissioner Page received letters from Representative Knute Hill and Marion A. Zioncheck about labor problems at Grand Coulee, and that Zioncheck urged the Bureau to fire Banks for anti-labor attitudes. In a letter to Banks, Page wrote that he had met with Interior Secretary Ickes to discuss organizing an impartial investigation, and he continued,

"Upon the advice of the Secretary I also called on Mr. Burr and Mr. Huxley of the Department of Labor, and they reaffirmed their assurance that the resignation of Mr. Sullivan and Mr. Beery was requested because of indiscretions in administration and not because of the union resolution. However, they said the public expression of the complaint probably hastened the administrative action. They seemed to have the same idea, although not expressed entirely, that the unions proposed to organize Grand Coulee, and that the removal of yourself and the NRS officers was proposed as a necessary step because of a pronounced feeling that the present administration was not sufficiently friendly to make this possible. We have had a flood of correspondence concerning this matter from the various labor organizations and from the Congressmen representing that district, and it apparently is assuming sufficient importance to justify the proposed investigation by the Secretary."

Beery found a job at Grand Coulee looking after non-union workers. Okanogan and Douglas county officials hired him in this capacity. Grant County refused to join in the group. County authorities claimed that the National Reemployment Service favored union employees from outside of the area around the dam and hired them over locals who were not union members. See Frank Banks to John Page, confidential, March 4, 1936; Walters to Zioncheck, March 6, 1936; Page to Banks, confidential, March 10, 1936, all in BOR Denver Mss.; *Wenatchee Daily World* (April 8, 1936), p. 1; (March 23, 1936), p. 1.
77. *Wenatchee Daily World* (June 4, 1936), p. 1; (June 6, 1936), p. 7; (June 8, 1936), p. 7.
78. *Ibid.* (August 6, 1936), p. 7.
79. *Ibid.* (February 3, 1936), p. 12; (February 4, 1936), p. 10.
80. *Ibid.* (September 5, 1936), p. 6; (October 27, 1936), p. 7; Project History—Grand Coulee Dam, 1936, p. 19.
81. Page to Seward, January 9, 1937, BOR Denver Mss.
82. *Wenatchee Daily World* (May 31, 1937), p. 7; Project History—Grand Coulee Dam, 1937, p. 70.
83. Banks to Walter, June 28, 1937, and June 26, 1937, BOR Denver Mss.; *Wenatchee Daily World* (June 29, 1937), p. 12.
84. *MWAK Columbian* (July 30, 1937), p. 1, Banks Mss. T. J. Walsh signed the letter.
85. *Wenatchee Daily World* (July 30, 1937), p. 1; Project History—Grand Coulee Dam, 1937, p. 14.
86. *Wenatchee Daily World* (July 31, 1937), p. 1.
87. Walter to Banks, July 31, 1937, BOR Denver Mss.; *Spokesman-Review* (August 1, 1937), O'Sullivan Mss.
88. *Wenatchee Daily World* (August 2, 1937), p. 1.

89. *Ibid.* (August 3, 1937), p. 7; (August 6, 1937), p. 1; *Spokesman-Review* (August 6, 1937), BOR Denver Mss. The fact that newspaper clippings concerning this event appeared in the files of the Bureau of Reclamation indicate how closely officials in Denver and Washington, D.C., followed the uproar at Grand Coulee.

90. *Wenatchee Daily World* (August 7, 1937), p. 1; *Spokane Chronicle* (August 7, 1937), BOR Denver Mss.

91. See Charles P. Larrowe, *Harry Bridges: The Rise and Fall of Radical Labor in the U.S.* (Westport, Conn.: Lawrence Hill & Company, 1972).

92. *Spokesman-Review* (August 9, 1937), BOR Denver Mss.

93. *Ibid.*

94. *Wenatchee Daily World* (August 10, 1937), p. 7.

95. Banks to Walter, August 13, 1937), BOR Denver Mss.

96. *Wenatchee Daily World* (August 24, 1937), p. 7; (August 25, 1937), p. 1; *MWAK Columbian* (August 27, 1937), p. 2.

97. In a private letter to the general counsel for the Public Works Administration, Reclamation Commissioner Page commented, "The contractor was probably hopeful that the agreement made with the A. F. of L. would solve his difficulties. It had the opposite effect but the agitation subsided after a few days, mainly, I think, because the contractor has never proceeded under the agreement made with the A. F. of L., and, for approximately two months, the work has proceeded without troubles. In another two or three months the work under the existing contract will be completed. I am, therefore, apprehensive that a submission of the matter to the national Labor Relations Board, as proposed, might result in disturbing the present peaceful situation. As you may know, the Labor Relations Board sent a representative to the project early in August but no action, so far as I am advised, was taken by the Board itself. The wholesale dismissal of employees, predicted by the telegram of August 1, failed to materialize." Page to Foley, September 22, 1937, BOR Denver Mss. Also see "Labor Controversy at Grand Coulee," *Western Construction News* 12 (September 1937): 340.

98. Banks to Page, September 29, 1937, BOR Denver Mss.

99. "When the Lower Coffer Broke At Coulee," *Pacific Builder and Engineer* 42 (June 5, 1937): 33.

100. "Serious Leak Checked in Cofferdam at Grand Coulee," *Engineering News Record* 118 (April 22, 1937): 595.

101. *Wenatchee Daily World* (March 19, 1937), p. 20; (March 20, 1937), p. 7; (April 6, 1937), p. 7; (November 12, 1961), p. 1; Project History—Grand Coulee Dam, 1937, p. 12.

102. "Contractors Win River Battle," *Engineering News Record* 119 (July 1, 1937): 13.

103. *MWAK Columbian* (April 2, 1937), p. 1; *Wenatchee Daily World* (March 27, 1937), p. 1; John H. D. Blanke, "Building Highlights of Grand Coulee," *The International Engineer* 73 (December 1938): 200. Years later, in 1983, Larry Roskowyk, who had worked at the dam in the 1930s, reported to the *Wenatchee World* that bentonite was not used at Grand Coulee to stop the leak in the cofferdam. See *Wenatchee World* (July 8, 1983), p. A34.

104. *Wenatchee Daily World* (May 29, 1937), p. 12; *Spokesman-Review* (May 29, 1937), O'Sullivan Mss.

105. Project History—Grand Coulee Dam, 1937, p. 14; *Wenatchee Daily World* (August 16, 1937), p. 1.

106. Project History—Grand Coulee Dam, 1937, p. 15; *Grant County Journal* (October 15, 1937), p. 1; Fred K. Ross, "CBI Prepares to Pour Concrete," *Pacific Builder and Engineer* 43 (June 4, 1938): 52.
107. Grant County Historical Society, *Memories of Grant County*, vol. 3 (March 1982), Hu Blonk interview, titled "Grand Coulee," recorded Nov. 10, 1981, p. 176.
108. *Wenatchee Daily World* (August 30, 1937), p. 7.
109. *Wenatchee Daily World* (August 16, 1937), p. 7; "Bids Close December 10 for Completion of Coulee Dam," *Pacific Builder and Engineer* 42 (November 6, 1937): 53.
110. Project History—Grand Coulee Dam, 1937, p. 16.
111. *Wenatchee Daily World* (September 15, 1937), p. 20.
112. *MWAK Columbian* (November 25, 1937), p. 1. Also see *Wenatchee Daily World* (November 26, 1937), p. 1.
113. *Wenatchee Daily World* (September 14, 1937), p. 7.
114. *Ibid.* (December 6, 1937), p. 7.
115. *Ibid.* (January 12, 1938), p. 7.
116. S. E. Hutton, "New Contract Covers Completion of the Grand Coulee Dam," *Reclamation Era* 27 (December 1937): 277; *Wenatchee Daily World* (December 3, 1937), p. 10; (December 18, 1937), p. 5.
117. *Spokesman-Review* (November 30, 1937), O'Sullivan Mss.
118. *Ibid.* (January 29, 1938), O'Sullivan Mss.
119. The sixty deaths total included all work done at the dam to this date, including overburden removal, railroad, and the MWAK contract. Total expenditures for this work by this time had exceeded $100 million. So the number of deaths for total work was lower than then "one death per $1 million" anticipation, although for MWAK work only the death figure slightly exceeded this rule-of-thumb figure. Forty-five men working with the MWAK Company died, one with Crick and Kuney, two with the Indian Service putting up telephone lines, and a number of others with various subcontractors. Twelve men died from falls, eight from falling objects hitting them, four due to machinery such as trucks or tractors, two from explosives, four were electrocuted, four were caught and crushed in machines, four drowned, two were run over by trains, three died in slides, two were caught in conveyor belts, and two died from burns. See *Wenatchee Daily World* (October 8, 1938), p. 12.
120. "On the Grand Coulee Dam Project the Contractor Emphasizes the Motto: Safety Pays," *Western Construction News* 12 (April 1937): 125.
121. "Aggregate Output Problems at Grand Coulee Project," *Western Construction News* 12 (November 1937): 431.
122. "Manufacturing 4,500,000 c. y. of Concrete for Coulee Dam," p. 34.
123. C. D. Riddle, "Construction Plant at Grand Coulee Dam," *Civil Engineering* 6 (October 1936): 639.
124. *Wenatchee Daily World* (February 21, 1938), p. 7. MWAK finished exactly twelve months and twelve days early. See Project History—Grand Coulee Dam, 1938, p. 39.
125. Project History—Grand Coulee Dam, 1938, p. 3.
126. "Consolidated Builders Tackle Work of Completing Coulee Dam," *Pacific Builder and Engineer* 43 (April 2, 1938): 38.
127. Project History—Grand Coulee Dam, 1938, p. 56.
128. *Ibid.*, 1940, p. 23; *Wenatchee Daily World* (January 11, 1940), p. 10.

## Notes for Chapter Ten

1. As quoted in Charles Henning, *The Wit and Wisdom of Politics* (Golden, Col.: Fulcrum, Inc., 1989), p. 11.
2. William L. Lang in Michael P. Malone, eds., *Historians and the American West* (Lincoln: University of Nebraska Press, 1983), p. 270.
3. Fred O. Jones, *Grand Coulee from "Hell to Breakfast," A Story of the Columbia River from Molten Lavas and Ice to The Grand Coulee Dam* (Portland: Binfords and Mort, 1947), p. 42.
4. Roosevelt may have welcomed the Supreme Court action that brought about the need for congressional approval for Grand Coulee. It allowed him to put more of the relief money directly into smaller projects with immediate results, as advocated by Harry Hopkins, rather than into the large long-term projects backed by Harold Ickes. The two advisors argued with each other and with the President over which avenue best worked to assuage the affects of the depression, and Ickes resigned (never accepted) more than once when he felt Hopkins had received more money than had the PWA. The action of the Congress allowed Congress to back the large projects while the administration funded the day-to-day relief efforts. See Robert E. Sherwood, *Roosevelt and Hopkins: An Intimate History* (New York: Grosset & Dunlap, 1948), pp. 68-71, 78-79, 202; T. A. Watkins, *Righteous Pilgrim: The Life and Times of Harold L. Ickes 1874-1952* (New York: Henry Holt and Company, 1990), pp. 410-429; *Wenatchee Daily World* (September 11, 1935), p. 1; (September 13, 1935), p. 1.
5. Richard Lowitt, *The New Deal in the West* (Bloomington: Indiana University Press, 1984), p. 91.
6. *Wenatchee Daily World* (January 6, 1936), p. 1. The *World* reported that $36 million had been allocated for Grand Coulee to this point. Of the entire $64 million request, the *World* reported that $30,810,000 would go to work in Utah, Oregon, Washington, and Idaho.
7. In a letter to James O'Sullivan, Representative Sam Hill wrote, "Some of our western Congressmen become rebellious to think they are not getting their share of the pot." Sam B. Hill to James O'Sullivan, March 3, 1936, O'Sullivan Mss.
8. Donald C. Swain, "The Bureau of Reclamation and the New Deal, 1933-1940," *Pacific Northwest Quarterly* 61 (July 1970): 142; *Wenatchee Daily World* (March 26, 1936), p. 1.
9. Murray Morgan, *The Dam* (New York: Viking Press, 1954), p. 29.
10. *Wenatchee Daily World* (March 27, 1936), p. 1; Bruce Mitchell, *Flowing Wealth: The Story of Water Resource Development in North Central Washington* (Wenatchee: *Wenatchee Daily World*, March 6, 1967), p. 44.
11. James O'Sullivan, "The Struggle for Grand Coulee Dam," p. 64, typed manuscript in O'Sullivan Mss.
12. K. P. Sexton to Rufus Woods, April 8, 1936, Woods Mss.
13. *Wenatchee Daily World* (May 20, 1936), p. 1; (May 21, 1936), p. 1.
14. *Grant County Journal* (June 19, 1936); *Wenatchee Daily World* (June 16, 1936), p. 5
15. "Grand Coulee Wins," *Pacific Builder and Engineer* 42 (July 4, 1936): 13, editorial.
16. Walter A. Averill to James O'Sullivan, July 8, 1936, O'Sullivan Mss.
17. *Spokesman-Review* (May 26, 1936), *ibid.*

18. O'Sullivan to Hill, May 12, 1936, *ibid.*
19. The denial became official on July 2. Project History—Grand Coulee Dam, 1936, p. 18.
20. Mitchell, *Flowing Wealth*, p. 38; Stoutmeyer to James O'Sullivan, April 3, 1937, BOR Denver Mss.
21. *Spokesman-Review* (September 15, 1936), O'Sullivan Mss.
22. *Wenatchee Daily World* (November 5, 1936), p. 1.
23. *Oregonian* (November 29, 1936), Dubuar Scrapbooks, #8, p. 41. Charles Leavy, as part of the Democratic sweep, won the fifth district seat in the House, replacing Sam Hill.
24. *Wenatchee Daily World* (January 8, 1937), p. 1.
25. Albert S. Goss to James O'Sullivan, January 11, 1937, Columbia Basin Commissions Mss.
26. *Wenatchee Daily World* (January 18, 1937), p. 1.
27. Kathleen Waugh, *Roll On, Columbia: Guide to the Records of the Columbia Basin Survey Commission and the Columbia Basin Commission 1919-1964* (Olympia, Wash.: Office of the Secretary of State, Division of Archives, 1987), p. 21.
28. O'Sullivan reports Fink as having said, "I must say that if any man in this State deserves a pension from the State, you deserve it on account of what you did for Grand Coulee Dam." O'Sullivan, "The Struggle for Grand Coulee Dam," p. 70, O'Sullivan Mss.; *Wenatchee Daily World* (August 13, 1937), p. 22.
29. *Wenatchee Daily World* (January 6, 1937), p. 1; *Spokesman-Review* (January 7, 1937), O'Sullivan Mss.
30. This is discussed in Chapter Sixteen.
31. *Wenatchee Daily World* (May 17, 1937), p. 1.
32. *Ibid.*
33. *Ibid.* (May 18, 1937), p. 1;
34. Columbia River Anti-Speculation Act, 50 *United States Statutes at Large*, p. 208-210.
35. *Wenatchee Daily World* (June 1, 1937), p. 12.
36. *Grant County Journal* (August 13, 1937), p. 1.
37. The American West has contributed more to the world's supply of myth and legend than any country or region since the Middle Ages. It was a fantasy land with a halo of legend leaving an enduring picture of what historian Robert Athearn called "a sort of sagebrush Shangri-La." It was only natural that the big dam at the ancient, and itself picturesque, Grand Coulee would jump from reality to a state of mind larger than life. For a quick background on the Paul Bunyan myth see Howard R. Lamar, ed., *The Reader's Encyclopedia of the American West* (New York: Harper & Row, 1977), p. 138. Also see Denis W. Brogan as quoted in Robert G. Athearn, *The Mythic West in Twentieth Century America* (Lawrence: University Press of Kansas, 1986), pp. 252, 257. For background see Henry Nash Smith, *Virgin Land: The American West as Symbol and Myth* (New York: Alfred Knopf, Random House, Vantage Books, 1950).
38. The Bureau of Reclamation went so far as to distribute selected pictures from construction of its various dams on a weekly basis. The best of these appeared in the *New York Times Magazine* and other Sunday supplements. See William E. Warne, *The Bureau of Reclamation* (New York: Praeger Publishers, 1973), p. 31.

39. Stewart H. Holbrook, *The Columbia* (New York: Rinehart and Company, Inc., 1956), pp. 308-309.

40. Gene Tollefson, *BPA and the Struggle for Power at Cost* (Portland: Bonneville Power Administration, 1987), p. 185.

41. When dam construction finally appeared imminent, Washington state Democratic representative Knute Hill proclaimed that it would be "the largest of man's structures." The Portland *Oregonian*, in reporting the speech, cautioned that perhaps he had overlooked the Great Wall of China. See George Sundborg, *Hail Columbia: The Thirty-Year Struggle for Grand Coulee Dam* (New York: The Macmillan Company, 1954), p. 251.

42. "Grand Coulee: Columbia River Billion Dollar 'Crazy Idea' Will Start with Giant Dam," *Newsweek* 4 (July 7, 1934): 6

43. *Seattle Daily Journal of Commerce* (June 10, 1935), Dubuar Scrapbooks, #10, p. 32.

44. "Grand Coulee," *Fortune* 16 (July 1937): 79.

45. *Seattle Times* (July 12, 1936), Dubuar Scrapbooks, #8, p. 32.

46. The *MWAK Columbian*, printed by the company building the first part of the dam, noted, "If all the cement used in the dam were placed on a freight train, it would stretch over 500 miles and 2,000 locomotives would be required to move it. Nothing ever made by man can be compared to its massive bulk. The gigantic Sphinx of Egypt would only be a sizable decoration for the top of one of the powerhouses." The *Wenatchee Daily World* had actually run a large front-page drawing of the finished dam with the great pyramid, looking rather small, sitting in front of it and the Sphinx sitting on top of the west powerhouse. See *MWAK Columbian* (October 30, 1936): 3; *Wenatchee Daily World* (August 13, 1935), p. 1.

47. Robert Ormond Case, "The Eighth World Wonder," *Saturday Evening Post* 208 (July 13, 1935): 23.

48. *Seattle Times* (August 4, 1935), Dubuar Scrapbooks, #10, p. 10.

49. "Grand Coulee Construction Spreads Work over 40 States," *Reclamation Era* 27 (March 1937): 52. Also see "Widespread Benefits from Grand Coulee Construction," *Reclamation Era* 29 (March 1939): 46.

50. See Owen P. White, "Spare that Desert!," *Collier's* 93 (June 16, 1934): 10-11, ff; Walter Davenport, "Power in the Wilderness," *Collier's* 94 (September 21, 1935): 10-11, ff.

51. *Wenatchee Daily World* (June 22, 1934), p. 1, ff.

52. *Ibid.* (September 13, 1935), p. 1, editorial.

53. *Ibid.* (April 26, 1939), p. 13.

54. In all, over 126,000 came in 1936; over 230,000 in 1937 and 1938; almost 300,000 in 1939; over 325,000 in 1940 and 1941. World War II diminished the flow by half or better as gas rationing and the rubber shortage limited travel. *Wenatchee Daily World* (July 18, 1936), p. 1: (October 5, 1940), p. 6; (January 4, 1943), p. 6; "Grand Coulee—Kibitzer's Paradise," *Reclamation Era* 29 (October 1939): 276.

55. *The Public Papers and Addresses of Franklin D. Roosevelt* (New York: Macmillan and Company, 1941), 1937 volume, p. 395.

56. *Wenatchee Daily World* (September 18, 1937), p. 1.

57. For lunch they fed the President chicken, served in Guy F. Atkinson's home where Roosevelt's daughter, Mrs. Anna Boettiger, (whose husband John published the Seattle *Post-Intelligencer*) and his daughter-in-law, the wife of James Roosevelt, joined him. A rather bizarre incident indirectly related to

that lunch involved Sidney Martin and George Crocker. While they cleaned and singed chickens, they attempted to refuel the alcohol stove they were using when it exploded. The blast killed Crocker immediately and Martin died from burns a day later. They would have been the forty-fifth and forty-sixth deaths at the dam, but officials agreed not to count them since theirs was an accident unrelated to construction. *Wenatchee Daily World* (October 2, 1937), p. 1; (October 4, 1937), p. 7; (October 6, 1937), p. 7; (January 14, 1938), p. 8.

58. *The Public Papers and Addresses of Franklin D. Roosevelt*, 1937 volume, p. 395.
59. *Ibid.*
60. Sundborg, *Hail Columbia*, p. 347.
61. John C. Page, "Also One-Third is Ill-Watered," *Reclamation Era* 28 (January 1938): 2.
62. See Steve Neal, ed., *They Never Go Back to Pocatello: The Selected Essays of Richard Neuberger* (Portland: Oregon Historical Society Press, 1988), especially Paul C. Pitzer's "The Writings of Richard L. Neuberger," pp. 338-363.
63. Richard L. Neuberger, "The Biggest Thing on Earth: Grand Coulee Dam," *Harper's Magazine* 174 (February 1937): 247-258; Richard L. Neuberger, "The Biggest Thing on Earth," *Scholastic* (March 20, 1937).
64. Richard L. Neuberger, "Colossus in the West," *New Republic* 97 (January 18, 1939): 310-311.
65. "America's Future—Pacific Northwest: The Story of a Vision and a Promised Land," *Life* 6 (June 6, 1939): 15-23. The Neuberger story appeared without byline. Also see *Wenatchee Daily World* (May 9, 1939), p. 8; (May 18, 1939), p. 6.
66. Richard L. Neuberger, "The Columbia Flows to the Land," *Survey Graphic* 28 (July 1939): 440-441.
67. "The Biggest Thing on Earth," *Popular Mechanics* 69 (April, 1938): 489.
68. "The Largest Thing Ever Built," *ibid.* 73 (April 1940): 546.
69. R. G. Skerrett, "Grand Coulee Progress," *Scientific American* 159 (December 1938): 296.
70. Garet Garrett, "Great Works," *Saturday Evening Post* 211 (April 8, 1939): 6.
71. "Grand Coulee Dam Project Will Affect Pacific Trade," *China Weekly Review* 96 (May 31, 1941): 406.
72. Morgan, *The Dam*, p. 4.
73. "Facts—Not Fancy," *Reclamation Era* 26 (December 1936): 285.
74. Fort Peck Dam in Montana is an earth-and-gravel structure two miles long and containing 135,000,000 cubic yards of material. It stands 241 feet high and is one-half mile thick at its base. It is, then, roughly ten times the size of Grand Coulee by volume. See Albert N. Williams, *The Water and the Power: Development of the Five Great Rivers of the West* (New York: Duell, Sloan and Pearce, 1951), p. 220.
75. Oscar Osburn Winther wrote in 1950, "The Grand Coulee Dam, as it is named, is now completed. By all odds the most massive structure ever made by man, this is, of course, the most stupendous undertaking of the Bureau of Reclamation." See Oscar Osburn Winther, *The Great Northwest: A History* (New York: Alfred Knopf, 1952), p. 317. In 1986 this appeared in another source: "As the decade wore on, the Coulee Dam progressed. When it was finished in 1941, it was the largest structure ever built." See Gordon Dodds, *The American Northwest: A History of Oregon and Washington*

(Arlington Heights, Ill.: Forum Press, 1986), p. 241. Donald Worster wrote, "Neuberger touted Grand Coulee Dam as 'the biggest thing on earth,' a boast that took in a lot of territory—the Pacific Ocean, Mount Everest, Antarctica, and the like." See Donald Worster, *Rivers of Empire: Water, Aridity, and The Growth of the American West* (New York: Pantheon Books, 1995), pp. 270-271.

76. *Wenatchee Daily World* (October 6, 1937), p. 12.

77. Henry Earle Riggs, "Waste in Ill-Considered Federal Public Works Projects: Part II, The Grand Coulee Project," *Public Utilities Fortnightly* 21 (April 14, 1938): 469.

78. *Wenatchee Daily World* (January 3, 1938), p. 1; (January 5, 1938), p. 4; (January 6, 1938), p. 7.

79. *Spokesman-Review* (January 22, 1938), O'Sullivan Mss.; *Wenatchee Daily World* (March 4, 1938), p. 1. For comment on the value of Grand Coulee Dam as an aid in flood control see Paul C. Pitzer, "The Mystique of Grand Coulee Dam and the Reality of the Columbia Basin Project," *Columbia: The Magazine of Northwest History* 4 (Summer 1990): 33.

80. As of the first of January 1938 the federal government had spent $58,110,000 on Grand Coulee Dam. *Spokesman-Review* (January 6, 1938); (January 14, 1938); (March 6, 1938), O'Sullivan Mss.

81. *Wenatchee Daily World* (April 28, 1938), p. 8.

82. *Ibid.* (May 12, 1938), p. 5; (May 14, 1938), p. 12.

83. *Spokesman-Review* (June 28, 1938), O'Sullivan Mss.; *Grant County Journal* (July 8, 1938).

84. "Basin War Bulwark," declared a headline in a later edition of the *Wenatchee Daily World* in one of a number of articles on this theme. A later issue called Grand Coulee one of the country's "outstanding national defense assets," and that in case of war it would provide both power and food. See *Wenatchee Daily World* (April 6, 1939), p. 10; (October 3, 1939), p. 1.

85. *Ibid.* (January 3, 1939), p. 1.

86. *Ibid.* (November 3, 1938), p. 1; (December 16, 1938), p. 1; *Spokesman-Review* (November 4, 1938), O'Sullivan Mss; Woods to Clapp and Matthews, November 8, 1938, Woods Mss.

87. *Wenatchee Daily World* (January 7, 1939), p. 10.

88. *Ibid.* (March 6, 1939), p. 1; (March 8, 1939), p. 1; *Spokesman-Review* (March 9, 1939), O'Sullivan Mss.

89. *Spokesman-Review* (March 11, 1939), O'Sullivan Mss.; *Wenatchee Daily World* (March 11, 1939), p. 10.

90. *Wenatchee Daily World* (March 16, 1939), p. 1; *Spokesman-Review* (March 17, 1939), O'Sullivan Mss.; *Seattle Post-Intelligencer* (March 18, 1939), editorial, O'Sullivan Mss.

91. *Wenatchee Daily World* (December 2, 1939), p. 7; (December 12, 1939), p. 10.

92. *Ibid.* (January 5, 1940), p. 1; (January 15, 1940), p. 7; (January 24, 1940), p. 10; (January 29, 1940), p. 1.

93. In 1940 Ernest R. Abrams wrote a book published in New York which said, in part, "That the vast area of rich and fertile land in Adams, Franklin and Grant counties, Washington, comprising the Columbia basin, must be irrigated if it is to be brought into production has never been questioned since the complete failure of the dry-farming experiment, nor has the productivity of this land when adequately watered ever been doubted. . . . These

lands can never carry the full burden of the required investment—[they] only can pay if all the power is sold at a firm rate." The author estimated the cost of irrigating each acre at about $333 as compared to the general cost of land around the state at $57.17. The author did admit that the power would draw some industry to the region, and that the Aluminum Company of America had, in fact, already announced its plans to build a plant at Vancouver, Washington. "Perhaps the worst that might be said of the Bonneville project is that it was constructed too far in advance of the demand for its power, but when Grand Coulee and other public power projects are developed in the same general area at a public cost of more than three-quarters of a billion dollars to produce far more power than the Northwest can absorb for generations, the entire program becomes ludicrous." See Ernest R. Abrams, *Power in Transition* (New York: Charles Scribner's Sons, 1940), pp. 50, 51, 63, 70.

94. *Wenatchee Daily World* (March 4, 1940), p. 1.
95. *Ibid.* (March 8, 1940), p. 7; (April 3, 1940), p. 10; (July 31, 1940), p. 12.
96. *Ibid.* (January 9, 1941), p. 12.
97. *Ibid.* (March 6, 1941), p. 1; George Sundborg, *Hail Columbia*, p. 376.
98. For background on the debate over the monopoly held in aluminum production by the Aluminum Company of America and emergence of the newcomer, Reynolds, see Gerald D. Nash, *World War II and the West: Reshaping the Economy* (Lincoln: University of Nebraska Press, 1990), pp. 94-98.
99. *Wenatchee Daily World* (April 14, 1941), p. 4, editorial, which reprinted the *Grand Rapids Herald* article and the *Post-Intelligencer* rebuttal; (April 16, 1941), p. 19.
100. *Wenatchee Daily World* (April 30, 1941), p. 1; (May 1, 1941), p. 9.

## Notes for Chapter Eleven

1. James Rorty, "Grand Coulee," *The Nation* 140 (March 20, 1935): 329.
2. "10,000 Montana Relief Workers Make Whoopee on Saturday Night," *Life* 1 (November 23, 1936): 9-11.
3. See Russell Martin, *The Story That Stands Like a Dam: Glen Canyon and the Struggle for the Soul of the West* (New York: Henry Holt and Company, 1989), pp. 132-158; Joseph E. Stevens, *Hoover Dam: An American Adventure* (Norman: University of Oklahoma Press, 1988), pp. 117-158.
4. *Wenatchee Daily World* (August 12, 1933), p. 9; (November 1, 1933), p. 1; (January 1, 1934), p. 6.
5. *Ibid.* (November 3, 1933), p. 2.
6. *Grant County Journal* (September 24, 1937), p. 1; (February 4, 1938), p. 1; (February 25, 1938), p. 1; *Wenatchee Daily World* (September 29, 1937), p. 1; (March 19, 1938), p. 1; (April 26, 1939), p. 62.
7. Banks to R. F. Walter, August 18, 1933, BOR Denver Mss.
8. *Wenatchee Daily World* (October 13, 1933), p. 1.
9. Ida Fleischman became Ida Bartels. Ida Bartels interview, March, 1989; *Wenatchee Daily World* (November 10, 1933), p. 17.
10. *Spokesman-Review* (April 29, 1934), Dubuar Scrapbooks, #9, p. 13.
11. *Wenatchee Daily World* (October 2, 1933), p. 1.
12. Seaton continued to live in the area of the dam for years and died on April 29, 1971 in an automobile accident. See *Wenatchee Daily World* (April 29, 1971), pp. 1, 2.

13. *Wenatchee Daily World* (November 4, 1933), p. 10; (November 14, 1933), p. 1.
14. "Ten Months' Construction Progress," *Engineering News Record* 115 (August 1, 1935): 145.
15. Henry W. Young, "A Dual Water Supply for Coulee's 'Model Town,'" *American City* 54 (April 1939): 50.
16. H. P. Bunger to Reclamation Commissioner, June 15, 1937; Banks to Walter, December 26, 1935. Both in BOR Denver Mss.
17. *Wenatchee Daily World* (May 3, 1934), p. 6.
18. Walter A. Averill, "U. S. Railroad for Grand Coulee Nears Completion," *Pacific Builder and Engineer* 40 (December 1, 1934): 23.
19. *Walla Walla Sunday Union Bulletin* (June 9, 1935), Dubuar Scrapbooks, #10, p. 12.
20. "Work Starts on Cofferdams on Coulee Project," *Pacific Builder and Engineer* 41 (January 5, 1935): 40; G. W. Hitchcock, "The World's First All-Electric City: Mason City," *Pacific Builder and Engineer* 41 (January 19, 1935): 3.
21. *Walla Walla Sunday Union Bulletin* (June 9, 1935), Dubuar Scrapbooks, #10, p. 11.
22. *MWAK Columbian* (December 1, 1938), p. 3.
23. George Weller, "Eighth Wonder of the World," *Esquire Magazine* 15 (May 1941): 45.
24. Grant County Historical Society, *Memories of Grant County* (March 1982), interview with Hu Blonk titled "Grand Coulee," p. 177.
25. *MWAK Columbian*, (October 16, 1936), p. 11.
26. *Wenatchee Daily World* (September 12, 1940), p. 8.
27. *Ibid.* (September 15, 1933), p. 1.
28. *Ibid.* (April 5, 1934), p. 8; (April 6, 1934), p. 6.
29. *Ibid.* (April 23, 1934), p. 10.
30. *Spokesman-Review* (April 27, 1934), O'Sullivan Mss.
31. *Spokesman-Review* (April 29, 1934), special edition, Dubuar Scrapbooks, #9, p. 19.
32. *Seattle Times* (August 19, 1934), Dubuar Scrapbooks, #9, p. 40.
33. *Walla Walla Sunday Union Bulletin* (August 19, 1934), Dubuar Scrapbooks, #9, p. 36.
34. *Wenatchee Daily World* (September 21, 1934), p. 20.
35. Banks to Walter, January 31, 1935; Walter to Banks, February 25, 1935. Both in BOR Denver Mss.
36. A. Gilbert Darwin, "Grand Coulee Dam and Power Plant Specifications," *Western Construction News* 9 (April 1934): 110.
37. *Grant County Journal* (September 28, 1934), p. 3; *Wenatchee Daily World* (November 15, 1934), p. 1; Project History—Grand Coulee Dam, 1934, p. 39.
38. The Continental Land Company owned the land, which it bought from John Purtee, the pioneer owner who settled there in 1907 and died in 1936. The company also held much of the adjoining land. Such companies promoted many of the settlements around Grand Coulee.
39. Hu Blonk interview, June 16, 1988.
40. *Wenatchee Daily World* (February 27, 1935), p. 12.
41. *Ibid.* (September 16, 1936), p. 7.
42. *Ibid.* (September 19, 1935), p. 7.
43. *Ibid.* (October 10, 1935), p. 12. Hu Blonk, who reported for the *Wenatchee Daily World* and the *Spokane Chronicle* during the construction period, stated

that there were no cases of syphilis during those years. He recalled that the local doctor told him that of the fifty-some prostitutes, none ever contracted disease. Hu Blonk interview, June 16, 1988. Frank Banks reported that the town of Grand Coulee held a population of 6,000 in the spring of 1936. Other estimates vary considerably. Banks to Commissioner on Schools, March 6, 1936, BOR Denver Mss.

44. Grand Coulee Dam Bicentennial Association (Vesta Seiler Chairman), *From Pioneers to Power: Historical Sketches of the Grand Coulee Dam Area* (Nespelem, Wash.: RIMA Printing and Graphics, 1976), pp. 296-298.
45. *Wenatchee Daily World* (June 11, 1937), p. 10.
46. *Ibid.* (June 24, 1937), p. 9.
47. *Oregonian* (July 25, 1937), magazine section, p. 1.
48. *Ibid.*
49. Richard L. Neuberger, *Our Promised Land* (New York: The Macmillan Company, 1938), pp. 372-378.
50. Dave Rawe interview, June 14, 1988; Ida Bartels interview, June 13, 1988.
51. This was many years before humorist Garrison Keillor used the similar slogan, "If we don't have it, you can get along without it," for one of his Lake Woebegone characters. *Gold Historian* 2 (August 31, 1981): 6; Dave Rawe interview, June 14, 1988.
52. Dave Rawe interview, June 14, 1988; Ida Bartels interview, June 13, 1988.
53. The first fire of note in the area occurred not in Grand Coulee but in the neighboring town of Coulee Heights. On July 1, 1934 a large fire in some buildings threatened the 100+ houses that existed in the settlement. See *Wenatchee Daily World* (July 1, 1934), p. 1.
54. *Ibid.* (August 3, 1936), p. 1; (August 10, 1936), p. 1; (July 27, 1937), pp. 6-7; (July 28, 1937), p. 12; (August 18, 1938), p. 8; (August 4, 1938), p. 7; (July 28, 1938), p. 1; (August 12, 1946), p. 1; (December 2, 1948), p. 1; *Grant County Journal* (July 29, 1938), p. 1.
55. *Wenatchee Daily World* (February 2, 1938), p. 12; (January 17, 1938), p. 7; (February 25, 1938), p. 7; (May 19, 1938), p. 11; (June 21, 1938), p. 10.
56. Murray Morgan, *The Dam* (New York: The Viking Press, 1954), p. 67.
57. The Continental Land Company of Spokane originally held the deed to what became Coulee Center as well as much of Grand Coulee. Higher up on the south side of the coulee, Alfred Alexander (A. A.) Elmore and Ida Fleischman (Mrs. Albert Fleischman, who later married Sidney Bartels) promoted Coulee Heights.
58. *Wenatchee Daily World* (May 6, 1935), p. 12.
59. *Grant County Journal* (June 7, 1935), p. 1.
60. *Wenatchee Daily World* (July 10, 1935), p. 6.
61. *Ibid.* (August 8, 1935), p. 7.
62. *Ibid.* (September 26, 1935), p. 9.
63. *Ibid.* (October 30, 1935), p. 14; *Spokesman-Review* (October 30, 1935), O'Sullivan Mss.
64. *Wenatchee Daily World* (June 25, 1936), p. 8.
65. *Ibid.* (June 28, 1939), p. 10; (November 16, 1939), p. 7; (May 3, 1940), p. 8; (May 14, 1940), p. 7; (June 6, 1940), p. 10.
66. *Ibid.* (December 2, 1937), p. 10.
67. *Ibid.*
68. Fred K. Ross, "CBI Prepares to Pour Concrete," *Pacific Builder and Engineer* 43 (June 4, 1938): 52; *Wenatchee Daily World* (July 6, 1938), p. 8.

69. *Wenatchee Daily World* (October 27, 1938), p. 16; (February 17, 1939), p. 8.
70. "The Dam Builders and Their Children," *School and Society* 55 (April 11, 1942): 419; *Wenatchee Daily World* (February 2, 1943), p. 6.
71. *Wenatchee Daily World* (August 19, 1941), p. 1; (September 27, 1941), p. 10; (February 4, 1942), p. 5.
72. *Ibid.* (July 29, 1943), p. 6. The consolidation became official on October 1, 1943. See *Wenatchee Daily World* (October 1, 1943), p. 6.
73. Murray Morgan, *The Dam*, p. 68.
74. Carl Abbott "American Planning Language," from a talk delivered at the Atomic West Conference, University of Washington, Seattle, Washington, September 25, 1992.

## Notes for Chapter Twelve

1. William E. Warne, *The Bureau of Reclamation* (New York: Praeger Publishers, 1973), pp. 111-112.
2. The so-called Central Valley Project in California is larger, but it is actually a combination of contiguous projects whereas the Columbia Basin Project is one single project.
3. The specifications, however, did not appear until November 2. Secretary of the Interior Ickes then announced the bid opening for December 10 at 1:30 in the afternoon and, despite complaints from Wenatchee businessmen who wanted the event in their city, he chose Spokane. *Wenatchee Daily World* (July 3, 1937), p. 1; (November 2, 1937), p. 1; (September 28, 1937), p. 5; (September 30, 1937), p. 1; Project History—Grand Coulee Dam, 1937, p. 16.
4. Walter to Banks, July 31, 1937, BOR Denver Mss.
5. Agreement between Interior Construction Company and various labor unions, December 9, 1937, copy found in the BOR Denver Mss.; George Sundborg, *Hail Columbia: The Thirty-Year Struggle for Grand Coulee Dam* (New York: The Macmillan Company, 1954), p. 337; *Wenatchee Daily World* (December 11, 1937), p. 1.
6. "Grand Coulee Dam: Plans for Completion Reveal Staggering Proportions of World's Greatest Concrete Structure Being Built on the Columbia River Washington," *Engineering News Record* 119 (December 23, 1937): 1022.
7. "America's Greatest Builders of Dams Merge to Bid Largest Concreting Contract," *Pacific Builder and Engineer* 42 (December, 1937): 36.
8. *Wenatchee Daily World* (December 10, 1937), p. 1. The Pacific Construction Company held seven firms: The Griffith Company, the American Concrete and Steel Pipe Company, the Metropolitan Construction Company, and L. E. Dixon Company—all of Los Angeles; Lawler and Maguire of Butte, Montana; D. W. Thurston of Pasadena, California; and Hunkin and Hinman of Denver, Colorado. See "America's Greatest Builders of Dams Merge," p. 36.
9. *Spokesman-Review* (December 11, 1937), O'Sullivan Mss.
10. *Spokesman-Review* (February 3, 1938), O'Sullivan Mss.; *Wenatchee Daily World* (February 8, 1938), p. 1.
11. "Consolidated Builders Tackle Work of Completing Coulee Dam," *Pacific Builder and Engineer* 43 (April 2, 1938): 38.
12. *Wenatchee Daily World* (March 18, 1938), p. 7; (March 22, 1938), p. 1; *CBI Columbian* (June 30, 1938), p. 2.

13. "New World's Record Set at Coulee Dam," *Pacific Builder and Engineer* 44 (June 3, 1939): 29.
14. O. G. F. Markhus, "Electric House Heating in Mason City," *Reclamation Era* 28 (August 1938): 149; *Wenatchee Daily World* (April 2, 1938), p. 5; (June 11, 1938), p. 7.
15. A. L. Osborne to Francis Perkins, May 18, 1938, BOR Denver Mss.
16. Page to Banks, May 9, 1938, *ibid.*
17. Williams to Glovan, May 18, 1938, *ibid.*
18. *Wenatchee Daily World* (April 14, 1938), p. 9; (April 20, 1938), p. 12.
19. Woods to Clapp, November 8, 1938, Woods Mss.
20. S. J. Lake, NAACP, to W. J. Trent, Department of the Interior, Adviser on Negro Affairs, May 10, 1939, BOR Denver Mss.
21. Charles H. Day, AFL, to Frank Maynard, Department of Labor Relations, Bureau of Reclamation, May 25, 1939, *ibid.*
22. W. B. Cook, Business Representative, Hod Carrier's Union No. 1151 to Maynard, Bureau of Reclamation, May 25, 1939, *ibid.*
23. S. J. Lake to Banks, May 26, 1939, *ibid.*
24. *Wenatchee Daily World* (September 6, 1940), p. 10.
25. R. S. Brabrook, "Erecting a 3,600-ft Steel Trestle for Placing Grand Coulee Concrete," *Western Construction News* 14 (February 1939): 42; Project History—Grand Coulee Dam, 1938, p. 32; *Wenatchee Daily World* (July 21, 1938), p. 1.
26. *Wenatchee Daily World* (September 6, 1938), p. 6; (September 10, 1938), p. 7.
27. *Ibid.* (July 21, 1938), p. 1; (July 23, 1938), p. 1; Fred K. Ross, "Moving a Mountain of Aggregate at Grand Coulee Dam," *Pacific Builder and Engineer* 43 (September 3, 1938): 52.
28. Project History—Grand Coulee Dam, 1938, p. 33; *Wenatchee Daily World* (August 19, 1938), p. 9.
29. *Wenatchee Daily World* (December 16, 1938), p. 1; (January 16, 1939), p. 5. By December 31, 1938 the government had spent an estimated $95 million on Grand Coulee Dam.
30. "How CBI Conquers the Columbia River," *Pacific Builder and Engineer* 44 (October 7, 1939): 30-32; "Steel Gates Close Gaps in Dam," *Engineering News Record* 121 (December 8, 1938): 733-734; "Closure Gates: Columbia Basin Project, Washington," *Reclamation Era* 30 (March 1940): 70-71.
31. "Refrigerating 11 Million Yards of Concrete," *Pacific Builder and Engineer* 43 (May 7, 1938): 40.
32. "Items to be Checked in Dam Construction," *Engineering News-Record* 124 (March 28, 1940): 367-375.
33. "New Set-up in Coulee Gravel Pit," *Pacific Builder and Engineer* 44 (April 1, 1939): 44; "New World's Record Set at Coulee Dam," p. 30; "Grand Coulee Dam," *Compressed Air Magazine* 46 (March 1941): 6397.
34. O. G. Patch, "Mixer Efficiency or Mortar-Mix Tests," *Journal of the American Concrete Institute* 35 (January 1939): 173-178, ff.
35. *Wenatchee Daily World* (May 25, 1939), p. 1; Project History—Grand Coulee Dam, 1939, p. 37; "20,684 Yards of Concrete in 24 Hours," *Pacific Builder and Engineer* 44 (July 8, 1939): 12; "Grand Coulee Breaks Concreting Record," *Engineering News-Record* 122 (June 8, 1929): 63.
36. *Wenatchee Daily World* (June 12, 1939), p. 7; unidentified clipping dated June 3, 1939, found in the scrapbooks in Terou Mss.
37. *Wenatchee Daily World* (June 28, 1939), p. 16.

38. *Ibid.* (July 11, 1939), p. 7; (August 4, 1939), p. 8; *Spokesman-Review* (August 4, 1939), O'Sullivan Mss.

39. *Spokesman-Review*, Progress Edition (January 28, 1940).

40. Project History—Grand Coulee Dam, 1939, p. 38.

41. S. E. Hutton, "The Grand Coulee Dam and the Columbia Basin Reclamation Project," *Mechanical Engineering* 62 (September 1940): 658.

42. *Spokesman-Review* (November 22, 1938), O'Sullivan Mss.

43. *Ibid.* (April 6, 1939); (August 20, 1939), O'Sullivan Mss.

44. *Ibid.* (November 1, 1939), O'Sullivan Mss.; *Wenatchee Daily World* (November 1, 1939), p. 1; Jean Nicholson interview, June 13, 1988, Electric City, Washington.

45. *Spokane Daily Chronicle*, January 3, 1940, Terou Mss.; *Wenatchee Daily World* (January 3, 1940), p. 12.

46. Tom A. Heatfield, "Grand Coulee: Big, Strong, and Straight," *Reclamation Era* 31 (September 1941): 239.

47. Mark S. Foster, *Henry J. Kaiser: Builder of the Modern American West* (Austin: University of Texas Press, 1989), especially chapter 13 and pages 214-215.

48. If the standard of one death per million dollars spent is accepted, then Grand Coulee construction, at $162 million, did well. Hu Blonk, "How Human Safety is Emphasized at Grand Coulee," *Pacific Builder and Engineer* 45 (April 6, 1940): 40.

49. *Grant County Journal* (December 9, 1938), p. 1; "New Set-up in Coulee Gravel Pit," p. 44; *Wenatchee Daily World* (July 16, 1938), p. 7; "Grand Coulee Generates First Power," *Reclamation Era* 31 (April 1941): 100.

50. "Steam Heat Aids Concreting as Winter Comes to Coulee," *Engineering News Record* 123 (December 14, 1939): 795; *Wenatchee Daily World* (November 3, 1939), p. 7.

51. *Wenatchee Daily World* (September 28, 1940), p. 6.

52. *Ibid.* (December 21, 1939), p. 12.

53. F. A. Latimer, "Here's What Will be Done in 1940 at Grand Coulee Dam," *Pacific Builder and Engineer* 45 (April 6, 1940): 40.

54. "To a Waterfall," *Reclamation Era* 30 (September 1940): 275.

55. Ivan E. Houk, "Twist Effects in Straight Gravity Dams: Part II; Twist Action In Grand Coulee Dam," *The Engineer* 164 (December 24, 1937): 702; "Grand Coulee Dam Nears Completion," *Engineering News Record* 127 (September 11, 1941): 91; L. Vaughn Downs, *The Mightiest of Them All: Memories of Grand Coulee Dam* (Fairfield, Wash.: Ye Galleon Press, 1986), p. 121; *Wenatchee Daily World* (April 22, 1940), p. 2; (June 28, 1941), p. 11; (July 9, 1940), p. 7.

56. *Wenatchee Daily World* (June 20, 1940), p. 13.

57. *Ibid.* (April 1, 1941), p. 1.

58. *Ibid.* (January 28, 1941), p. 6; (February 27, 1941), p. 10.

59. *Grant County Journal* (March 21, 1941).

60. "Coulee Ready," *Business Week* (March 15, 1941): 22; Richard L. Neuberger, "Mightiest Man-Made Thing," *New York Times Magazine* (March 16, 1941): 10.

61. "Grand Coulee Powerhouse Goes to Work," *Reclamation Era* 31 (May 1941): 133.

62. Project History—Grand Coulee Dam, 1941, p. 31.

63. *Wenatchee Daily World* (July 20, 1940), p. 7; (October 16, 1940), p. 7.

64. *Ibid.* (January 15, 1941), p. 3.
65. Project History—Grand Coulee Dam, 1941, p. 33.
66. "Grand Coulee Dam Nears Completion," p. 92.
67. Project History—Grand Coulee Dam, 1942, p. 30.
68. *Spokesman-Review*, Progress Edition (January 31, 1943), p. 3.
69. *Ibid.*, p. 8.
70. *Wenatchee Daily World* (October 24, 1941), p. 7.
71. "The Dam Builders and Their Children," *School and Society* 55 (April 11, 1942): 419; *The Star Newspaper* of Grand Coulee, Washington, 50th Anniversary edition, 1983, p. 22.
72. "Power for Defense," *Time* 37 (March 31, 1941): 10.
73. Hubert Blonk interview, June 16, 1988. Blonk's reporting accounts for a great deal that has gone into these chapters. He was paid by the amount of material he filed with each newspaper. The *Wenatchee Daily World* carried his stories regularly, usually in a column on page seven. Blonk worked to guarantee the accuracy of the material he sent out and, taken together, the material in the *World* chronicles the construction period.
74. *Wenatchee Daily World* (January 2, 1940, p. 7.
75. "Refrigerating 11 Million Yards of Concrete," p. 40.
76. Project History—Columbia Basin Project, 1946, p. 31.
77. As quoted in Vera Springer, *Power and the Pacific Northwest: A History of the Bonneville Power Administration* (Washington, D.C.: U.S. Government Printing Office, 1976), p. 33; "Roll on Columbia," 1941, words by Woody Guthrie, music based on "Goodnight Irene" by Huddie Ledbetter and John Lomax, TRO© 1957, 1963 Ludlow Music Inc., New York.

# Notes for Chapter Thirteen

1. Acting regional director to the Commissioner, Bureau of Reclamation, memo dated May 15, 1946, subject: Columbia Basin Fisheries Development Association Publication *Wealth of the River*, BOR-FRS Mss.
2. *Wenatchee Daily World* (January 12, 1935), p. 10; (July 13, 1935), p. 7. In his autobiography, Charles Chase Pearsons confirms the problems with rattlesnakes during the surveying. See Charles Chase Pearsons, "An Autobiography, 82 Years and Growing," *The Gold Historian* 4 (August 1983): 20. *The Gold Historian* is a publication put out locally at Grand Coulee Dam.
3. Project History—Grand Coulee Dam, 1935, p. 81.
4. *Wenatchee Daily World* (April 13, 1936), p. 7.
5. This will be detailed later in this chapter.
6. *Wenatchee Daily World* (November 1, 1937), p. 1; (December 21, 1937), p. 1.
7. *Spokesman-Review* (December 22, 1937), O'Sullivan Mss.; *Wenatchee Daily World* (December 23, 1937), p. 7.
8. *Wenatchee Daily World* (February 10, 1938), p. 7; (June 11, 1938), p. 1; (June 13, 1938), p. 1.
9. Project History—Grand Coulee Dam, 1938, p. 33.
10. *Wenatchee Daily World* (August 13, 1938), p. 1; (August 15, 1938), p. 6; (September 7, 1938), p. 7; (October 1, 1938), p. 7; (October 4, 1938), p. 12; Project History—Grand Coulee Dam, 1938, p. 34.
11. Figures on exactly how many worked on the clearing are difficult to determine as the sources disagree and because the crews were transient, moving

from place to place due to the work pattern. In the early summer of 1939 the *Wenatchee Daily World* estimated that 2,500 men worked on the project. In October it reported that the total would be increased to 3,500. One year later the paper reported 2,336 men on the job. *Wenatchee Daily World* (June 24, 1939), p. 7; (October 5, 1939), p. 8; (October 3, 1940), p. 18.

12.  *Ibid.* (March 6, 1939), p. 7.

13.  "Clearing Grand Coulee Reservoir Site," *International Engineer* 80 (November 1941): 152.

14.  *Wenatchee Daily World* (April 26, 1939), p. 14.

15.  H. M. Sheerer, "Columbia River Reservoir Clearing Project," *Reclamation Era* 31 (January 1941): 24.

16.  *Spokesman-Review* (February 26, 1939), p. 1 (includes picture); "New Set-up in Coulee Gravel Pit," *Pacific Builder and Engineer* 44 (April 1, 1939): 44; *Wenatchee Daily World* (October 2, 1939), p. 6.

17.  *Wenatchee Daily World* (August 12, 1940), p. 6. Along with the barges, the WPA built and launched a power boat named the *Paul Bunyan*. The men wanted to name the craft the *Banks*, but Frank Banks refused the honor. See Project History—Grand Coulee Dam, 1939, pp. 36, 45 and *Wenatchee Daily World* (January 5, 1939), p. 1.

18.  Bashore to Miller, February 8, 1941, BOR Denver Mss.; *Wenatchee Daily World* (February 7, 1941), p. 6.

19.  *Wenatchee Daily World* (April 26, 1941), p. 7.

20.  *Ibid.* (September 6, 1941), p. 7; (March 2, 1939), p. 10.

21.  *Ibid.* (November 6, 1941), p. 15. The *World* reported that exactly 53,860 acres had been cleared, with 11,426 acres cleared of stumps. Also see *Wenatchee Daily World* (September 6, 1941), p. 7. The nearly 54,000 acres did not represent the full size of the reservoir.

22.  Bureau of Reclamation data states that the Grand Coulee reservoir covers 82,300 acres. That would indicate that 28,000 acres were not cleared. Some of the land may simply not have needed clearing and other acres represented the river bottom itself, which also needed no clearing. From the information available, it was not possible to determine how much land, if any, was left uncleared. See "Columbia Basin Project," an excerpt from *Project Data Book*, Bureau of Reclamation, April 1983, p. 11.

23.  The reservoir had not yet been named Franklin D. Roosevelt Lake. Project History—Grand Coulee Dam, 1941, p. 32; L. Vaughn Downs, *The Mightiest of Them All: Memories of Grand Coulee Dam* (Fairfield, Wash.: Ye Galleon Press, 1986), p. 55.

24.  *Wenatchee Daily World* (September 6, 1941), p. 7; (November 6, 1941), p. 15. The *World* noted that the final report on the clearing had been filed on October 31, 1941, which is true as far as it went. Additional work followed, however, repairing damage to the shore, and the final report on the clearing and shore reconstruction effort which followed was not finished until 1952. See Bert A. Hall, "Final Construction Report on Lake Roosevelt Reservoir, 1952," BOR Denver Mss. This report gives details on the initial clearing effort. Also see "Clearing Grand Coulee Reservoir Site," p. 152.

25.  Grand Coulee Dam Bicentennial Association, Vesta Seiler, Chairman, *From Pioneers to Power: Historical Sketches of the Grand Coulee Dam Area*, (Nespelem, Wash.: RIMA Printing and Graphics, 1976), p. 345; *Wenatchee Daily World* (July 30, 1935), p. 7.

26. Other sources dispute the amount of the acreage and set it at 840.28 acres. See Final Construction Report, Grand Coulee Pump Plant, 1956-1977, chronology. The *Wenatchee Daily World* reported that the land in question was 837 acres. See *Wenatchee Daily World* (December 29, 1933), p. 3; (August 4, 1934), p. 4. The *Spokesman-Review* listed the amount at 1,102 acres. See *Spokesman-Review* (August 3, 1935), O'Sullivan Mss. The debate hinged on whether private companies (which would have to negotiate independently with each owner, thereby raising the land values) could build the dam or if only the federal government could do the job. If only the government could do the job, the government argued that land values would not go up. The government's lawyers took the latter position and maintained that then the government needed to pay only the assessed value before the advent of the dam.

27. Project History—Grand Coulee Dam, 1935, p. 21.

28. *Spokesman-Review* (August 3, 1935), O'Sullivan Mss. William Rath received $2,507 on 191 acres, The Continental Land Company received $1,261 on 169 acres, and Julius Johnson received $624 for 164 acres. See *Wenatchee Daily World* (August 5, 1935), p. 2.

29. In making their case, Seaton and the others pointed to the $100,000 the government paid Mrs. Sam Hill for thirty-five acres at Bonneville Dam. On February 15, 1937 the circuit court upheld the jury award and later that court denied a rehearing. *Spokesman-Review* (August 3, 1935), O'Sullivan Mss.

30. "Final Construction Report, Grand Coulee Dam Pump Plant, 1956-1977," reports on file at Grand Coulee Dam; Project History—Grand Coulee Dam, 1935, p. 69.

31. *Wenatchee Daily World* (September 14, 1934), p. 20; (June 18, 1937), p. 8.

32. *Ibid.* (July 6, 1938), p. 8. A later article in the *Wenatchee Daily World* reported that most of the moving had been done with good grace and free from hard feelings. See *Wenatchee Daily World* (September 5, 1940), p. 14.

33. *Ibid.* (December 27, 1938), p. 1; (September 5, 1940), p. 14; Stewart H. Holbrook, *The Columbia* (New York: Rinehart and Company, 1956), p. 309; Murray Morgan, *The Dam* (New York: The Viking Press, 1954), p. 32; "Paul Bunyan's Pond," *Popular Mechanics* 75 (February 1941): 104A.

34. *Wenatchee Daily World* (March 21, 1939), p. 7; (May 4, 1939), p. 13.

35. *Ibid.* (April 16, 1941), p. 17.

36. *Ibid.* (June 21, 1939), p. 7.

37. *Ibid.* (July 20, 1933), p. 14.

38. *Ibid.* (July 14, 1937), p. 5.

39. Project History—Grand Coulee Dam, 1938, p. 56.

40. Typical of the many letters of this sort found in the records of the Bureau of Reclamation was one from Erma Gunther, Director of the University of Washington's Washington State Museum, which was addressed to Frank Banks, October 12, 1938, found in BOR Office Mss.

41. Page to J. C Stevens, President, Oregon Museum of Natural History, Portland, July 1, 1939, BOR Office Mss.

42. Floyd E. Dotson, chief clerk, to acting commissioner of Reclamation, February 15, 1939, *ibid.*

43. Mrs. J. W. Dunning to J. H. Miner, acting supervising engineer, Bureau of Reclamation, July 12, 1939, *ibid.*

44. Page to Office of Indian Affairs, March 15, 1939, *ibid.*

45. *Wenatchee Daily World* (October 12, 1939), p. 13.
46. "Preliminary Report on the Coulee Dam Archaeological Project," July 24, 1939, BOR Office Mss.
47. *Wenatchee Daily World* (May 18, 1939), p. 7; (July 29, 1939), p. 7.
48. Project History—Grand Coulee Dam, 1939, p. 65.
49. Memorandum signed by Torkelson, February 6, 1940, BOR Office Mss.
50. Project History—Grand Coulee Dam, 1939, p. 65.
51. *Wenatchee Daily World* (September 4, 1940), p. 11.
52. *Ibid.* (July 25, 1939), p. 7.
53. *Ibid.* (October 7, 1939), p. 7.
54. *Ibid.* (June 7, 1940), p. 8; (September 12, 1940), p. 8.
55. *Spokesman-Review* (March 23, 1994), p. A8.
56. *Wenatchee Daily World* (April 21, 1978), p. 3. The film, *The Price We Paid*, is available on request from the Colville Confederated Tribes, through their library in Nespelem, Washington.
57. *Wenatchee World* (July 13, 1980), p. 2. Of the tribes' 8,000 members, 2,191 voted for the agreement and 154 opposed the settlement. That left only congressional approval to conclude the deal. *Spokesman-Review* (March 23, 1994), p. A1, ff.; (March 24, 1994), p. A1, ff.; (March 27, 1994), p. 5B; (April 15, 1994), p. B1; (April 17, 1994), p. B4; *Lewiston Morning Tribune* (March 23, 1994), p. 5A; (March 27, 1994), p. 5B; (April 17, 1994), p. 7C; *Portland Oregonian* (April 17, 1994), p. D9.
58. Project History—Grand Coulee Dam, 1935, p. 81.
59. *Pacific Builder and Engineer* estimated the railroad work at $1.5 million and the total for road and railroad rebuilding at about $5 million. See "Here's What Will Be Done at Grand Coulee Dam in 1941," *Pacific Builder and Engineer* 46 (May 1941): 44; *Wenatchee Daily World* (May 8, 1940), p. 8.
60. Hall, "Final Construction Report on Lake Roosevelt Reservoir, 1952," BOR Denver Mss.
61. Anthony Netboy, *The Columbia River Salmon and Steelhead Trout: Their Fight for Survival* (Seattle: University of Washington Press, 1980), pp. 72-102.
62. *Ibid.*, p. 75, and with special note to footnote 2 from Chapter 6 on p. 161.
63. Project History—Grand Coulee Dam, 1933, p. 62. This study accompanied the study by Major John S. Butler, completed for the Army Corps of Engineers, and sometimes called the "308 Report."
64. *Wenatchee Daily World* (July 31, 1933), p. 6; (November 27, 1933), p. 1.
65. Frank Bell to Elwood Mead, March 7, 1934, BOR-FRS Mss.
66. Frank Banks to Elwood Mead, April 20, 1934, BOR Denver Mss.
67. George Sundborg, *Hail Columbia: The Thirty Year Struggle for Grand Coulee Dam* (New York: The Macmillan Company, 1954), p. 281.
68. H. H. Wrong, Canadian Legation, Washington, D. C., to Minister of Fisheries, Water Power Branch, Ottawa, Canada, October 16, 1934, Document F 30495; and William A. Found to U. D. Skelton (Undersecretary of State for External Affairs), October 27, 1934. These letters, from the Canadian government files in Ottawa, provided by Mr. Ron McCleod of Surrey, British Columbia.
69. Commissioner Elwood Mead died suddenly in January, 1926. Page became the actual Commissioner in January, 1937. Harper to Page, April 17, 1936, BOR Denver Mss.
70. *Wenatchee Daily World* (October 29, 1937), p. 8; (December 27, 1937), p. 1.

71. *Wenatchee Daily World* (February 1, 1938), p. 1; (April 9, 1938), p. 1; (April 13, 1938), p. 1.
72. Fish hatcheries were not experimental in the 1930s. The first salmon hatchery had been built by the U. S. Fish Commission on the McCloud River in northern California in 1872 and the first in the Columbia Basin on the Clackamas River in Oregon in 1877. See Netboy, *The Columbia River Salmon and Steelhead Trout*, p. 105. The whole idea of such large fish ladders and of transplanting salmon runs was experimental in the late 1930s. See Richard L. Neuberger, *Our Promised Land* (New York: The Macmillan Company, 1938), pp. 132-133.
73. "Migratory Fish Consultants," *Reclamation Era* 28 (December 1938): 243; Neuberger, *Our Promised Land*, p. 133. Also see Harold L. Ickes, *The Secret Diary of Harold L. Ickes: The First Thousand Days* (New York: Simon and Schuster, 1953), pp. 183-184.
74. *Wenatchee Daily World* (January 19, 1939), p. 1; (January 21, 1939), p. 7; "Salmon Control at Grand Coulee Dam," *Reclamation Era* 29 (March 1939): 46.
75. Project History—Grand Coulee Dam, 1939, p. 36; *Wenatchee Daily World* (March 10, 1939), p. 8; (March 25, 1939), p. 1; (March 31, 1939), p. 1; "Plans for Control of Migratory Fish at Grand Coulee Dam Approved," *Reclamation Era* 29 (April 1939): inside front cover.
76. Project History—Grand Coulee Dam, 1939, p. 37; *Wenatchee Daily World* (May 3, 1939), p. 1; (July 19, 1939), p. 1; Sterling B. Hill, "Transplanting the Migratory Salmon and Steelhead of the Upper Columbia River," *Reclamation Era* 30 (June 1940): 175. Also see "Uncle Sam to Nurse Forty Million Little Fish," *Compressed Air Magazine* 45 (June 1940): 6174-6178.
77. Note that these were 1940 dollars. *Wenatchee Daily World* (April 24, 1940), p. 72.
78. Project History—Grand Coulee Dam, 1940, p. 33.
79. *Wenatchee Daily World* (April 10, 1939), p. 1; (May 29, 1939), p. 6; (May 1, 1940), p. 1.
80. Rufus Woods, "Dam Destroyed Kettle Falls Fishery," *Wenatchee World* (July 8, 1983), p. A22-A23.
81. *Spokesman-Review* (June 14, 1940); (June 18, 1940); *Wenatchee Daily World* (June 4, 1940), p. 6; (June 15, 1940), p. 6.
82. Morgan, *The Dam*, pp. 32-34.
83. "It has been recommended that this [ending the transport of fish] be done and that the fish not be captured and transported any more as that caused much loss of life among the fish." Monthly Report of Progress of Biological Work in Connection with the Grand Coulee Fish-Salvage Program, January, 1944, report found in the BOR-FRS Mss.
84. Netboy, *The Columbia River Salmon and Steelhead Trout*, p. 74.
85. Kai N. Lee, Donna Lee Klemka, and Marion E. Marts, *Electric Power and the Future of the Pacific Northwest* (Seattle: University of Washington Press, 1980), p. 100. The work cites as its source a seven-volume environmental impact study done in 1977 by the Bonneville Power Administration which assessed its role in the Pacific Northwest Power Supply System. Netboy records the downstream loss at 15 percent per dam and quotes the Corps of Engineers as indicating that it had made progress in dealing with but had been unable to solve the problem. See Netboy, *The Columbia River Salmon and Steelhead Trout*, p. 85.

86. Herbert S. Lampman, "$10,000,000 Fish Story: With the Grand Coulee Dam Blocking the Fall Run of Spawning Salmon a Whole Industry is At Stake," *New York Times Magazine* (September 14, 1941): 16; *Wenatchee Daily World* (August 6, 1941), p. 1.
87. *Wenatchee Daily World* (October 20, 1943), p. 1; Netboy, *The Columbia River Salmon and Steelhead Trout*, p. 74.
88. Banks to Page, July 3, 1943, BOR-FRS Mss.
89. Commissioner Page to Harry Bashore, September 14, 1944, *ibid.*
90. Project History—Grand Coulee Dam, 1944, p. 38.
91. Anthony Netboy wrote that until discovery of the so-called Oregon pellet revolutionized salmon culture after 1960, the fish hatcheries suffered large losses, problems, and disease. See Netboy, *The Columbia River Salmon and Steelhead Trout*, p. 106.
92. Project History—Grand Coulee Dam, 1949, p. 55.
93. Meeting Announcement, January 26, 1944, BOR-FRS Mss. The salmon industry amounted to a $15-$20 million dollar annual business and its lobby remained most active and vociferous in any discussion of development on the river. See Albert N. Williams, *The Water and the Power: Development of the Five Great Rivers of the West* (New York: Duell, Sloan and Pearce, 1951), p. 281.
94. Memorandum to Chief Engineer from D. C. McConaughy, January 3, 1944, BOR-FRS Mss.
95. *Wenatchee Daily World* (April 20, 1944), p. 1. Anthony Netboy disagrees and wrote that stragglers headed for Grand Coulee Dam until 1946. See Netboy, *The Columbia River Salmon and Steelhead Trout*, p. 74. The *Wenatchee Daily World* reported stunned salmon at the base of the Grand Coulee spillway in May, 1943. *Wenatchee Daily World* (May 4, 1943), p. 13.
96. Project History—Grand Coulee Dam, 1946, p. 73. Actually, despite this overly optimistic prediction the Bureau had little hope that the runs would greatly increase and it attempted to explain why. In the long letter refuting charges made in a magazine article critical of the Columbia Basin Project, the director wrote, "It is not to be denied that water resource development on the Columbia River system has had an adverse effect on the fishing industry. Similarly indisputable is the fact that fishing practices of the industry have also contributed significantly to the reduction in size of Columbia River salmon runs. Coastal trolling for Columbia River fish, for example, has expanded in recent years to the point where its annual catch totals nearly 12,000,000 pounds. The size of the catch taken by sports fishermen in the Columbia system recently has increased, moreover, to an estimated 2,000,000 pounds annually. Also, changes in fishing equipment have increased the proportion of the total annual runs taken by the commercial river fisheries. A dip net fishery recently established at Celilo Falls is one example of such changes. The Department of the Interior is interested in preserving and conserving fish resources just as in conserving and developing all other natural resources. Intensive study of the whole problem of fish depletion in the Columbia with a view toward restoring and maintaining the heavy runs of the past is being continued by the Department." See acting regional director to Commissioner Page, May 15, 1946, BOR-FRS Mss.
97. Project History—Grand Coulee Dam, 1948, p. 49.
98. Netboy, *The Columbia River Salmon and Steelhead Trout*, p. 57.

99. *Wenatchee Daily World,* series of articles from February 27 through March 10, 1988, by Bob Woods.

100. Oral Bullard, *Crisis on the Columbia* (Portland: Touchstone Press, 1968), p. 111; Ivan Bloch, "The Columbia River Salmon Industry," *Reclamation Era* 28 (February 1938): 26; Netboy, *The Columbia River Salmon and Steelhead Trout,* p. 143.

101. Marc Reisner and Sarah Bates, *Overtapped Oasis: Reform or Revolution for Western Water* (Covelo, Cal.: Island Press, 1990), pp. 36-38. Reisner and Bates claim that hatchery fish are inferior to wild stock.

## Notes for Chapter Fourteen

1. Secretary of the Interior Harold L. Ickes as quoted in *Spokesman-Review* (August 20, 1941), O'Sullivan Mss.

2. Note that early CVA proposals called for a Columbia Valley Authority, but that later proposals named the agency the Columbia Valley Administration. Albert N. Williams, *The Water and the Power: Development of the Five Great Rivers of the West* (New York: Duell, Sloan and Pearce, 1951), p. 292.

3. No comprehensive history has yet been written about electric power in the Pacific Northwest, the struggles to dominate it, and the debates and manipulations involved concerning its development, distribution, and use. It is a complex subject that will be touched on but hardly explored in depth in this chapter. There is a place for a study of considerable significance.

4. Arthur Maass, *Muddy Waters: The Army Engineers and the Nation's Rivers* (Cambridge, Mass.: Harvard University Press, 1951), pp. 189, 198.

5. Daniel Miller Ogden, "The Development of Federal Power Policy in the Pacific Northwest" (unpublished Ph.D. dissertation, University of Chicago, 1949), p. 305.

6. As an example see *Wenatchee Daily World* (October 16, 1937), p. 16. Roosevelt had the notion that cheap power produced at the federal dams would set the standard for the cost of electricity throughout the Northwest and perhaps even nationally. He called it the concept of the "yardstick," where all other rates were measured against the measure established by the government. He hoped that this would break the hold of the so-called power trust.

7. These became HR 2790 and S 869 respectively.

8. Herman C. Voeltz, "Genesis and Development of a Regional Power Agency in the Pacific Northwest, 1933-1943," *Pacific Northwest Quarterly* 53 (April 1962): 66.

9. Bruce Mitchell, *Flowing Wealth: The Story of Water Resource Development in North Central Washington* (Wenatchee: *Wenatchee Daily World,* March 6, 1967), p. 55.

10. With Bonneville and Grand Coulee being built simultaneously, they faced another problem. An October 1933 article from the *Oregon Evening Journal* cautioned, "A great engineer warns friends of Bonneville not to permit electric energy from that project to be combined or be mixed in any way with Grand Coulee power. He is D[avid] C. Henny, consulting engineer for Boulder Dam. . . . The friends of Grand Coulee want such an arrangement because Bonneville power will be much lower in cost. The low-cost power there, and the transportation advantages, will be a powerful lure to attract to the Bonneville area vast metallurgical and chemical industries. And the

low cost of the power at Bonneville will place electric light and power within the reach of cities and rural communities in Oregon and Washington. Certain interests want Bonneville tied to Grand Coulee, as a service to Grand Coulee. Among them are the people who intrigued for a low dam at Bonneville, and nearly succeeded. At one time it was a conspiracy to construct the Bonneville dam solely for navigation purposes and not for power at all. Failing in that, the plotters now want the low-cost power at Bonneville tied to Grand Coulee, with power distribution placed by both projects at the same figure." *Oregon Evening Journal* (October 24, 1933), typed manuscript of the article found in BOR Denver Mss.

11. Charles McKinley, *Uncle Sam in the Pacific Northwest: Federal Management of Natural Resources in the Columbia River Valley* (Berkeley: University of California Press, 1952), pp. 157-158; Mitchell, *Flowing Wealth*, p. 42.

12. Maass, *Muddy Waters*, pp. 74-75, 101; McKinley, *Uncle Sam in the Pacific Northwest*, pp. 158-160.

13. Historian William E. Leuchtenburg argues that despite the publicity about anywhere from seven to fifteen additional TVA-like agencies, the President was never very warm toward more TVAs. His goal was unified planning, and he was willing to get it the best way possible. Roosevelt's aim, claims Leuchtenburg, was for regional planning, not TVA clones and it was the President himself who, in the end, killed the CVA idea. That may be true, but it is certain that the Bureau of Reclamation, the Army Corps of Engineers, and the people of Oregon and Washington, especially those like editor Rufus Woods, helped considerably. And it must also be noted that in June of 1937, two months before he signed the Bonneville Power Act, Roosevelt again called for the Congress to create seven regional authorities modeled on TVA and that the President continued to push for such agencies for the rest of his time in office. See William E. Leuchtenburg, "Roosevelt, Norris and the 'Seven Little TVAs,'" *Journal of Politics* (August 1952): 430-439. Also see Mitchell, *Flowing Wealth*, p. 55; *Wenatchee Daily World* (June 3, 1937), p. 1; Voeltz, "Genesis and Development of a Regional Power Agency in the Pacific Northwest," p. 69.

14. The conclusions of the Pacific Northwest Regional Planning Council had been strongly shaped by a report issued by Reed College Professor Charles McKinley. McKinley rejected any need for a TVA-like agency in the Northwest, an argument that he later elaborated in his lengthy 1952 book, *Uncle Sam in the Pacific Northwest*. See Gene Tollefson, *BPA and the Struggle for Power at Cost*, (Portland: Bonneville Power Administration, 1987), p. 127.

15. Chaired by Harold Ickes, the group included Frederick Delano of the National Resources Committee, Robert H. Healey of the Securities and Exchange Commission, Morris L. Cooke of the Rural Electrification Association, and Frank R. McNinch of the Federal Power Commission. Mitchell, *Flowing Wealth*, pp. 49-50.

16. *Ibid.*, p. 50.

17. McKinley, *Uncle Sam in the Pacific Northwest*, p. 160.

18. Voeltz, "Genesis and Development of a Regional Power Agency in the Pacific Northwest," p. 68.

19. *Wenatchee Daily World* (August 21, 1937), p. 12.

20. Homer Bone to Clarence Martin, August 17, 1937, letter found in the Governor Clarence Martin papers, Papers of the Governors.

21. Tollefson, *BPA and the Struggle for Power at Cost*, p. 128.

22. Kirby Billingsley quoted in "The Rivers of the West," *Electrical West* 129 (August 1962): 369.
23. Woods to editors, July 14, 1937, Woods Mss.
24. *Wenatchee Daily World* (October 30, 1937), p. 1.
25. *Ibid.* (February 2, 1938), p. 6; (February 8, 1938), p. 1.
26. With interruptible power, an industry could obtain cheaper blocks of power with the understanding that in times of heavy use their flow of electricity could be cut off or interrupted. See Murray Morgan, *The Dam* (New York: Viking Press, 1954), pp. 32, 39, 47; Gus Norwood, *Columbia River Power for the People: A History of Policies of the Bonneville Power Administration* (Washington, D.C.: U.S. Government Printing Office, 1981), pp. 66-68, 79-86.
27. For the last month of Ross's life, while he was ill, Charles E. Carey acted as BPA administrator.
28. Craig Wollner, *Electrifying Eden: Portland General Electric 1889-1965* (Portland: Oregon Historical Society Press, 1990), p. 208.
29. As something of a reward for his efforts, Banks's alma mater, the University of Maine, awarded him an honorary Doctorate of Engineering a few months later. *Wenatchee Daily World* (August 21, 1939), p. 1; Project History—Grand Coulee Dam, 1939, p. 37.
30. George C. Tenney, "What the Administrator Says about Bonneville," *Electrical West* 84 (January 1940): 23.
31. The BPA became Grand Coulee's marketing agent through executive order #8526. The Columbia Basin Project Act of March 10, 1943 reaffirmed that order. "President Roosevelt Coordinates Marketing of Power at Bonneville and Grand Coulee Dams," *The Commercial and Financial Chronicle* 151 (August 31, 1940): 1211.
32. As quoted in Bruce Mitchell, *The Story of Rufus Woods and the Development of North Central Washington* (Wenatchee: *Wenatchee World*, May 25, 1965), p. 26. Also see Mitchell, *Flowing Wealth*, p. 52, and Charles C. Kerr, *The World of the World: From Frontier to Community, 1905-1980* (Wenatchee: *Wenatchee Daily World*, 1980), p. 108.
33. W. T. Prosser, "Puget Sound—Hotspot of Public vs. Private Ownership," *Western Advertising* (November 5, 1940): 40; Francis X. Welch, "The Washington Outlook for Utilities—1941," *Public Utilities Fortnightly* 27 (January 2, 1941): 3.
34. This was a part of a much larger fight on the part of Ickes. The Secretary of the Interior wanted a reshaping of the Interior and Agriculture departments. Ickes would control what would become a Department of Conservation. It would include as one of its bureaus all power production at public facilities and would have a hand in regulating private power producers. See Thomas H. Watkins, *Righteous Pilgrim: The Life and Times of Harold L. Ickes 1874-1952* (New York: Henry Holt and Company, 1990), pp. 447-595.
35. Harold L. Ickes, *The Secret Diary of Harold L. Ickes: The Lowering Clouds* (New York: Simon and Schuster, 1954), p. 78.
36. Herman C. Voeltz "Proposals for a Columbia Valley Authority: A History of Political Controversy" (unpublished Ph.D. dissertation, University of Oregon), p. 63.
37. *Wenatchee Daily World* (August 14, 1941), p. 14.
38. On April 20, 1941, Harold Ickes confided to his diary that he felt McNary would back his plan for an agency with one administrator. Ickes, *The Secret Diary of Harold L. Ickes: The Lowering Clouds*, pp. 478-479. Also see Voeltz,

"Genesis and Development of a Regional Power Agency in the Pacific Northwest, p. 68; William Fred Bohrensen, "A History of the Grand Coulee Dam and the Columbia Basin Reclamation Project" (unpublished masters thesis, University of Washington, 1942), p. 94.

39. Mitchell, *Flowing Wealth*, p. 55; *Spokane Daily Chronicle*, (June 23, 1941), O'Sullivan Mss.; *Wenatchee Daily World* (June 23, 1941), p. 1.
40. Richard L. Neuberger, "Kilowatt Battlefront," *Nation* 153 (September 6, 1941): 199.
41. David E. Lilienthal, *The Journals of David E. Lilienthal, Volume I, The TVA Years, 1939-1945* (New York: Harper & Row, 1964), pp. 476-477.
42. Ogden, "The Development of Federal Power Policy in the Pacific Northwest," p. 318.
43. *Ibid.*, p. 306. Also see Marshall T. Jones, "Columbia River Basin: Study of the Sale of Electricity and the Resulting Revenues as Applied to the Repayment of Construction Costs," typed study dated August 1938 in BOR Denver Mss. The study predicted that in twenty-seven years, or before 1968, the entire cost of the power aspect of the dam would be repaid with 3 1/2 percent interest. By 1983, including the eighty cents per acre to be paid by water users, the full cost of the entire irrigation system would be repaid.
44. The entire Columbia Basin Project, including Grand Coulee Dam and all irrigation, was then estimated at $487,000,000: $342,000,000 for irrigation, $114,000,000 for power, $30,000,000 allocated to downstream benefits, and $1,000,000 to flood control and navigation. Water users on the Columbia Basin Project were, by contract, obligated to only $85 an acre in repayment and that would total, eventually, assuming it was all paid, $138,000,000. This left $349,000,000 to be repaid by power sales with $204,000,000 of that as the subsidy to irrigation. Add interest, and power users would have to repay $419,000,000. See Mitchell, *Flowing Wealth*, p. 52; Paul C. Pitzer, "Visions, Plans, and Realities: A History of the Columbia Basin Project" (unpublished Ph.D. dissertation, University of Oregon, 1990).
45. Ogden, "The Development of Federal Power Policy in the Pacific Northwest." p. 318.
46. McKinley, *Uncle Sam in the Pacific Northwest*, p. 210.
47. Mitchell, *The Story of Rufus Woods*, p. 32; Ogden, "The Development of Federal Power Policy in the Pacific Northwest," pp. 319-322.
48. Ogden, "The Development of Federal Power Policy in the Pacific Northwest," pp. 319-322. This is a fairly simplistic rendering of a complicated decision. There is not room here to detail the complete debate, which would require a book of its own.
49. Michael C. Robinson, *Water for the West: The Bureau of Reclamation, 1902-1977* (Chicago: Public Works Historical Society, 1979), p. 77.
50. As quoted in Ogden, "The Development of Federal Power Policy in the Pacific Northwest," pp. 319-322.
51. Mitchell, *Rufus Woods*, p. 32.
52. House Document 172.
53. George Macinko, "Types and Problems of Land Use in the Columbia Basin Project Area, Washington" (unpublished Ph.D. dissertation, University of Michigan, 1961), pp. 67-68.
54. Mitchell, *Flowing Wealth*, p. 57; McKinley, *Uncle Sam in the Pacific Northwest*, p. 562.

55. Ogden, "The Development of Federal Power Policy in the Pacific Northwest," p. 324; Harlan H. Edwards, "River Basin Authorities Bitterly Denounced by Reclamation," *Pacific Builder and Engineer* 51 (December 1945): 50.

56. George William Scott, "Arthur B. Langlie: Republican Governor in a Democratic Age" (unpublished Ph.D. dissertation, University of Washington, 1971), p. 366; "The Columbia Basin Inter-Agency Commission," *Pacific Builder and Engineer* 52 (June 1946): 45, editorial.

57. "CVA Boys are At it Again," *Pacific Builder and Engineer* 52 (October 1946): 46, editorial.

58. McKinley, *Uncle Sam in the Pacific Northwest*, p. 217; *Wenatchee Daily World* (May 27, 1946), p. 1; Ogden, "The Development of Federal Power Policy in the Pacific Northwest," pp. 337-338; George Sundborg, *Hail Columbia: The Thirty-Year Struggle for Grand Coulee Dam* (New York: The Macmillan Company, 1954), p. 419.

59. McKinley, *Uncle Sam in the Pacific Northwest*, p. 562.

60. "Congressman Horan Proposes Interstate Commission for Columbia Basin," *Pacific Builder and Engineer* 53 (August 1947): 47.

61. "Commission for the Basin?" *Pacific Builder and Engineer* 53 (August 1947): 44.

62. HR 1886, HR 2873, and HR 2874 all attempted to overturn the solicitor's opinion by making interest paid on power facilities unavailable to help offset irrigation costs. For details see Magnuson to O'Sullivan, April 7, 1947, Columbia Basin Commissions Mss. As Magnuson pointed out, removing the ability or the interest to be applied to reclamation would make many borderline reclamation projects in the West too expensive to be feasible.

63. As quoted in "Bureau of Reclamation Proposed Columbia Basin Plan: 238 Projects to Cost $5,600,000,000," *Pacific Builder and Engineer* 53 (June 1947): 96.

64. "Columbia Basin—Bureau Urges Consolidation Program," *Western Construction News* 22 (April 1947): 95; *Wenatchee Daily World* (December 2, 1946), p. 1; (April 18, 1947), p. 9; Hu Blonk, "New Horizons in the Columbia River Basin," *Reclamation Era* 33 (May 1947): 107.

65. "Bureau of Reclamation Proposed Columbia Basin Plan," p. 96; Maass, *Muddy Waters*, p. 119.

66. Morgan, *The Dam*, p. 73; Theron D. Weaver, "Optimum Development of Columbia Basin's Unmatched Water Resources," *Pacific Builder and Engineer* 54 (November 1948): 67; "Grand-Scale Plan Proposes Spending $3,000,000,000 for Columbia Basin," *Pacific Builder and Engineer* 54 (December 1948): 58

67. *Spokesman-Review* (June 11, 1948), O'Sullivan Mss.

68. McKinley, *Uncle Sam in the Pacific Northwest*, p. 643; "Proposed Legislation for Columbia Valley Administration," *Reclamation Era* 35 (June 1949): 139.

69. Mitchell, *Rufus Woods*, p. 34; *Wenatchee Daily World* (January 27, 1949), p. 18; (January 29, 1949), pp. 1, 10.

70. *Wenatchee Daily World* (February 17, 1949), p. 1.

71. Williams, *The Water and the Power*, p. 294.

72. Historian Leonard Arrington points out the traditional Western dislike for big government combined with a willingness to call on Washington, D.C., to finance development, build railroads, forts, and fund reclamation. See

Leonard J. Arrington, "The Sagebrush Resurrection: New Deal Expenditures in the Western States, 1933-1939," *Pacific Historical Review* 52 (February 1983): 1-16.

73. Elmo Richardson, *Dams, Parks and Politics: Resource Development and Preservation in the Truman-Eisenhower Era* (Lexington: University of Kentucky Press, 1973), pp. 37-38.

## Notes for Chapter Fifteen

1. Murray Morgan, *The Dam* (New York: The Viking Press, 1954), p. 58.
2. From the song "Grand Coulee Dam," as heard in an actual recording sung by Woody Guthrie on an album titled "Woody Guthrie: Columbia River Collection" (Rounder Records, Cambridge, Mass.), ©1958, 1963, 1976, and 1987. For a different version see: "Columbia Epic—Grand Coulee's Third Powerhouse," *Pacific Builder and Engineer* 76 (December 4-18, 1970): 6.
3. Ernest R. Abrams, *Power in Transition* (New York: Charles Scribner's Sons, 1940), p. 18.
4. Donald C. Swain, "The Bureau of Reclamation and the New Deal, 1933-1940," *Pacific Northwest Quarterly* 61 (July 1970): 145.
5. *Wenatchee Daily World* (September 24, 1940), p. 1; (September 28, 1940), p. 12.
6. Quoted in a pamphlet by Joseph G. McMacken, *Grand Coulee of Washington: The Grand Coulee Dam in Picture and Story* (Spokane: C. W. Hill Printing Company, 1948), p. 29.
7. "'Electric Army' Represented by Generators at Grand Coulee," *Scientific American* 167 (October 1942): 173.
8. Frank J. Taylor, "The White Elephant Comes Into Its Own," *Saturday Evening Post* 215 (June 5, 1943): 27.
9. "Firm Power Spells Mastery of the Air," *Pacific Builder and Engineer* 49 (July 1943): 14, editorial.
10. *Wenatchee Daily World* (January 25, 1944), p. 3. The *World* added later, "without question, no other power system in the United States, public or private, is contributing so much of its output to the winning of the war." (February 14, 1944), p. 1.
11. Quoted in *Wenatchee Daily World* (December 30, 1942), p. 5.
12. Ralph E. Dyar, *News for an Empire: The Story of the Spokesman-Review of Spokane, Washington, and of the Field it Serves* (Caldwell, Ida.: The Caxton Printers, 1952), pp. 419-420.
13. George Sundborg, *Hail Columbia: The Thirty Year Struggle for Grand Coulee Dam* (New York: Macmillan Company, 1954), pp. 430-431.
14. As quoted in *Wenatchee Daily World* (October 14, 1948), p. 20.
15. Marc Reisner, *Cadillac Desert: The American West and Its Disappearing Water* (New York: Viking, 1986), p. 165.
16. In commenting about Hoover Dam, historian Joseph E. Stevens wrote, "Hoover Dam played a vital role in this speedy defense build up. The electricity it generated flowed to Los Angeles, where it powered steel and aluminum mills and the Douglas, Lockheed, and North American aircraft plants, which built approximately sixty-two thousand fighters, bombers, and cargo planes—a fifth of the nation's entire aircraft production between 1941 and 1945." See Joseph E. Stevens, *Hoover Dam: An American Adventure*

(Norman: University of Oklahoma Press, 1988), p. 260. In his evaluation of the Tennessee Valley Authority, North Callahan wrote, "The tremendous production of TVA power helped defeat the Axis just as surely and effectively as if its great thrust had originated on the battlefields of Europe and Asia." See North Callahan, *TVA—Bridge Over Troubled Waters: A History of the Tennessee Valley Authority* (New York: A. S. Barnes and Company, 1980), p. 131.

17. Gerald D. Nash, *World War II and the West: Reshaping the Economy* (Lincoln: University of Nebraska Press, 1990), p. 1.

18. To be technically correct, there are eighteen generators rated at 108,000 kilowatts and three small-station service units of 10,000 kilowatts each, making a total of 2.28 million kilowatts for the left and right powerhouses.

19. "Aid for Bonneville: Supplemental Power from Coulee is Due Soon, and Need for it Emphasizes How Defense has Industrialized the West," *Business Week* (August 2, 1941): 18.

20. William Ira Davisson, "The Impact of Electric Power on the Economic Development of the Pacific Northwest" (unpublished Ph.D. dissertation, Cornell University, 1961), p. 83; Gene Tollefson, *BPA and the Struggle for Power at Cost* (Portland: Bonneville Power Administration, 1987), p. 216; "More Power for America," *Popular Mechanics* 77 (May 1942): 19.

21. Below the surface, however, Interior Secretary Ickes identified a potential problem. As Ickes wrote in his diary on February 2, 1941, "The desire of the Aluminum Company of America to add materially to its plant at Bonneville [actually, Vancouver, Washington] presents a grave issue. There is no doubt that we will need a great deal more aluminum for defense purposes, but if we give the company all of the power that it wants, it is bound to cause criticism later because the Aluminum Company will have what might be charged as being a monopoly of the power that is supposed to be manufactured for the people of the Northwest. The situation is all the more embarrassing because only a day or two ago the Federal Government indicted the Aluminum Company and it principal officers, along with other concerns and individuals, for conspiracy in restraint of trade." Ickes reported that he and President Roosevelt hoped to break the "aluminum trust," perhaps by diverting some power to a new venture by Henry Kaiser or some other company. After World War II, the government canceled Alcoa's wartime leases and offered to sell the aluminum plants, which the government actually built, as war surplus. Eventually, under Ickes's direction, three competing companies began producing aluminum. Morgan, *The Dam*, pp. 42-46; Harold L. Ickes, *The Secret Diary of Harold L. Ickes: Volume III, The Lowering Clouds, 1939-1941*, (New York: Simon and Schuster, 1954), entry for February 2, 1941, p. 420; Charles P. McKinley, *Uncle Sam in the Pacific Northwest: Federal Management of Natural Resources in the Columbia River Valley* (Berkeley: University of California Press, 1952), p. 182.

22. McKinley, *Uncle Sam in the Pacific Northwest*, p. 184.

23. A. L. Walker, et. al., *The Economic Significance of the Columbia Basin Project*, Washington Agricultural Experiment Station, College of Agriculture, Washington State University, Bulletin 669, September, 1966, p. 48.

24. *Wenatchee Daily World* (April 16, 1942), p. 1.

25. Gus Norwood, *Columbia River Power for the People: A History of Policies of the Bonneville Power Administration* (Washington, D. C.: U. S. Government

Printing Office, 1981), p. 221; Stewart H. Holbrook, *The Columbia* (New York: Rinehart and Company, 1956), p. 315; Morgan, *The Dam*, pp. 52-53.

26. Vernon M. Murray, "Grand Coulee and Bonneville Power in the National War Effort," *Journal of Land and Public Utility Economics* 18 (May 1942): 139.

27. *Spokesman-Review* (April 30, 1941), O'Sullivan Mss.

28. "Second 183-mile Line Allotted $2,319,400," *Electrical World* 117 (January 10, 1942): 84; *Wenatchee Daily World* (February 2, 1942), p. 6; S. E. Schultz, "Grand Coulee Linked to Bonneville by 230 Kv.," *Electrical World* 117 (April 4, 1942): 1168.

29. Project History—Grand Coulee Dam, 1942, pp. 27, 29; "Contracts Let for Three More Grand Coulee Units," *Electrical West* 87 (September 1941): 74.

30. *Wenatchee Daily World* (April 8, 1942), p. 8.

31. "House Votes $6,000,000 to Speed Grand Coulee," *Electrical World* 116 (October 25, 1941): 1337.

32. *Wenatchee Daily World* (August 18, 1942), p. 6.

33. McKinley, *Uncle Sam in the Pacific Northwest*, p. 176; *Wenatchee Daily World* (April 29, 1942), p. 10; Project History—Grand Coulee Dam, 1942, p. 30.

34. *Wenatchee Daily World* (May 4, 1942), p. 6; Memorandum from L. N. McClellan, May 7, 1942, BOR Denver Mss.; "Shasta Units Shifted to Grand Coulee," *Electrical West* 88 (June 1942): 128; *Spokesman-Review*, Progress Edition, (January 31, 1943), p. 4.

35. *Wenatchee Daily World* (February 26, 1943), p. 3; (May 7, 1943), p. 10; (November 30, 1945), p. 3; Project History—Grand Coulee Dam, 1943, p. 38.

36. Project History—Grand Coulee Dam, 1942, p. 33; *Wenatchee Daily World* (October 28, 1942), p. 2; (November 7, 1942), p. 1.

37. Project History—Grand Coulee Dam, 1943, pp. 39, 40; Project History— Grand Coulee Dam, 1944, p. 40

38. *Wenatchee Daily World* (October 11, 1943), p. 8.

39. *Ibid.* (January 4, 1943), p. 10.

40. Grand Coulee and Bonneville dams did heroic jobs producing large blocks of power during the war. From 1939 through 1946, the two together generated 33.8 billion kilowatt hours of electricity. Their installed capacity in 1945 equaled all the 150 hydroelectric and steam plants built in the Northwest during the previous fifty years. Over the war years power consumption in the Pacific Northwest rose 190 percent with industrial use rising 290 percent (most of that going to aluminum). In its first four years, Grand Coulee Dam alone produced $37,000,000 worth of electricity, or about 15,000,000,000 kilowatt hours. That would have been comparable to one million men working eight-hour days every day for seventy-eight years. The percentage of power sold by the BPA supplied by Grand Coulee Dam was as follows: 1941— .08 percent; 1942—29.1 percent; 1943—50.1 percent; 1944—62.2 percent; 1945—62.5 percent. By 1969, Grand Coulee Dam had produced $646 million worth of power. See Walker, et. al, *The Economic Significance of the Columbia Basin Project*, p. 49. Edwin J. Cohn, Jr. *Industry in the Pacific Northwest and the Location Theory* (New York: King's Crown Press, Columbia University, 1954), p. 115; *Wenatchee World* (January 29, 1970), p. 2; (June 23, 1945), p. 1; Morgan, *The Dam*, p. 58.

41. *Wenatchee Daily World* (October 16, 1944), p. 6; (October 24, 1944), p. 10.

42. Tollefson, *BPA and the Struggle for Power at Cost,* p. 228.

43. Kinsey M. Robinson, "In Union There is Strength," *Electrical West* 86 (May 1941): 27; "Public and Private Power in the Northwest," *Public Utilities Fortnightly* 27 (May 22, 1941): 643-646.
44. "Aid for Bonneville," p. 22.
45. "Strange Marriage: Bonneville-Grand Coulee Project Seems to Have Reversed a Policy in Transmission Deal with Two Utility Companies," *Business Week* (May 2, 1942): 32.
46. Herman C. Voeltz, "Genesis and Development of a Regional Power Agency in the Pacific Northwest, 1933-1943," *Pacific Northwest Quarterly* 53 (April 1962): 74; Don Campbell, "The Pacific Northwest Power Pool," *Reclamation Era* 34 (October 1948): 190.
47. McKinley, *Uncle Sam in the Pacific Northwest*, p. 178.
48. Cohn, *Industry in the Pacific Northwest and the Location Theory,* p. 109; "Power to Burn," *Fortune* 31 (February 1945): 141, 143.
49. "New Grand Coulee Units to Replace Two from Shasta," *Electrical World* 124 (July 14, 1945): 10; *Wenatchee Daily World* (August 6, 1945), p. 1.
50. Project History—Grand Coulee Dam, 1946, p. 71.
51. Schultz to McClelland, February 21, 1946, BOR-FRS Mss.; Project History—Grand Coulee Dam, 1946, pp. 47, 48.
52. When a generator produces less power than is being requested of it, due to having reached its capacity, then the frequency of sixty cycles per second drops. See *Wenatchee Daily World* (October 22, 1946), p. 1.
53. *Ibid.* (October 20, 1947), p. 10.
54. Project History—Columbia Basin Project, 1948, p. 47; Project History—Grand Coulee Dam, 1948, p. 39.
55. "Power Shortage Looms in P. N. W. Experts tell ASCE," *Pacific Builder and Engineer* 54 (August 1948): 70.
56. *Wenatchee Daily World* (December 3, 1948), p. 1; (December 16, 1948), p. 1; (January 27, 1949), p. 1.
57. Project History—Grand Coulee Dam, 1949, p. 36; "Grand Coulee's 14th Generator," *Reclamation Era* 36 (October 1950): 199.
58. "The Pacific Northwest: Land of the Big Blue River," *Time* 55 (May 15, 1950): 23.
59. Richard L. Neuberger, "Northwest Giant: President Truman, Dedicating the Grand Coulee Dam, Will Unleash a Vast Reclamation Project," *New York Times Magazine* (April 30, 1950): 18.
60. *Wenatchee Daily World* (April 18, 1950), p. 6; (May 9, 1950), p. 1; (May 10, 1950), p. 1; (May 11, 1950), p. 1; "The Pacific Northwest: Land of the Big Blue River," p. 23.
61. Project History—Grand Coulee Dam, 1950, n. p.; Project History—Grand Coulee Dam Division, v. 18, 1950, p. 36; *Wenatchee Daily World* (April 18, 1950), p. 6; (June 5, 1950), p. 2; (July 17, 1950), p. 6.
62. It is notable that just over one year later, in December 1952, the Bureau of Reclamation proposed a huge twenty-two state power grid and development of facilities that could transmit energy up to 660 miles. *Wenatchee Daily World* (June 7, 1951), p. 7; (December 30, 1952), p. 6.
63. O. E. Walsh, "The Relationship of Hydroelectric Power to the Columbia River Basin Development," *Electrical Engineering* 70 (October 1951): 861.
64. Jacob E. Warnock, "Models Guide Work on Western Dams: Experiments Aid in Design at Grand Coulee," *Civil Engineering* 6 (November 1936): 737.

65. Acting Supervising Engineer to Chief Engineer, May 2, 1952, BOR-FRS Mss.
66. *Wenatchee Daily World* (October 17, 1942), p. 6; (January 14, 1947), p. 1; Project History—Grand Coulee Dam, 1943, pp. 37, 45.
67. Banks to Chief Engineer, December 18, 1943, BOR-FRS Mss.
68. W. Nalder memorandum dated September 22, 1948, *ibid.*; W. I. Morgan, "The Floating Caisson of Grand Coulee Dam," *Reclamation Era* 34 (May 1948): 90-91.
69. Kenneth B. Keener, "Spillway Erosion at Grand Coulee Dam," *Engineering News-Record* 133 (July 13, 1944): 95.
70. Project History—Grand Coulee Dam, 1949, p. 31; *Wenatchee Daily World* (January 21, 1949), p. 6.
71. Project History—Grand Coulee Dam, 1950, unnumbered page and p. 36; *Wenatchee Daily World* (September 16, 1950), p. 6.
72. William E. Warne, *The Bureau of Reclamation*, (New York: Praeger Publishers, 1973), p. 145; *Wenatchee Daily World* (November 29, 1951), p. 6.
73. Richard Lowitt, *The New Deal and the West* (Bloomington: Indiana University Press, 1984), p. 164.
74. Banks to Commissioner, February 6, 1937, BOR Denver Mss.
75. Harold L. Ickes, "Grand Coulee Dam, a National Development," *Reclamation Era* 28 (January 1938): 1.
76. Stuart Chase, "Great Dam," *Atlantic Monthly* 162 (November 1938): 594.
77. Jerome Hugo Johnson, "Development of By-Product Power in the Columbia Basin Reclamation Project" (unpublished Ph.D. dissertation, Oregon State College, 1953), p. 19.
78. *Wenatchee Daily World* (June 1, 1948), p. 2.
79. Project History—Grand Coulee Dam, 1948, p. 40.
80. J. W. Ball, "Field Trip Report on Flow Conditions at Grand Coulee Dam During the 1948 Flood," August 5, 1948, BOR-FRS Mss.
81. *Wenatchee Daily World* (November 13, 1948), p. 1.
82. S. O. Harper to Chief Engineer, January 25, 1949, BOR-FRS Mss.
83. Darland to Boise, July 9, 1948, *ibid.*
84. Reclamation Memorandum, October 6, 1948, *ibid.*
85. R. J. Newell to Commissioner, May 18, 1949, *ibid.*
86. Theron D. Weaver was the Division Engineer for the Army Corps of Engineers, North Pacific Division. Regional Director A. N. Nelson to District Manager, April 22, 1949, *ibid.* Also see William Whipple, Jr., "Comprehensive Plan for the Columbia Basin," *American Society of Civil Engineers—Papers* 76 (November 1950): 11.
87. *Wenatchee Daily World* (April 27, 1950), p. 6; (June 9, 1950), p. 1.
88. "Conference on Grand Coulee," minutes, September 30, 1952, BOR-FRS Mss; *Columbia River and Tributaries, Northwestern United States*, 81st Congress, 2nd Session, House Document 531, Army Corps of Engineers, Vol. VII, March 20, 1950, (Washington, D.C.: U.S. Government Printing Office, 1952), p. 2824.
89. *Wenatchee Daily World* (May 12, 1954), p. 1; "Hungry Horse and Grand Coulee Will Cut High Flood Crest on Columbia River," *Engineering News-Record* 152 (May 27, 1954): 24.
90. Floyd Dominy to Henry M. Jackson, February 23, 1961, BOR Office Mss.
91. Walker, et. al, *The Economic Significance of the Columbia Basin Project*, p. 51.

92. *Ibid.*, p. 52.
93. *Wenatchee Daily World* (May 2, 1951), p. 6. Also, for a full telling of the story of the Grand Coulee Dam flood, see Morgan, *The Dam*. Morgan tells the story with considerable detail and analysis. The information for this episode told here comes largely from Morgan unless otherwise noted.
94. Frank Banks, "Report on Outage of March 14, 1952," undated report, BOR Denver Mss.
95. Holmdahl and two other men received medals in Washington, D.C., from Interior Secretary Oscar Chapman. *Wenatchee Daily World* (September 23, 1952, p. 5.
96. The internal flood cost the Bonneville Power Administration $13,650 in lost revenue, $50,000 for power from other sources, and it spoiled 85,000 pounds of aluminum.
97. *Wenatchee Daily World* (June 10, 1954), p. 1.
98. A novel by Christopher Hyde, titled *The Wave*, published in the 1970s, was built around just such a theme. Also see *Wenatchee World* special edition (July 8, 1983), p. A51.
99. *Wenatchee Daily World* (February 5, 1957), p. 1.
100. *Grant County Journal* (June 2, 1958), clipping found in the collection of Coulee Dam, Washington, City Hall.
101. *Wenatchee Daily World* (April 29, 1957), p. 9; "Grand Coulee in Color," *Reclamation Era* 42 (August 1957): 58.
102. As quoted in Tollefson, *BPA and the Struggle for Power at Cost*, p. 118.
103. Quoted in Vera Springer, *Power and the Pacific Northwest: A History of the Bonneville Power Administration* (Portland: Bonneville Power Administration, 1976), frontispiece [quoted there by permission of Huddie Ledbetter and John Lomax, TRO © 1957, 1963, Ludlow Music, Inc., New York].

## Notes for Chapter Sixteen

1. Arthur Maass, *Muddy Waters: The Army Engineers and the Nation's Rivers* (Cambridge, Mass.: Harvard University Press, 1951), p. 61.
2. For Franklin Roosevelt's commitment to and concern for planning see Frank Freidel, *Franklin D. Roosevelt: Launching the New Deal* (Boston: Little Brown & Co., 1973), pp. 78-82, 164-166, 351-353. Also see Frank Freidel, *Franklin D. Roosevelt: A Rendezvous with Destiny* (Boston: Little Brown & Co., 1990), pp. 81, 85.
3. For additional information on M. L. Wilson see William D. Rowley, *M. L. Wilson and the Campaign for the Domestic Allotment* (Lincoln: University of Nebraska Press, 1970).
4. Paul K. Conkin, *Tomorrow a New World: The New Deal Community Program* (Ithica, N. Y.: Cornell University Press, for the American Historical Association, 1959), pp. 37-39.
5. Richard Lowitt, *The New Deal and the West* (Bloomington: Indiana University Press, 1984), pp. 138-171.
6. Project History—Columbia Basin Project, 1961, Appendix, pp. 6, 8.
7. Eric Ernest Elder, "Economic Impacts of Irrigation Development in Washington" (unpublished Ph.D. dissertation, Washington State University, 1985), p. 14. Elder's dissertation, with an agricultural/economic theme, deals with all irrigation within Washington state.

8. In 1924 and 1925, congressional adjustment and amendatory acts mandated such studies to determine feasibility of projects before federal development and later control by an organized water users' association. Money for the surveys came at first from the Public Works Administration fund. Project History—Columbia Basin Project, 1935, p. 84; *Wenatchee Daily World* (August 6, 1935), p. 1; Roy M. Robbins, *Our Landed Heritage: The Public Domain, 1776-1936* (New York: Peter Smith, 1950, reprinted from New Jersey: Princeton University Press, 1942), pp. 404-405.

9. Project History—Grand Coulee Dam, 1938, p. 72; *Wenatchee Daily World* (April 5, 1937), p. 6; *Grant County Journal*, (September 3, 1937).

10. *Wenatchee Daily World* (June 17, 1935), p. 7.

11. Grant Gordon, Associate Engineer to Construction Engineer, Department of the Interior, October 24, 1935, General Correspondence Engineering files, BOR Denver Mss.

12. The work took over four years and involved 45,000 soil borings, hundreds of test pits, and detailed examination of over 8,000 soil samples. William Fred Bohrnsen, "A History of the Grand Coulee Dam and the Columbia Basin Project" (unpublished masters thesis, University of Washington, 1942), p. 53; W. W. Johnston, "Preparing for Irrigation," *Reclamation Era* 38 (April 1952): 80.

13. Project History—Grand Coulee Dam, 1941, p. 65; William E. Warne, *The Bureau of Reclamation* (New York: Praeger Publishers, 1973), p. 127. Class One was the best, followed by Class Two and Three, all of which were arable. Classes Four and Five were to remain unused for various reasons. Class Six land was not arable. The project area in total encompassed 2,500,000 acres and previous estimates of irrigable acreages usually ran to about 1,200,000 acres. The surveys showed that fewer acres could actually be served due to elevation, soil quality, and other reasons.

14. Project History—Grand Coulee Dam, 1934, p. 24; "Settlement of the Columbia Basin Reclamation Project," Department of the Interior, Bureau of Reclamation, 1944, revised and printed 1946, from a brochure found in BOR Denver Mss.

15. *Spokesman-Review* (June 8, 1935), O'Sullivan Mss.

16. Project History—Columbia Basin Project, 1946, 27; Clayton R. Koppes, "Public Water, Private Land: Origins of the Acreage Limitation Controversy, 1933-1953," *Pacific Historical Review* 47 (November 1978): 612; Mel A. Hagood, "Training New Agents," *Reclamation Era* 38 (April 1952): 81; Chester F. Cole, "Factors Relative to Settlement in the Columbia Basin Reclamation Project," *Journal of Geography* 47 (January 1948): 1-8; *Wenatchee Daily World*, (July 30, 1937), p. 20.

17. Koppes, "Public Water, Private Land," p. 613.

18. Warne, *The Bureau of Reclamation*, pp. 63, 123; Richard L. Berkman and W. Kip Viscusi, *Damming the West: Ralph Nader's Study Group Report on the Bureau of Reclamation* (New York: Grossman Publishers, 1973), pp. 132-133. The Reclamation Act of 1939 was also known as the Wheeler-Case Act.

19. Lowitt, *The New Deal and the West,* p. 90; Craig Lynn Infanger, "Income Distributional Consequences of Publicly Provided Irrigation: The Columbia Basin Project" (unpublished Ph.D. dissertation, Washington State University, 1974), p. 58; Robert Albert Weinkauf, "The Columbia Basin Project, Washington: Concept and Reality, Lessons for Public Policy" (unpublished Ph.D. dissertation, Oregon State University, 1973), p. 8.

20. Weinkauf, "The Columbia Basin Project," p. 8.
21. *Wenatchee Daily World* (July 29, 1937); John C. Page to Jim Ford, Spokane Chamber of Commerce, July 1, 1937, BOR Denver Mss.
22. Frank Banks to Bernard E. Stoutmeyer, July 2, 1937, BOR Denver Mss.; James O'Sullivan, "The Struggle for Grand Coulee Dam," p. 69, typed unpublished manuscript held in O'Sullivan Mss.; Bohrnsen, "A History of the Grand Coulee Dam and the Columbia Basin Project," p. 61.
23. *Grant County Journal* (August 28, 1938). Fear of Spokane dominating the project came from the competition between the pumpers and the gravity people in the 1920s. That fear had prevented formation of one large irrigation district in 1930. See *Wenatchee Daily World* (December 23, 1930). p. 1.
24. James O'Sullivan to John C. Page, November 12, 1937, BOR Denver Mss.
25. John C. Page to Frank Banks, October 30, 1937, BOR Denver Mss.; *Grant County Journal* (August 6, 1937); George Sundborg, *Hail Columbia: The Thirty-Year Struggle for Grand Coulee Dam,* (New York: The Macmillan Company, 1954), p. 328.
26. *Wenatchee Daily World* (December 8, 1937), p. 1; Sundborg, *Hail Columbia,* p. 331.
27. Of the approximately 2,000 landowners in the proposed Quincy-Columbia Basin Irrigation District, only about 150 lived there. Nearly 600 were in Seattle. To inform them, and other absentee owners, district backers created the West Columbia Basin Land Owners League, headquartered in Seattle. Members of the old Quincy District insisted that the name Quincy be somehow retained. Project History—Grand Coulee Dam, 1938, p. 56; Sundborg, *Hail Columbia,* p. 346.
28. *Wenatchee Daily World* (February 20, 1939), p. 1; (February 21, 1939), p. 1; *Spokane Spokesman-Review* (February 20, 1939), clipping in BOR Denver Mss.
29. *Wenatchee Daily World* (February 21, 1939), p. 1.
30. John C. Page to Raymond F. Walter, Bureau of Reclamation engineer, May 10, 1939, BOR Denver Mss.
31. *Wenatchee Daily World* (August 11, 1939), p. 9.
32. *Ibid.* (December 11, 1939), p. 1; Bernard E. Stoutmeyer to John C. Page, December 11, 1939, BOR Denver Mss. The vote in the east was 617 to 67. In the south it was 735 to 49.
33. Sundborg, *Hail Columbia,* p. 359; *Wenatchee Daily World* (October 21, 1939), p. 6.
34. "Joint Investigations, Columbia Basin Project," *Reclamation Era* 30 (August 1940): 219.
35. Barrows's field coordinator was Dr. Edward N. "Nat" Torbert, who had been one of his students. Torbert quickly opened offices in Ephrata. The Department of Agriculture named Marion Clawson, then with its Bureau of Agricultural Economics, as its field representative for the work. The Bureau of Agricultural Economics was the one remaining planning agency left in the Department of Agriculture after 1935. The Department of Agriculture and the Department of the Interior often competed with each other, especially when Secretary of the Interior Ickes moved to reform his department and include in it some agencies held by Agriculture, particularly the Forest Service. Cooperation with the Department of the Interior on the Joint Investigations caused difficulty for Clawson and others in the Agriculture Department. For additional information see Richard S. Kirkendall, *Social Scientists and Farm Policies in the Age of Roosevelt* (Columbia: University of Missouri Press,

1966), pp. 223-225; Hagood, "Training New Agents," p. 81; Project History—Grand Coulee Dam, 1939, p. 53; *Wenatchee Daily World* (July 11, 1939), p. 7.

36. Carl C. Taylor, "The Sociologists' Part in Planning the Columbia Basin," *American Sociological Review* 11 (November 1946): 321.

37. "Joint Investigations, Columbia Basin Project," p. 219; Weinkauf, "The Columbia Basin Project," p. 52.

38. Department of the Interior Bureau of Reclamation, *Columbia Basin Joint Investigations; Character and Scope* (Washington, D.C.: U.S. Government Printing Office, 1941), passim.

39. James Eugene Brooks, "Settlement Problems Related to Farm Size in the Columbia Basin Project, Washington" (unpublished Ph.D. dissertation, University of Washington, 1957), p. 28; Warne, *The Bureau of Reclamation*, 124. "It is history's greatest major technical planning job. The area is as large as Delaware," wrote the Spokane *Spokesman-Review*. See *Spokesman-Review*, Progress Edition (January 25, 1942), part 2, 4.

40. "Columbia Basin: Test for Planning," *New Republic* 107 (September 7, 1942): 279.

41. *Ibid.*

42. Taylor, "The Sociologists' Part in Planning the Columbia Basin," p. 331. The emphasis is his.

43. Alvin F. Darland to John C. Page, March 20, 1943, BOR Denver Mss. The Larson Air Base at Moses Lake and the Ephrata Air Base each brought between 3,000 and 5,000 people into the area. See Elbert E. Miller, "Economic and Social Changes in the Columbia Basin, Washington," *Land Economics* p. 46 (November 1965): 336.

44. *Wenatchee Daily World* (March 29, 1941), p. 6.

45. *Ibid.* (February 3, 1942), p. 1.

46. *Spokesman-Review* (March 7, 1942), O'Sullivan Mss. For an influential, widely read, similar criticism of planning see the intellectual polemic against centralized state planning by the Central European emigré, Friedrich A. Hayek: *The Road to Serfdom* (Chicago: University of Chicago Press, 1944). Political scientist Herbert Finer countered with an impassioned rebuttal to Hayek in his *Road to Reaction* (Boston: Little Brown and Company, 1945). In the post-New Deal climate of the middle and late 1940s, public opinion was moving with the anti-planning philosophy of Hayek.

47. "Settlement of the Columbia Basin Reclamation Project," *Wenatchee Daily World* (March 11, 1943), p. 7.

48. Infanger, "Income Distributional Consequences of Publicly Provided Irrigation," p. 63; Yahaya Doka, "Policy Objectives, Land Tenure, and Settlement Performance: Implications for Equity and Economic Efficiency in the Columbia Basin Irrigation Project" (unpublished Ph.D. dissertation, Washington State University, 1979), pp. 63-64; Harold L. Ickes to Bureau of Reclamation, March 18, 1943, BOR Denver Mss.

49. Subbarayan Prasanna, "Morphogenesis of the Settlement Landscape in the Columbia Basin Project Area" (unpublished Ph.D. dissertation, University of Washington, 1976), p. 118; Warne, *The Bureau of Reclamation*, p. 127.

50. George Macinko, "The Columbia Basin Project: Expectations, Realizations, Implications," *Geographical Review* 53 (April 1963): 187.

51. Warne, *The Bureau of Reclamation*, p. 127.

52. Brooks, "Settlement Problems Related to Farm Size in the Columbia Basin Project, Washington," p. 55.

Grand Coulee: Harnessing a Dream

53. Otis Graham argues that national planning ended in 1935 when the Supreme Court declared parts of the National Industrial Recovery Act unconstitutional. Regional and local efforts continued, however. See Otis L. Graham, *Toward a Planned Society: From Roosevelt to Nixon* (New York: Oxford University Press, 1976), pp. 30, 32, 83. The death of Elwood Mead in 1936 further removed another advocate for planning and for the Columbia Basin Project. See Conkin, *Toward a New World*, p. 55.

54. Roy F. Bessey, *Pacific Northwest Regional Planning—A Review*, Bulletin No. 6, Division of Power Resources (Olympia: State of Washington Printing Plant, 1978), p. 19.

55. It would have been impossible to involve resident farmers in pre-planning for the project because those farmers did not materialize, for the most part, until after the delivery of water, at which point most of the planning had been completed. Lowitt argues that in the Pacific Northwest, planning was done more on a regional and state level than in any other part of the country. See Lowitt, *The New Deal and the West*, p. 138.

56. *Wenatchee Daily World* (January 30, 1943), p. 1.

57. Its seven members included a representative from each irrigation district, three members appointed by the governor, and the director of the State Department of Conservation and Development as chairman. The original members were Edward "Deke" Davis as chairman, J. A. Weber of Quincy, Don Damon of Cunningham for the east district, W. C. Marshall of Connell for the south district, and Thomas D. Welborn, Fred Cunningham, and Rufus Woods appointed by the governor. Members changed frequently over the years. *Wenatchee Daily World* (April 15, 1943), p. 12; Minutes of the Columbia Basin Commission, April 12, 1943, Columbia Basin Commissions Mss.

58. John C. Page to Chief Engineer, Denver, March 13, 1940, BOR Denver Mss.

59. Project History—Grand Coulee Dam, 1940, p. 53; *Wenatchee Daily World* (October 1, 1940), p. 1.

60. The first year of water delivery was an experimental year. The next ten years were the developmental period. Charges would start in the twelfth year. Not all land was to be assessed $85.00 an acre, which was actually only an average. High quality land would pay more, and low quality land would pay less. However, from the start there would be assessments for day-to-day operation and maintenance of the project applied equally to all land. Project History—Grand Coulee Dam, 1944, p. 25; *Wenatchee Daily World* (March 7, 1944), p. 1; (March 17, 1944), p. 1; Harry W. Bashore to Harold L. Ickes, March 6, 1945, BOR-FRS Mss.

61. *Wenatchee Daily World* (March 23, 1945), p. 1. The 25 percent is figured on the assumption that no interest is charged on the money. The percentage is considerably lower if an interest component is included. This percentage also assumed that 100 percent of the project would be developed and that all of the landowners of all of the land would make payments.

62. *Wenatchee Daily World* (February 6, 1945), p. 1.

63. Hal Holmes to James O'Sullivan, March 23, 1945, Columbia Basin Commissions Mss. For legal reference to repayment contracts, see 59 *United States Statutes at Large*, Public Law 39, April 24, 1945, p. 75.

64. Ninety-eight percent of those voting approved the repayment contracts. Project History—Columbia Basin Project, 1946, 44; *Wenatchee Daily World* (July 23, 1945), p. 1; (August 2, 1945), p. 1.

65. *Wenatchee Daily World* (September 1, 1945), p. 1.

66. *Ibid.* (October 10, 1945), p. 1.
67. *Ibid.* (August 13, 1945), p. 1.
68. Alfred R. Golzé, *Reclamation in the United States* (New York: McGraw-Hill Book Company, Inc., 1952), p. 83.
69. Warne, *The Bureau of Reclamation*, p. 240.
70. For a brief description of Commissioner of Reclamation Michael Straus see Marc Reisner, *Cadillac Desert: The American West and Its Disappearing Water* (New York: Viking Penguin Inc., 1986), esp. pp. 143-146; Koppes, "Public Water, Private Land," p. 620.
71. *Wenatchee Daily World* (July 1, 1948), p. 1.
72. George William Scott, "Arthur B. Langlie: Republican Governor in a Democratic Age" (unpublished Ph.D. dissertation, University of Washington, 1971), p. 159.
73. *Wenatchee Daily World* (February 15, 1949), p. 8.
74. *Ibid.* (July 24, 1948), p. 6. In 1946, *Business Week* wrote, "Construction on the Columbia Basin Project to be the Single Biggest Construction Job in the Pacific Northwest and Single Biggest Job ever Contemplated [bigger even than Grand Coulee Dam]." See "Big Diggings," *Business Week* (May 18, 1946), p. 21.
75. Project History—Grand Coulee Dam, 1945, p. 42.
76. *Wenatchee Daily World* (January 14, 1946), p. 1.
77. George Sundborg, "Narrative of Grand Coulee Dam from James O'Sullivan Records," p. 52, unpublished manuscript in O'Sullivan Mss.
78. Project History—Grand Coulee Dam, 1950, opening unnumbered page; *Wenatchee Daily World* (July 20, 1949), p. 1. The road relocation cost the State of Washington $4.5 million; see *Wenatchee Daily World* (July 18, 1949), p. 1.
79. Project History—Columbia Basin Project, 1946, p. 16.
80. "Lytle-Amis-Green Works New Equipment on Fourth Longest Dam," *Pacific Builder and Engineer* 53 (August 1947): 54. Also see Sundborg, *Hail Columbia*, p. 428.
81. *Moses Lake Herald* (February 6, 1948), O'Sullivan Mss.
82. Sundborg, *Hail Columbia*, p. 428; *Wenatchee Daily World* (June 30, 1949), p. 1.
83. *Wenatchee Daily World* (September 9, 1948), p. 1; (September 24, 1948), p. 1; (September 27, 1948), p. 1.
84. *Ibid.* (February 16, 1949), p. 1; Sundborg, *Hail Columbia*, p. 449.
85. William "Billy" Clapp died on May 16, 1965 at the age of eighty-seven. *Wenatchee Daily World* (May 16, 1965), p. 1.
86. Department of the Interior information release to the press, September 3, 1958, BOR Denver Mss.; *Wenatchee Daily World* (December 16, 1957), p. 1; (June 17, 1958), p. 4; (July 31, 1958), p. 1.
87. Carl Blom, "Development of the Hydraulic Design for the Grand Coulee Pumps," *Transactions of the American Society of Mechanical Engineers* 72 (January 1950): 53.
88. "Bids Called on Columbia Basin Pumps: World's Largest," *Pacific Builder and Engineer* 52 (January 1946): 52; "Grand Coulee Pumps—1600 cfs Each," *Power Plant Engineering* 50 (March 1946): 76; *Wenatchee Daily World* (June 10, 1946), p. 1.
89. Jerome Hugo Johnson, "Development of By-Product Power in the Columbia Basin Reclamation Project" (unpublished Ph.D. dissertation, Oregon State College, 1953), p. 25.

90. *Wenatchee Daily World* (May 17, 1951), p. 6; Charles P. McKinley, *Uncle Sam in the Pacific Northwest: Federal Management of Natural Resources in the Columbia River Valley* (Berkeley: University of California Press, 1952), p. 217.
91. *Wenatchee Daily World* (March 26, 1946). p. 1; H. T. Nelson, regional director for the Bureau of Reclamation to Glenn C. Lee, Publisher of the *Tri-City Herald*, January 3, 1947, letter in the files of the CBD League Mss.
92. Project History—Columbia Basin Project, 1948, p. 59. For a fuller discussion of the control zones see Michele Stenehjem Gerber, *On the Home Front: The Cold War Legacy of the Hanford Nuclear Site* (Lincoln: University of Nebraska Press, 1992), pp. 107-112.
93. Project History—Columbia Basin Project, 1949, p. 35.
94. Nelson to Lee, January 3, 1967, CBD League Mss.; Project History—Columbia Basin Project, 1958, p. 7.
95. Richard White, from a paper delivered at the Atomic West Conference, University of Washington, September 25, 1992.
96. Taylor, "The Sociologists' Part in Planning the Columbia Basin," p. 321; Project History—Columbia Basin Project, 1946, p. 26. The population of the project area in 1985 was under 80,000. See "Department of the Interior, Bureau of Reclamation, Columbia Basin Project Population, information sheet of 1985." The population figures include the military complexes of Moses Lake, Larson Air Force Base, and also the Moses Lake Job Corps facility.
97. Murray Morgan, *The Dam* (New York: Viking Press, 1954), p. 62; Stewart H. Holbrook, *The Columbia* (New York: Rinehart and Company, 1956), p. 320.
98. W. J. Granberg, "How Much Land Can a Man Farm Best?" *Evergreen* 1 (September 1946): 16; *Planning Farm Units*, brochure produced by the Department of the Interior, 1949, pp. 2-3; McKinley, *Uncle Sam in the Pacific Northwest*, p. 141.
99. *Seattle Times* (March 7, 1946), O'Sullivan Mss.; press release dated May 22, 1946 from the Department of the Interior, Columbia Basin Commissions Mss.
100. Although there were around 8,000 total owners of project lands, those who had withdrawn were no longer included.
101. Project History—Grand Coulee Dam, 1947, p. 60, 71; *Wenatchee Daily World* (March 14, 1947), p. 6; *Columbia Basin Herald* (August 1, 1947), O'Sullivan Mss.
102. *Farming Opportunities, Columbia Basin Project*, booklet issued by the Bureau of Reclamation, June 23, 1949, BOR Denver Mss.
103. *Farmer's Handbook*, booklet issued by the Bureau of Reclamation and the Agricultural Science Department of the State College of Washington, circa 1950-1951, Columbia Basin Commissions Mss.
104. Project History—Grand Coulee Dam, 1945, p. 41; *Wenatchee Daily World* (January 7, 1947), p. 9.
105. Project History—Columbia Basin Project, 1951, p. 44.
106. *Wenatchee Daily World* (January 27, 1949), p. 8; (October 2, 1948), p. 1; (October 14, 1950), p. 1; (July 28, 1952), p. 6; (October 5, 1953), p. 1.
107. The Pasco Pump Unit was powered with Grand Coulee Dam electricity. Project History—Columbia Basin Project, 1946, p. 15; *Wenatchee Daily World* (April 11, 1946), p. 1.

108. *Wenatchee Daily World* (February 6, 1948), p. 2.
109. "Pasco Unit to Get Water," *The Northwest* 22 (April 1948): 8. *The Northwest* was a monthly publication of the Northern Pacific Railroad.
110. Project History—Columbia Basin Project, 1948, p. 161; "First Water to Flow in Columbia River Basin Project," *Newsweek* 31 (May 3, 1948): 36; *Wenatchee Daily World* (May 15, 1948), p. 1.
111. Today Block 1 (the Pasco Unit) receives water from the Potholes Canal.
112. In all, veterans qualified for about 75 percent of the farm units sold by the government. From 1952 through 1960, veterans had preference in drawings for units. In 1960 the law giving this preference expired. However, veterans, as well as others, had to prove two years experience in farm work, and had to have $4,500 in liquid assets before they could apply. See Warne, *The Bureau of Reclamation*, p. 132.
113. In 1956 the Bureau of Reclamation announced that the number of veteran applicants exceeded the number of farms offered for sale through the system of drawings by fifty-six times. See *Progress and Prospects: Columbia Basin Project, Washington*, brochure produced by the Bureau of Reclamation, March 1, 1956, p. 23. Also see: *Wenatchee Daily World*, (January 24, 1951), p. 1; (January 8, 1951), p. 10; (June 18, 1951), p. 1; Project History—Columbia Basin Project, 1951, p. 44.
114. Project History—Columbia Basin Project, 1951, p. 46; *Wenatchee Daily World* (June 4, 1946), p. 1; Milo W. Hoisveen, "Sprinkler Tests on Columbia Basin Project," *Reclamation Era* 35 (August 1949), p. 18.
115. *Wenatchee Daily World* (October 15, 1951), p. 6.
116. Project History—Columbia Basin Project, 1951, p. 44.
117. *Wenatchee Daily World* (January 3, 1952), p. 1.
118. *Ibid.* (September 1, 1957), p. 1; Project History—Columbia Basin Project, 1957, p. 10e.
119. Today George, Washington, is located near Ephrata at a junction with Interstate Highway 90. It consists of two or three small buildings and an undetermined population not exceeding twenty to thirty.
120. The lift depends on the level of Franklin D. Roosevelt Lake, the reservoir behind Grand Coulee Dam. This lift can be anywhere from 292 to 310 feet or more.
121. Bureau of Reclamation, *The Story of the Columbia Basin Project* (Washington, D. C.: U. S. Government Printing Office, 1978), p. 16; *Wenatchee Daily World* (May 24, 1951), p. 1; (June 15, 1951), pp. 1, 7; (June 12, 1951), p. 2; (June 8, 1951). p. 1; "World's Largest Pump Starts to Fill Coulee Reservoir for Basin Project," *Electrical World* 135 (June 25, 1951): 6. Also see Paul C. Pitzer, "A 'Farm-in-a-Day' The Publicity Stunt and Celebrations That Initiated the Columbia Basin Project," *Pacific Northwest Quarterly* 82 (January 1991): 2-7.
122. Typed report on the "Water-of-All-States" Ceremony dated June 14, 1951, Columbia Basin Commissions Mss.
123. *Wenatchee Daily World* (October 31, 1951), p. 6.
124. *The Big Search*, a sixteen-page pamphlet by the Veterans of Foreign Wars of Kansas City, circa 1952, Columbia Basin Commissions Mss.
125. *Wenatchee Daily World* (May 1, 1952), p. 6; Hugh H. Moncrieff, "Columbia Basin's Showplace," *Reclamation Era* 38 (September 1952): 214.
126. Donald Dunn left the Columbia Basin Project after only a few years. The rest of his story is part of the next chapter.

127. Morgan, *The Dam*, p. 61; *Spokane Daily Chronicle* (May 28, 1952), p. 1; *Spokesman-Review* (May 30, 1952), p. 1.
128. Columbia Basin Commission, *Biennial Report of the Columbia Basin Commission of the State of Washington* (Olympia: Washington State Printing Office, September 30, 1952), p. 3.
129. Scott, "Arthur B. Langlie," p. 239. The Columbia Valley Administration, considered off and on since 1934 and earlier called the Columbia Valley Authority, was patterned after the Tennessee Valley Authority. It would have superseded the Bonneville Power Administration, the Army Corps of Engineers, the Bureau of Reclamation, and other government agencies as the federal planning and governing agency in the Columbia Basin.
130. Elmo Richardson, *Dams, Parks and Politics: Resource Development and Preservation in the Truman-Eisenhower Era* (Lexington: University of Kentucky Press, 1973), pp. 74, 115.
131. *Wenatchee Daily World* (December 1, 1952), p. 1; (December 31, 1952), p. 1. Also see *Spokane Daily Chronicle* (December 31, 1952), p. 1.
132. Richardson, *Dams, Parks and Politics*, p. 115; Michael C. Robinson, *Water For the West: The Bureau of Reclamation, 1902-1977* (Chicago: Public Works Historical Society, 1979), pp. 79-80.
133. Robinson, *Water for the West,* p. 81; Geoffrey Wandesforde-Smith and Robert Warren. *A Comparative Analysis of American and Canadian Government Arrangements for the Development of Regional Water Policy in the Columbia River Basin* (Seattle: Washington Water Resources Research Center, Office of Water Resources, Research Completion Report for the University of Washington, September 1970), pp. 19, 77. McCarthy and others seeing planning and reclamation as subversive is ironic. In the 1930s, Roosevelt's efforts at planning met resistance because they were seen as something done by Fascists. See Graham, *Toward a Planned Society*, p. 296.
134. *Wenatchee Daily World* (April 23, 1953), p. 1; (May 16, 1953), p. 1.
135. *Ibid.* (May 29, 1956), p. 8. Eisenhower proposed putting forty million acres into the "soil bank" which meant that at least crops in surplus would not be grown on that land. Proposed January 9, 1956, the controversial and expensive program finally passed Congress the following May.
136. CH2M Hill, *Draft Environmental Impact Statement: Continued Development of the Columbia Basin Project, Washington* (Boise, Ida.: Bureau of Reclamation, September 20, 1989), p. I-6. Although nearly half a million acres could technically receive water by 1960, in most years only between 45 percent to 60 percent of project lands were actually irrigated. Land unit development was slower than development of irrigation facilities and at all times some farmers held land out either for fallow or for other reasons. See E. R. Franklin, W. U. Fuhriman, and B. D. Parrish, *Economic Problems and Progress of Columbia Basin Project Settlers* Washington Agricultural Experiment Stations, Institute of Agricultural Sciences, State College of Washington, Bulletin 597, January 1959, p. 2.
137. Project History—Columbia Basin Project, 1952, p. 3. If 1940 is considered the base, prices rose 230 percent by January 1, 1952 and 240 percent by December 1952.
138. Rudolph Ulrich, "Relative Costs and Benefits of Land Reclamation in the Humid Southwest and the Semiarid West," *Journal of Farm Economics* 35 (February 1953): 67. This amount did not include the power subsidy.

139. Murray A. Straus and Bernard D. Parrish. *The Columbia Basin Settler: A Study of Social and Economic Resources in New Land Settlement*: Washington Agricultural Experiment Stations Institute of Agricultural Sciences, State College of Washington, Bulletin 566, May 1956, pp. 3, 4, 7, 9.

140. The Bureau of Reclamation and the Columbia Basin Commission received occasional letters from farmers complaining that obtaining loans through the Farmer's Home Administration was not as easy as they had been led to believe. For example, see Mrs. Earl Halverson to Hubert Walters of the Columbia Basin Commission, May 3, 1956, Columbia Basin Commissions Mss. The letter is typical of a number that appeared in the files. Also see *Wenatchee Daily World* (October 26, 1956), p. 1.

141. Straus and Parrish, *The Columbia Basin Settler*, pp. 9, 11, 27; *Wenatchee Daily World* (April 10, 1953), p. 6.

142. *Wenatchee Daily World* (March 19, 1953), p. 1.

143. "From Waste Land to Farm Land," *Business Week* (June 7, 1952): 98.

144. Project History—Columbia Basin Project," 1955, p. 7; Kenneth Clyde Scott, "The Redistributional Consequences of Public Recreation Provision at the Potholes Reservoir—Columbia Basin Project, Washington" (unpublished Ph.D. dissertation, Washington State University, 1975), p. 37; *Wenatchee Daily World* (May 2, 1953), p. 6; "A Preliminary Report on the Fish and Wildlife Resources and Development Plans Recommended for Columbia Basin Project, Columbia River, Washington," typed report dated August 1953, BOR Denver Mss.

145. *Wenatchee Daily World* (May 1, 1957), p. 5. For comment on the growth of Moses Lake see Rita G. Seedorf and Martin F. Seedorf, "Runways and Reclamation: The Influence of the Federal Government on Moses Lake," *Columbia: The Magazine of Northwest History* 8 (Summer 1984): 30-37.

146. *Ibid.* (May 2, 1962), p. 7. Note that the population in the project area reached a peak early in the 1970s and that it has been dropping since then. In 1989 it was about 80,000.

147. There is a problem in confusing the term "farm" and the term "unit." A unit was an area designed by the Bureau of Reclamation and deemed sufficient land for a family to make a living by farming. No one owner was legally able to possess more than one unit. But owners of one unit could, by making arrangements with the Bureau, rent and work additional units. In looking at Bureau statistics and newspaper reports, it is often unclear whether the term farm means only one unit or multiple units, or if the terms unit and farm are being used interchangeably. Considerably more will be said about renting and about the size of the units in the next chapter. It is sufficient to point out here that the practice of extensive renting may not have contravened the letter of the law, but it certainly stood in opposition to its original intent.

148. Department of the Interior, *Progress and Prospects: Columbia Basin Project*, brochure dated March 1, 1956, p. 3, Banks Mss.

149. *Wenatchee Daily World* (January 3, 1955), p. 6; (February 17, 1959), p. 3.

150. *Progress and Prospects: Columbia Basin Project*, p. 9.

151. Some, such as historian Donald Worster, would ask whether they investigated the right questions in the first place—items such as the possible long-term effect on the environment. See Donald Worster, *Under Western Skies: Nature and History in the American West* (New York: Oxford University Press, 1992), esp. pp. 64-78.

# Notes for Chapter Seventeen

1. Project History—Columbia Basin Project, 1959, p. 9.
2. T. J. Fatherree and Cecil Hurst, "The Spa Treatment of Thromboangitis Obliterans at Soap Lake, Washington," *American Heart Journal* 35 (August 1941): 180; George W. Fuller, *A History of the Pacific Northwest: With Special Emphasis on the Inland Empire* (New York: Alfred A. Knopf, 1952 edition), p. 18.
3. Chief Design Engineer to District Manager in Ephrata, November 21, 1951, BOR-FRS Mss.
4. A. J. Davidson, Chief of Design Branch to Supervising Engineer, January 30, 1953, as contained in "Pertinent Data, Soap Lake Problem, Columbia Basin Project," Volume I, Ephrata, Washington, June 1, 1953, a typed document found in BOR Denver Mss.; Assistant Secretary of the Interior Fred G. Aandahl to Senator Henry M. Jackson, April 6, 1953, Jackson Mss. Aandahl pointed out in his letter that there was little evidence that the United States was responsible for the rise in the level of Soap Lake.
5. "Report on Conditions at Soap Lake, Washington," February 1954, typed manuscript, BOR Denver Mss.
6. "Proceedings of Soap Lake Conference, Chief Engineers Office, Confidential," typed report dated June 2-3, 1954, 2, 13, 62, BOR-FRS Mss; *Wenatchee Daily World* (May 30, 1954), p. 10.
7. H. T. Nelson, Regional Director to Commissioner of Reclamation Wilbur Dexheimer, February 15, 1954, BOR Denver Mss.
8. Warren G. Magnuson to James E. Murray, January 12, 1956; Magnuson to Wilbur A. Dexheimer, January 16, 1956, both in BOR-FRS Mss.
9. *Wenatchee Daily World* (April 27, 1956), p. 1.
10. *Ibid.* (June 10, 1956), p. 8.
11. *Ibid.* (August 20, 1970), p. 9. In 1975 the Bureau of Reclamation announced in the environmental impact study of that year that its action to save Soap Lake had worked, and that the area had no more worries. See *Wenatchee World* (February 10, 1975), p. 3. In 1984, three years of heavy rainfall again caused Soap Lake to rise and it reached 1956 levels. However it stabilized and dropped again in 1985. See Project History—Columbia Basin Project, 1984, p. 24; 1985, p. 21.
12. There had been some warnings about potential drainage problems in general. In 1913 Frederick Newell wrote, "It frequently happens that excess water accumulates on the low lands, reducing the value of these. It is sometimes possible, under such conditions, to utilize cheap power transmitted by electrical devices in draining these lands, this waste water being used for the reclamation of additional areas. Where drainage waters are used for irrigation, attention must be given to their quality, as they frequently contain large quantities of harmful alkali salts in solution." See Frederick Haynes Newell and Daniel William Murphy, *Principles of Irrigation Engineering: Arid Lands, Water Supply, Storage Works, Dams, Canals, Water Rights and Products* (New York: McGraw-Hill Book Company, Inc., 1913), p. 129.
13. Project History—Columbia Basin Project, 1948, pp. 59-60; George Macinko, "Types and Problems of Land Use in the Columbia Basin Project Area, Washington" (unpublished Ph.D. dissertation, University of Michigan, 1961), p. 108.

14. Ronald Albert Weinkauf, "The Columbia Basin Project, Washington: Concept and Reality, Lessons for Public Policy" (unpublished Ph.D. dissertation, Oregon State University, 1973), pp. 125-126. Weinkauf quoted William O. Watson, Chief of the Drainage Branch of the Project, who commented on drainage problems in 1962.
15. *Wenatchee Daily World* (February 11, 1953), p. 9.
16. *Ibid.* (March 11, 1955), p. 9; (March 29, 1955), p. 6.
17. Macinko, "Types and Problems of Land Use in the Columbia Basin Project Area, Washington," p. 115.
18. Murray A. Straus and Bernard D. Parrish, *The Columbia Basin Settler: A Study of Social and Economic Resources in New Land Settlement* (Pullman: Washington Agricultural Experiment Station Institute of Agricultural Sciences, State College of Washington, Bulletin 566, May 1956), p. 36.
19. *Wenatchee Daily World* (August 8, 1956), p. 10; (September 16, 1956), p. 1.
20. Project History—Columbia Basin Project," 1955, p. 10.
21. *Ibid.*, 1956, p. 2d.
22. *Wenatchee Daily World* (June 6, 1957), p. 26.
23. *Ibid.* (April 5, 1961), p. 9.
24. Assistant Secretary of Interior Fred G. Aandahl to Senator Henry M. Jackson, February 7, 1958, Columbia Basin Commissions Mss.
25. Project History—Columbia Basin Project, 1957, p. 10f.
26. Under the law, the farmers had eleven years to begin their farms free from making the repayment charges on the basic irrigation. Macinko, "Types and Problems of Land Use in the Columbia Basin Project Area, Washington," p. 98.
27. Weinkauf, "The Columbia Basin Project, Washington: Concept and Reality, Lessons for Public Policy," pp. 99-100.
28. Straus and Parrish, *The Columbia Basin Settler*, p. 8; George Macinko, "The Columbia Basin Project: Expectations, Realizations, Implications," *Geographical Review* 53 (April 1963): 193.
29. E. R. Franklin, W. U. Fuhriman, and B. D. Parrish, *Economic Problems and Progress of Columbia Basin Project Settlers* (Pullman: Washington Agricultural Experiment Stations, State College of Washington, Bulletin 597, January 1959), p. 16.
30. *Wenatchee Daily World* (March 6, 1955). p. 1.
31. In this respect, Dunn was less unique than he maintained as other farmers throughout the project, including many on developing farms, frequently complained that getting loans, either from the Farmer's Home Administration or from banks, was very difficult. The files of the Columbia Basin Commission in the Washington State Archives in Olympia contain many letters from farmers voicing such complaints.
32. *Wenatchee Daily World* (February 13, 1958). p. 1.
33. *Wenatchee World* (October 8, 1976), p. 3; (July 8, 1983), p. A53.
34. William E. Warne, *The Bureau of Reclamation* (New York: Praeger Publishers, 1973), p. 132.
35. Subbarayan Prasanna, "Morphogenesis of the Settlement Landscape in the Columbia Basin Project Area" (unpublished Ph.D. dissertation, University of Washington, 1976), p. 124.
36. United States Army Corps of Engineers, *Columbia Basin Water Withdrawal Environmental Review. Appendix B—Social Effects—Part II—Community*

*Social Effects* (Washington, D.C.: U.S. Government Printing Office, May 1980), p. 41.

37. *Wenatchee Daily World* (November 17, 1955), p. 12.
38. *Ibid.* (March 23, 1956), p. 6.
39. Prasanna, "Morphogenesis of the Settlement Landscape in the Columbia Basin Project Area," p. 113; *Wenatchee Daily World* (November 17, 1955), p. 12.
40. Press release dated July 16, 1956, Columbia Basin Commissions Mss.
41. Project History—Columbia Basin Project, 1957, p. 5b.
42. *Wenatchee Daily World* (January 3, 1957), p. 1; (January 4, 1957). p. 1; Project History—Columbia Basin Project, 1957, p. 4.
43. Prasanna, "Morphogenesis of the Settlement Landscape in the Columbia Basin Project Area," p. 119; also see Project History—Columbia Basin Project, 1957, p. 3a.
44. The Boise Project in Idaho was 65.3 acres, on the Umatilla Project in Oregon it was 38.7 acres, on the Owyhee Project in Oregon it was 61 acres, and on the Yakima Project right next door it was only 35.9 acres. Prasanna, "Morphogenesis of the Settlement Landscape in the Columbia Basin Project Area," p. 117.
45. Fred G. Aandahl to James E. Murray, March 8, 1957, BOR-FRS Mss.
46. *Wenatchee Daily World* (March 6, 1957). p. 6.
47. *The Wall Street Journal* (March 27, 1957), p. 1; *Wenatchee Daily World* (March 29, 1957), p. 8.
48. *The Wall Street Journal* (March 27, 1957), p. 1. In 1973, William Warne wrote, "Some Joint Investigation recommendations are rendered obsolete by technology—for example—all fences do not need to be 'horse-high, bull-strong, or hog-tight.' Some farmers have no fences—electric fences now control most livestock." Warne, *The Bureau of Reclamation*, p. 137.
49. 74 *United States Statutes at Large*, Public Law 85-264, p. 590-591. Commenting a few years later, Senators Magnuson and Jackson wrote, "In 1956 we introduced legislation to lift some of the land ownership limitations which were ham-stringing Basin farmers and pushed this bill through Congress in 1957 despite opposition from the Bureau of Reclamation." Magnuson and Jackson to Earl Coe, Director of the Washington State Department of Conservation, telegram dated September 1, 1961, Columbia Basin Commissions Mss.
50. Weinkauf, "The Columbia Basin Project, Washington: Concept and Reality, Lessons for Public Policy," p. 150.
51. *Columbia Basin Water Withdrawal Environmental Review. Appendix B*, p. 7.
52. Raymond Moley, *What Price Federal Reclamation?* (New York: American Enterprise Association, Inc., 1955), p. 44. The figure includes not only irrigation, but also the costs of Grand Coulee Dam and its power facilities.
53. Macinko, "Types and Problems of Land Use in the Columbia Basin Project Area, Washington," p. 98; Project History—Columbia Basin Project, 1957, p. 10f.
54. Project History—Columbia Basin Project, 1957, p. 10g; Fred G. Aandahl to Henry Jackson, February 7, 1958, Columbia Basin Commissions Mss.
55. In 1930, Rufus Woods had stated that power would pay for the entire project. In 1932 he claimed that the project would pay itself off in thirty years through power revenues. See *Wenatchee Daily World* (October 30, 1930), p. 4; (May 27, 1932), p. 1.

56. Murray Morgan, *The Dam* (New York: The Viking Press, 1954), 63; Walker R. Young, "World's Greatest Pumps For Grand Coulee Project," *Pacific Builder and Engineer* 51 (May 1945): 51; Macinko, "Types and Problems of Land Use in the Columbia Basin Project Area, Washington," p. 105.

57. Note that this is almost exactly the contract which the Bureau and the farmers finally signed six years later. Had the farmers agreed in 1957, the cost to them would have been considerably lower.

58. Macinko, "Types and Problems of Land Use in the Columbia Basin Project Area, Washington," p. 96; Project History—Columbia Basin Project, 1957, p. 10h.

59. Earl Coe, Chairman of the Columbia Basin Commission to Joel D. Wolfson, Commission lawyer, September 18, 1958, Columbia Basin Commissions Mss.

60. *Wenatchee Daily World* (September 14, 1958). p. 2; (December 19, 1958). p. 7.

61. Macinko, "Types and Problems of Land Use in the Columbia Basin Project Area, Washington," p. 106.

62. Macinko, "The Columbia Basin Project: Expectations, Realizations, Implications," p. 90; Macinko, "Types and Problems of Land Use in the Columbia Basin Project Area, Washington," p. 105.

63. Macinko, "The Columbia Basin Project: Expectations, Realizations, Implications," p. 192.

64. Both sides used statistics to their advantage. While the farmers might have been able to pay more in dollars, because of inflation, their standard of living may well have dropped. On the other hand, the $85 repayment cost was based on 1940 dollars, and had that cost been adjusted for inflation, something not allowed in the contract, the farmers would have been obligated to considerably more per acre. Macinko, "Types and Problems of Land Use in the Columbia Basin Project Area, Washington," p. 104; *Wenatchee Daily World* (May 4, 1958), p. 6.

65. Macinko, "Types and Problems of Land Use in the Columbia Basin Project Area, Washington," p. 101. Senator Magnuson unsuccessfully tried to have the percentage of the project attributed to flood control increased, which would have lowered repayment costs. Earl Coe, Director of the Columbia Basin Project to Joel Wolfsohn, January 30, 1959, Columbia Basin Commissions Mss. More recently, in the 1970s, in light of added upstream storage capacity resulting from new dams in the United States and Canada, and connected with the third powerhouse at Grand Coulee Dam, the flood control component has been recalculated and significantly increased.

66. *Spokane Daily Chronicle* (May 5, 1959), O'Sullivan Mss.; *Wenatchee Daily World* (May 5, 1959). p. 2.

67. *Wenatchee Daily World* (June 9, 1959), p. 7.

68. *Ibid.* (July 2, 1959). p. 9; (July 14, 1959). p. 12.

69. *Ibid.* (July 29, 1959), p. 1; Annual Project History—Columbia Basin Project, 1959, p. 7.

70. *Tri-City Herald* (July 29, 1959), p. 1, editorial.

71. Operation and maintenance charges, like the repayment charges, varied with the quality of the land. While some farmers paid more, and some less, the averages are used here as often as possible. *Wenatchee Daily World* (September 21, 1959), p. 1; (October 2, 1959), p. 10; Project History—Columbia Basin Project, 1959, p. 7; Max E. Kirkpatrick to Henry M. Jackson, February 15, 1960, Jackson Mss.; *Tri-City Herald* (April 16, 1959), Magnuson Mss.

72. *Tri-City Herald* (November 18, 1959), p. 1.
73. *Wenatchee Daily World* (November 19, 1959), p. 1; (November 20, 1959), p. 1. In April 1960 the East and South districts reaffirmed that they would not pay the drainage assessments for 1960. See William Clapp to Willis T. Batcheller, April 20, 1960, Batcheller Mss.
74. There was precedent for such action on the part of the Bureau of Reclamation. In 1921 the government had withheld water from irrigated land on the Yakima Indian Reservation. That land was farmed by white renters who refused to pay five-dollar water charges. See William Stuart Forth, "Wesley L. Jones: A Political Biography" (unpublished Ph.D. dissertation, University of Washington, 1962), p. 474.
75. *Wenatchee Daily World* (November 27, 1959), p. 2.
76. *Ibid.* (December 14, 1959), p. 9.
77. *Ibid.* (December 22, 1959), p. 10; (December 23, 1959), p. 2.
78. *Ibid.* (February 9, 1960), p. 1. Nalder resigned in order to become the project manager of the Helmand Valley Project in Afghanistan. That project was advised and administered by the Bureau of Reclamation. In his absence, L. Vaughn Downs served for about the next three months as acting Columbia Basin Project Manager. See Project History—Columbia Basin Project, 1960, p. 1.
79. "Four-flusher" is a poker term indicating that the person identified is bluffing. *Wenatchee Daily World* (February 10, 1960), p. 1.
80. *Ibid.* (February 10, 1960), p. 7.
81. Magnuson and Jackson to James E. Murray, February 16, 1960, Columbia Basin Commissions Mss.
82. Magnuson and Jackson included in their letter to James E. Murray, Chairman of the Committee on Interior and Insular Affairs, "It is incredible that the Interior Department, with the vast resources of personnel and records at its disposal, should produce this delay. If the officials of the Department do not have readily available to them the fundamental facts which you requested, we must seriously question the quality of their decisions. Their failure in this instance is consistent, however, with the failures and ineptitude which have characterized the handling of amendatory contract negotiations with the water users." *Ibid.*
83. *Wenatchee Daily World* (March 13, 1960), p. 2.
84. *Ibid.* (April 19, 1960). p. 7; (April 24, 1960), p. 2.
85. Project History—Columbia Basin Project, 1960, p. 2b.
86. *Ibid.*, 1961, appendix, p. 12.
87. *Wenatchee Daily World* (May 17, 1960), p. 1; Project History—Columbia Basin Project, 1960, p. 2b.
88. *Wenatchee Daily World* (May 19, 1960), p. 2; (May 20, 1960), p. 1.
89. *Ibid.* (July 20, 1960), p. 6; Project History—Columbia Basin Project, 1960, p. 2b; Earl Coe to Magnuson, July 28, 1960, Magnuson Mss.
90. Project History—Columbia Basin Project, 1960, p. 2b.
91. *Wenatchee Daily World* (October 6, 1960), p. 11; Project History—Columbia Basin Project, 1960, p. 2c.
92. John F. Kennedy to Albert D. Rosellini, November 4, 1960, Batcheller Mss.
93. *Wenatchee Daily World* (November 6, 1960), p. 2.
94. In Grant County, the heart of the Columbia Basin Project, Nixon, with 6,785 votes, defeated Kennedy, who drew 6,554 votes. *Wenatchee Daily World* (November 9, 1960), p. 11.

95. Walter Le Page to Warren Magnuson, December 4, 1960, Magnuson Mss.
96. *Wenatchee Daily World* (January 8, 1961), p. 8; Undated memo, c. 1963, Jackson Mss.
97. *Wenatchee Daily World* (April 18, 1961), p. 12; Project History—Columbia Basin Project, 1961, p. 10. The board consisted of its chairman, William R. Gianelli, former assistant to the California Director of Water Resources; LaSelle E. Coles, manager of the Ochoco Irrigation District, Prineville, Oregon, and President of the National Reclamation Association; and Dr. Roy E. Huffman, Dean of Agriculture and Director of the experimental station at Montana State University.
98. *Wenatchee Daily World* (September 1, 1961), p. 3.
99. Columbia Basin Project Board of Consultants, "Report to the Secretary of the Interior Re Repayment Problems Columbia Basin Project," mimeographed report, July 28, 1961, copy located in North Central Regional Library, Wenatchee; Project History—Columbia Basin Project, 1961, p. 10; appendix, pp. 6-32.
100. The Quincy board recommended fast action, the South board looked favorably at new overtures from the Bureau, and the East District remained reluctant. When negotiations revealed that the repayment costs in the first ten years of the new contract would drop from $2.12 per acre to one dollar and that overpayments made by farmers in 1961 would return over $300,000 to farmers on their greatly reduced operation and maintenance charges in 1962, the atmosphere warmed appreciably, especially among the Quincy people. *Wenatchee Daily World* (September 6, 1961), p. 10; (September 15, 1961), p. 12.
101. *Ibid.* (February 8, 1962), p. 3.
102. The vote was 2,263 to 297. See Project History—Columbia Basin Project, 1962, p. 11; Assistant Secretary of the Interior Kenneth Holum to Senator Magnuson, March 15, 1962, Magnuson Mss.
103. Magnuson and Jackson to Columbia Basin Commission, September 1, 1961, telegram in Columbia Basin Commissions Mss.
104. 76 *United States Statutes at Large*, Public Law 87-728, pp. 677-679; *Wenatchee Daily World* (July 19, 1962), p. 1; (October 3, 1962), 1; Project History— Columbia Basin Project, 1962, p. 11. Senator Henry C. Dworshak of Idaho protested that the deal gave Columbia Basin farmers lower rates than those on Idaho projects. Before the final vote in the Senate, Dworshak died of a sudden heart attack.
105. Secretary Stewart Udall to George R. Locke, Chairman of the Joint Columbia Basin Irrigation District Boards, September 1, 1961, Batcheller Mss.; Project History—Columbia Basin Project, 1962, p. 15; *Wenatchee Daily World* (June 14, 1963), p. 8.
106. The Quincy District favored joint operation, the East District favored separate operation and the South District remained undecided. See Project History—Columbia Basin Project, 1965, pp. 12-13.
107. The districts agreed to establish a joint committee to handle problems that affected all aspects of the project. See Project History—Columbia Basin Project, 1965, p. 14.
108. The contract signed was the sixth draft of a document painfully negotiated over hundreds of hours. See Spokane *Spokesman-Review* (December 24, 1968), CBD League Mss.
109. *Wenatchee Daily World* (December 19, 1968), p. 1.

110. Weinkauf, "The Columbia Basin Project, Washington: Concept and Reality, Lessons for Public Policy," p. 155.
111. A study in 1973 done under the direction of Ralph Nader claimed that while the Bureau of Reclamation frequently comments on the virtue of repayment on reclamation projects, actually the facts and figures offered are contrived to hide enormous subsidies. See Richard L. Berkman and W. Kip Viscusi, *Damming the West: Ralph Nader's Study Group Report on the Bureau of Reclamation* (New York: Grossman Publishers, 1973), p. 131.
112. *Wenatchee Daily World* (March 13, 1961), p. 5.
113. See *Tri-City Herald* (September 23, 1970), CBD League Mss.
114. Clayton R. Koppes, "Public Water, Private Land: Origins of the Acreage Limitation Controversy, 1933-1953," *Pacific Historical Review* 47 (November 1978): 609.
115. *Wenatchee Daily World* (July 6, 1967), p. 2.
116. "Irrigation Blocks, Acreages and Farm Units, Columbia Basin Project— 1948-1987," fact sheet provided by the Bureau of Reclamation, Columbia Basin Project Office, Ephrata, Washington.

## Notes for Chapter Eighteen

1. *Seattle Post-Intelligencer* (October 28, 1985), p. A5.
2. Project History—Columbia Basin Project, 1964, p. 13.
3. Typed membership list for the Columbia Basin Development League for 1968 and 1969, CBD League Mss. The League's budget for 1968 was $15,111.93 and in 1969 it was $18,480. David A. Gallant, a Realtor and insurance company owner in Pasco, became president of the League in 1968.
4. Project History—Columbia Basin Project, 1964, p. 13; *Wenatchee Daily World* (December 3, 1964), p. 3.
5. Project History—Columbia Basin Project, 1965, p. 16.
6. *Wenatchee Daily World* (February 20, 1970), p. 1; (April 27, 1970), p. 7; (October 4, 1970), p. 1; (February 19, 1971), p. 1; Interview with James V. Cole, Project Manager, Columbia Basin Project, Ephrata, Washington, March 23, 1989. Also see *Wenatchee World* (February 23, 1973), p. 2; (March 30, 1973), p. 13; (December 12, 1973), p. 1.
7. William E. Warne, *The Bureau of Reclamation* (New York: Praeger Publishers, 1973), p. 128.
8. Project History—Columbia Basin Project, 1965, p. 16.
9. *Wenatchee Daily World* (December 1, 1966), p. 6.
10. Project History—Columbia Basin Project, 1966, p. 57.
11. *Ibid.*, p. 10.
12. *Spokane Spokesman-Review* (November 10, 1967), CBD League Mss.
13. *Columbia Basin Herald* (December 28, 1965).
14. *Ibid.* (September 5, 1968).
15. *Ibid.* (September 5, 1968).
16. Floyd Dominy to Regional Director, Boise, July 1968, memo in BOR Office Mss.
17. Lee D. Dumm to Chief Engineer, July 28, 1969, *ibid.*
18. Commissioner of Reclamation to Regional Director, Boise, April 17, 1974, *ibid.*
19. *Wenatchee World* (November 9, 1980), p. 2. An article in the *Seattle Times* in 1971 indicated that 50 percent of the project would require expensive drainage

works. However, it pointed out that unlike central California, the Columbia Basin was not plagued with salts and alkali. See *Seattle Times*, magazine section (January 31, 1971), p. 12.

20. Warne, *The Bureau of Reclamation*, p. 143.
21. *Tri-City Herald* (October 20, 1969), BOR Ephrata Mss.
22. *Wenatchee Daily World* (October 28, 1963), p. 1.
23. C. E. Brown and E. M. Tomsic, "Pump-Generating Units for the Grand Coulee Pumping-Generating Plant," *I. E. E. Transactions on Power Apparatus and Systems* 92 (June 1973): 1057; *Wenatchee World* (August 24, 1971), p. 6; (September 6, 1973), p. 2; (October 31, 1973), p. 13; Project History—Columbia Basin Project, 1970, p. 1.
24. *Wenatchee World* (May 27, 1979), p. 3; Project History—Grand Coulee Project Office, vol. V, 1979, p. 2.
25. *Wenatchee Daily World* (May 23, 1968), p. 1; (September 20, 1968), p. 8.
26. *Seattle Daily Journal of Commerce* (March 21, 1969); *Tri-City Herald* (April 1, 1969), CBD League Mss. Also see *Wenatchee Daily World* (April 4, 1969), p. 1.
27. *Grant County Journal* (April 7, 1969), CBD League Mss. Otis Graham argues that the Nixon administration was an active agent for planning, and that it used the Bureau of the Budget as its main controller. Assuming that this is accurate, as will be seen below, the purpose of the Bureau of the Budget in this case was to control, if not even curtail, reclamation. See Otis L. Graham Jr., *Toward a Planned Society: From Roosevelt to Nixon* (New York: Oxford University Press, 1976), pp. xii, 69, 206.
28. *Wenatchee Daily World* (June 11, 1970), p. 7; (January 29, 1971), p. 1.
29. Project History—Columbia Basin Project, 1971, p. 6.
30. *Wenatchee World* (May 1, 1974), p. 1.
31. In 1972 the Washington state voters had approved $75 million in bonds to finance public water facilities. The legislation that passed in 1975 simply allocated $15 million of that to the second Bacon Siphon and Tunnel. See Eric Ernest Elder, "Economic Impacts of Irrigation Development in Washington" (unpublished Ph.D. dissertation, Washington State University, 1985), p. 28.
32. *Wenatchee World* (February 21, 1975), p. 1; (February 26, 1975), p. 1.
33. *Wenatchee World* (May 11, 1975), p. 2; (May 18, 1975), p. 2; (May 29, 1975), p. 2; (June 29, 1975), p. 3; (November 25, 1975), p. 3; (December 17, 1975), p. 1; "Columbia Basin Project Second Bacon Siphon and Tunnel, State-Local-Federal Cost Sharing Report," August 1975, typed report, BOR Office Mss.
34. Alan Sanford Kezis, "An Examination of Economies of Size and Revenues on Columbia Basin Farmers: Implications for Acreage Limitation Policy" (unpublished Ph.D. dissertation, Washington State University, 1978), p. 8.
35. CH2M Hill, *Draft Environmental Impact Statement: Continued Development of the Columbia Basin Project, Washington* (Boise, Ida.: Bureau of Reclamation, September 1989), p. I-14.
36. *Wenatchee World* (May 11, 1975), p. 2.
37. The charge of $1,560 was the portion to be paid by the water users. The Bureau tentatively figured the full cost of development at $2,153 per acre. See "Proposed Contractual Arrangements for the Initial Development of the Columbia Basin Project Extension," typed report dated April 5, 1976, BOR Office Mss.

38. *Wenatchee World* (May 18, 1975), p. 2.
39. Kezis, "An Examination of Economies of Size and Revenues on Columbia Basin Farmers: Implications for Acreage Limitation Policy," p. 9. The farmers paid the $131.60 over a fifty-year period on a sliding scale. Each year the amount rose gradually. This repayment fee was in addition to the yearly operation and maintenance fees.
40. *Ibid.*, p. 10.
41. Professor of Agricultural Economics Norman K. Whittlesey represented one position to emerge from the faculty at Washington State University. His appears to have conflicted with the opinions of E. E. Weeks and Arthur W. Peterson, also agricultural economists at the school. Generally the Agricultural Extension Station of the university, another department, supported the project and many of their bulletins are also cited here. The various Ph.D. dissertations which have come out of Washington State University, from the Agricultural Economics Department and from other departments, also cited here, cover the full range of opinions about the Columbia Basin Project. None of those dissertations was directed by Professor Whittlesey, although he sat on the committees for Eric Elder and Craig Infanger. Whittlesey telephone interview, 1989.
42. *Wenatchee World* (January 9, 1976), p. 1; (January 14, 1976), p. 1; (February 18, 1976), p. 1.
43. *Ibid.* (May 21, 1976), p. 2.
44. *Ibid.*, p. 3; (August 5, 1976), p. 1.
45. *Ibid.* (April 6, 1976), pp. 1, 2; (April 13, 1976), p. 1; (May 16, 1976), p. 1.
46. *Ibid.* (June 29, 1976), p. 1.
47. *Ibid.* (October 3, 1976), p. 11; (October 7, 1976), p. 1; (October 11, 1976), p. 1; Project History—Columbia Basin Project, 1976, p. 2.
48. *Seattle Post-Intelligencer* (June 27, 1978), BOR Ephrata Mss. Carter's action may have been influenced in small part by the fact that in the 1976 election he lost every state west of Missouri except Texas.
49. *Wenatchee World* (January 6, 1977), p. 3; (March 9, 1977), p. 1; (September 28, 1978), p. 2.
50. *Grant County Journal* (October 5, 1978), BOR Ephrata Mss.
51. *Seattle Post-Intelligencer* (June 27, 1978), *ibid.*
52. Norman K. Whittlesey, *Benefits and Costs of Irrigation Development in Washington, Volume I, Executive Summary* (Olympia: State Printing Plant for the House of Representatives and the Legislative Transportation Committee, Washington State Legislature, October 1976), pp. 3, 4, 8, 9, 17.
53. *Seattle Post-Intelligencer* (June 27, 1978), BOR Ephrata Mss.
54. About 20 percent of the water pumped for irrigation eventually returns to the Columbia River.
55. *Wenatchee World* (September 26, 1977), p. 1; (June 3, 1977), p. 1; (May 1, 1978), p. A-39; (May 27, 1977), p. 6. Carter agreed to fund such projects (or refused to veto them) reluctantly after realizing the political gridlock powerful congressional representatives like Jackson and Magnuson could bring to all of his programs if he did not. Of course, it did not help his cause that he had appointed a Secretary of Interior—Cecil Andrus—from an irrigation state—Idaho—who took to Washington with him an old Idaho irrigation ally, R. Keith Higgenson, as Bureau of Reclamation Commissioner. In this struggle, Carter did not even have the complete backing of two of his most important staffers on water policy.

56. Project History—Columbia Basin Project, 1979, p. 51.
57. *Wenatchee World* (May 23, 1978), p. 8; (November 27, 1979), p. 3; (December 1, 1978), p. 3; Project History—Columbia Basin Project, 1979, p. 1; 1980, p. 1.
58. *Wenatchee World* (March 5, 1981), p. 1; (March 12, 1981), p. 14; (April 1, 1981), p. 10; (July 22, 1981), p. 2.
59. *Ibid.* (October 11, 1981), p. 14; (October 17, 1982), p. 18.
60. Cole interview.
61. Project History—Columbia Basin Project, 1982, pp. 1, 9; *Draft Environmental Impact Statement*, p. vi; *Wenatchee World* (December 15, 1982), p. 1.
62. *Wenatchee World* (December 15, 1982), p. 1; (January 6, 1983), p. 3.
63. Cole interview. For information on further development of plans allowing farmers to build their own access lines in order to get water out of canals provided by the Bureau of Reclamation see *Wenatchee World* (April 18, 1982), p. 3.
64. *Wenatchee World* (April 13, 1977), p. 1; (February 27, 1978), p. 2.
65. Richard Lowitt, *The New Deal and the West* (Bloomington: Indiana University Press, 1984), pp. 96-97.
66. *Wenatchee World* (November 11, 1962), p. 3.
67. Carlos A. Schwantes, *The Pacific Northwest: An Interpretive History* (Lincoln: University of Nebraska Press, 1989), p. 349.
68. *Wenatchee World* (November 22, 1978), p. 2.
69. *Ibid.* (December 19, 1978), p. 1; (January 5, 1979), p. 1; (January 19, 1979), p. 1.
70. *Ibid.* (January 28, 1979), p. 1; (February 5, 1979), p. 2; (February 14, 1979), p. 7; (March 1, 1979), p. 1; (March 11, 1979), p. 1.
71. *Ibid.* (March 15, 1979), p. 2; (March 19, 1979), p. 1; (March 20, 1979), p. 1.
72. Project History—Columbia Basin Project, 1979, pp. 3, 50.
73. A year later nature contributed a new problem when, on May 18, 1980, Mt. St. Helens erupted and covered much of eastern Washington with a heavy coat of volcanic ash. The amount varied from a trace at Pasco to fifteen centimeters in Ritzville. It ruined much of the early alfalfa crop and clogged or broke down machinery. However, project farmers noted that the ash killed most of the grasshoppers that season, and the Bureau of Reclamation found that here and there it sealed sections of the leakier irrigation canals. The ash produced no long-term damage, and some speculated that it increased soil fertility. Cole interview; Project History—Columbia Basin Project, 1980, 1; 1981, p. 15.
74. Yahaya Doka, "Policy Objectives, Land Tenure, and Settlement Performance: Implications for Equity and Economic Efficiency in the Columbia Basin Irrigation Project" (unpublished Ph.D. dissertation, Washington State University, 1973), p. 48. The case in question was *Yellen vs. Hickel*, 352 F. supp. 1972.
75. Clayton R. Koppes, "Public Water, Private Land: Origins of the Acreage Limitation Controversy, 1933-1953," *Pacific Historical Review* 47 (November 1978): 607. Also see Donald Worster, *Rivers of Empire: Water, Aridity, and the Growth of the American West* (New York: Pantheon Books, 1985), p. 299.
76. Carter people were encouraged by a group called National Land for the People, Incorporated. That organization was a reform group that looked largely at holdings in California and advocated strict adherence to the law as a way to revive family farms. Worster, *Rivers of Empire*, pp. 294, 299.

77. Two studies, done through the Department of Agriculture at Washington State University, argued to the contrary, maintaining that larger farms were not more efficient, and that farms of 160 to 320 acres reached maximum economies. Doka, "Policy Objectives, Land Tenure, and Settlement Performance," p. 133; Kezis, "An Examination of Economies of Size and Revenues on Columbia Basin Farmers: Implications for Acreage Limitation Policy," p. 91. Also see *Wenatchee World* (November 16, 1977), p. 1; (March 17, 1978), p. 2; Worster, *Rivers of Empire*, pp. 304-305.

78. Marc Reisner argues that Carter's attack on reclamation and other water projects cost him his re-election in 1980. See Reisner, *Cadillac Desert: The American West and Its Disappearing Water* (New York: Viking Penguin, Inc., 1986), pp. 324-344. Also see, Worster, *Rivers of Empire*, pp. 301-302.

79. Richard L. Berkman and W. Kip Viscusi, *Damming the West: Ralph Nader's Study Group Report on the Bureau of Reclamation* (New York: Grossman Publishers, 1973), p. 148.

80. *Wenatchee World* (July 18, 1982), p. 1; (September 30, 1982), p. 5; (October 1, 1982), p. 1; (October 14, 1982), p. 1; Project History—Columbia Basin Project, 1982, p. 12. Also see 92 *United States Statutes at Large*, Public Law 97-293, October 12, 1982, p. 1263.

81. W. H. Keating for Ellis L. Armstrong to Timothy Atkenson, General Counsel, Council on Environmental Quality, May 27, 1971, BOR Office Mss. This means a finished project would draw 5 percent of the flow of the Columbia River at Grand Coulee, and return later about 2 1/2 percent. This does not necessarily agree with other figures.

82. Cole interview; interview with Craig Sprankle, Public Information Officer, Grand Coulee Dam, Bureau of Reclamation, March 24, 1989, Grand Coulee Project Office, Grand Coulee, Washington. Cole pointed out that under existing law, irrigation and the power to pump irrigation water has first call on the Columbia River. He admitted that the law could change, and in order to forestall that, the Bureau and other government agencies coordinate the pumping of irrigation water so that it will not interfere with power generation or other uses of the river, such as navigation and fisheries. Cole believed that the greatest challenge to use of the water might come from Native American groups. They are making legal challenges concerning use of the water for recreation, maintenance of fish runs, and irrigation of their lands. For an article quoting Bonneville Power Administrator Sterling Monro pointing out that power for irrigation is set aside and guaranteed by law see *Wenatchee World* (April 6, 1979), p. 2.

83. *Wenatchee World* (January 12, 1976), p. 2.

84. *Ibid.* (July 25, 1973), p. 1; (October 30, 1975), p. 1; (April 6, 1979), p. 2.

85. Bruce W. Cone, et. al., *An Analysis of Agricultural Potential in the Pacific Northwest with Respect to Water and Energy*, Report Number 2311103556 from the Northwest Agricultural Development Project, Sponsored by the Pacific Northwest Regional Commission, Northwest Economic Associates, Vancouver, Washington, December 1979, p. vii.

86. *Tri-City Herald* (May 28, 1981), BOR Ephrata Mss. Much of this was probably based on the work of Professor Norman K. Whittlesey.

87. *Wenatchee World* (April 5, 1979), p. 1.

88. Washington State Legislature, *Report on Proposed Completion of the Columbia Basin Project: Policy Question and Findings, Report No. 84-18* (Olympia: State Printing Plant, December 1984), p. 9.

89. *Wenatchee World* (February 17, 1983), p. 1.
90. Elder, "Economic Impacts of Irrigation Development in Washington," p. 28. Elder alludes to the doctrine of "first user" adopted in most Western states and argues that this influenced concern on the part of irrigators.
91. *Tri-City Herald* (October 26, 1985), BOR Ephrata Mss.
92. *Draft Environmental Impact Statement*, p. I-17. It is significant to note here that since construction of the third powerhouse at Grand Coulee Dam, little water has flowed over the spillway there. The combination of increased power generation and greatly enlarged upstream storage has evened the flow of the river so that today little water is wasted. During the tourist season, the Bureau of Reclamation agreed to spill water twice daily for thirty minutes sessions as a public relations gesture. In the summer of 1989, part of that spill was replaced with a laser light show. The estimated savings in lost power potential amount to about $250,000 yearly, which will more than repay the $100,000 cost of the lighting. Sprankle interview.
93. Elder, "Economic Impacts of Irrigation Development in Washington," p. 16.
94. *Wenatchee World* (April 5, 1979), p. 2.
95. Carol Prochaska spoke for the Bureau of Reclamation. *Ibid.* (July 8, 1983), p. A56.
96. *Ibid.*
97. *Columbia Basin Herald* (January 19, 1984), BOR Ephrata Mss. In the article, project manager James Cole was quoted as saying he could not argue with Whittlesey's figures.
98. *Wenatchee World* (April 5, 1979), p. 2; (January 15, 1984), p. 1.
99. *Ibid.* (January 20, 1984), p. 1.
100. *Ibid.* (January 20, 1984), p. 1; (February 10, 1984), p. 2; (February 22, 1984), p. 1; (March 7, 1984), p. 1; (March 9, 1984), p. 12. Other project backers said that Whittlesey's work was "phony and incomplete." See *Grant County Journal* (July 12, 1984), BOR Ephrata Mss.
101. *Wenatchee World* (April 15, 1984), p. 3.
102. Project History—Columbia Basin Project, 1985, 172; *Draft Environmental Impact Statement*, pp. IV-3-4.
103. *Draft Environmental Impact Statement*, pp. 18-19.
104. In his Ph.D. dissertation, done at Washington State University in 1985, Eric Ernest Elder calculated that every year the direct costs of the project to Washington state exceeded its direct benefits by $63 million. Elder claimed that the United States as a whole lost significantly more. In fact, he added, "The project farmers were the only group which showed direct benefits in excess [of] costs." Elder pointed out that one economic outcome of the project would be the redistribution of income from the urban/industrial west side of Washington state to the east side, where agriculture predominated. The result would be a situation where "a few people stand to gain quite a bit while a much larger number of people stand to lose a little." Finally, he calculated the energy loss resulting from each acre of land irrigated at between $100 and $220. Elder, "Economic Impacts of Irrigation on Development in Washington," pp. iv, 19, 132, 191.
105. *Draft Environmental Impact Statement*, p. vii.
106. James V. Cole to Whom it May Concern, April 30, 1989, letter mailed to all patrons of the Columbia Basin Project, a copy in the possession of the author. Also see *Draft Environmental Impact Statement*, p. iv-4.
107. *Draft Environmental Impact Statement*, p. viii, 4-5.

108. *Ibid.*, p. viii, 4-10.
109. *Ibid.*, p. iii, 78-79.
110. For a discussion on the nature of the West and its role as part of a larger capitalist world economy historically see William G. Robbins, "Western History, A Dialectic on the Modern Condition," *The Western Historical Quarterly* 20 (November 1989): 429-449.
111. Cole interview; David M. Dornbush & Company, *Columbia Basin Water Withdrawal Environmental Review, Appendix B—Social Effects, Part II— Community Social Effects*, (Portland: United States Army Corps of Engineers, January 1980), pp. 12, 14.
112. *Columbia Basin Water Withdrawal Environmental Review, Appendix B— Social Effects, Part II*, p. 48.
113. Interview with Wilfred Woods, Editor, *Wenatchee World*, March 28, 1989, Wenatchee, Washington.
114. Cole interview.
115. *Ibid.*

# Notes for Chapter Nineteen

1. As quoted in Edward Gross, "Grand Coulee: Heading Toward First Place," *Science News* 97 (March 28, 1970): 329.
2. When completed in the fall of 1954 it increased Grand Coulee Dam's prime power capacity from 1,117,000 Kw to 1,280,000 Kw. *Wenatchee Daily World* (February 4, 1944), p. 1; "Backlog of Reclamation Construction in Pacific Northwest Totals $1,389,899,400," *Pacific Builder and Engineer* 51 (June 1945): 62.
3. Neil A. Swainson, *Conflict Over the Columbia: The Canadian Background to an Historic Treaty* (Montreal: McGill-Queens University Press, 1979), p. 40; John V. Krutilla, *The Columbia River Treaty: The Economics of an International River Basin Development* (Baltimore, Md.: Resources for the Future, Johns Hopkins Press, 1967), p. 8.
4. Swainson, *Conflict over the Columbia*, p. 5.
5. Strauss to Taylor, December 22, 1947, BOR Denver Mss.
6. "Second Progress Report to the Federal Inter-Agency River Basin Committee: Measurement Aspects of Benefit-Cost Practices," November 1948, BOR Denver Mss.; *Wenatchee Daily World* (January 29, 1948), p. 1; (December 16, 1948), p. 1.
7. Gus Norwood, *Columbia River Power For the People: A History of Policies of the Bonneville Power Administration* (Washington, D.C.: U.S. Government Printing Office, 1981), p. 229.
8. C. E. Webb to Canadian Section, International Columbia River Engineering Committee, memorandum, January 3, 1949, BOR Denver Mss. For details of Dill's involvement see Kerry E. Irish, "Clarence Dill: The Life of a Western Politician" (unpublished Ph.D. dissertation, University of Washington, 1994), pp. 346-361.
9. Minutes of the Columbia Basin Commission, January 9, 1949, Columbia Basin Commissions Mss.; Webb to Hewitt, January 26, 1949, BOR Denver Mss.
10. "Preliminary Report on Arrow Lakes Storage to the International Commission by the International Columbia River Engineering Board," April 1, 1949, BOR Denver Mss.; Swainson, *Conflict over the Columbia*, p. 42.

11. "Libby Dam, Kootenai River, Montana Discussion of Problems Presented to the Corps of Engineers by Local Residents," report dated March 13, 1950, BOR Denver Mss.

12. Ralph W. Johnson, "The Canada-United States Controversy Over the Columbia River," *University of Washington Law Review* 41 (August 1966): 713; *Great Falls Tribune* (March 20, 1951), BOR Denver Mss.

13. Swainson, *Conflict Over the Columbia*, p. 45.

14. *Ibid.*, p. 24.

15. Norwood, *Columbia River Power For the People*, p. 230; Krutilla, *The Columbia River Treaty*, p. 12.

16. *Wenatchee Daily World* (April 27, 1954), p. 9.

17. Norwood, *Columbia River Power for the People*, p. 230.

18. Swainson, *Conflict Over the Columbia*, pp. 20, 53, 68.

19. Norwood, *Columbia River Power For the People*, p. 230; Johnson, "The Canada-United States Controversy Over the Columbia River," p. 715; Swainson, *Conflict Over the Columbia*, pp. 57-64.

20. *Wenatchee Daily World* (October 21, 1954), p. 1.

21. Howard Jacoby, "Impasse Over Columbia River Power," *Engineering News-Record* 157 (August 23, 1956): 21; Johnson, "The Canada-United States Controversy Over the Columbia River," p. 716.

22. Bruce Hutchison, "The Coming Battle for the Columbia," *Maclean's* (September 29, 1956): 11.

23. *Wenatchee Daily World* (October 19, 1960), p. 1; "Negotiations Agree on Treaty Terms," *Pacific Northwest Public Power Bulletin* 14 (December 1960): 4.

24. "Columbia Treaty Goes to Senate," *Engineering News-Record* 166 (January 26, 1961): 29.

25. *Wenatchee Daily World* (January 9, 1961), p. 1; (January 17, 1961), p. 1.

26. *Ibid.* (March 8, 1961), p. 1.

27. Swainson, *Conflict Over the Columbia*, p. 80; Krutilla, *The Columbia River Treaty*, p. 11.

28. Bruce Hutchison, "The Great Columbia River Foul-up," *Maclean's* (June 3, 1961): 15.

29. Swainson, *Conflict Over the Columbia*, pp. 248, 251, 256; *Wenatchee Daily World* (December 16, 1964), p. 3.

30. The agreement ends in the year 2003. "Final Taming of the Columbia Expected," *Electrical West* 131 (April 1964): 24; "Columbia River Power Contracts Signed," *Electrical West* (September 1964): 24.

31. Norwood, *Columbia River Power For the People*, p. 235; *Wenatchee Daily World* (June 11, 1964), p. 1; Kai N. Lee, Donna Lee Klemka, and Marion E. Marts, *Electric Power and the Future of the Pacific Northwest* (Seattle: University of Washington Press, 1980), p. 55; "Columbia Pact Ok'd, Work Starts," *Engineering News-Record* 172 (June 18, 1964): 66.

32. The Columbia River Treaty had a significant ramification. It discredited the so-called Harmon Doctrine which, since 1909, had said that a sovereign nation could do as it pleased with the portion of any international river found in its borders regardless of the downstream impact. This would affect the Rio Grande and Colorado Rivers. See Johnson, "The Canada-United States Controversy Over the Columbia River," p. 681.

33. Although unaltered since its final acceptance, the controversial treaty generated concern some years later. In 1972 British Columbia Premier David

Barrett stated that the United States had "skinned" the Canadians in the 1964 negotiations. Inflation in the years after the treaty raised the anticipated $500 million cost of the Canadian dams to well over $765 million. Since the treaty fixed the United States' portion of that cost at $344.4 million, this increased by more than $300 million Canada's portion. Barrett wanted the 1964 contract renegotiated and for a few weeks the issue made headlines, mostly in Canada. Then it faded away. *Wenatchee World* (December 12, 1972), p. 2; (December 18, 1972), p. 8.

34. Project History—Grand Coulee Dam Third Powerhouse, Volume VII, 1973, p. 2.

35. Bonneville Power Administration, *The Treaty with Canada for Joint Development of the Columbia River Handbook* (Portland: Bonneville Power Administration), p. 1, found in BOR Denver Mss.

36. Norwood, *Columbia River Power for the People,* p. 237; Gene Tollefson, *BPA and the Struggle for Power at Cost* (Portland: Bonneville Power Administration, 1987), p. 131.

37. For further background on the intertie, see Floyd E. Dominy, "A New Power Giant Materializes on the West Coast," *Reclamation Era* 51 (August 1965): 63-67.

38. "Pacific Intertie Plan Submitted to Congress," *Electrical West* 131 (August 1964): 26.

39. *Wenatchee Daily World* (August 14, 1964), p. 16; (January 26, 1965), p. 3; Norwood, *Columbia River Power for the People*, p. 244.

40. "Grand Coulee Completed—or is it? New Powerhouse Plans are Studied," *Public Power* 9 (October 1951): 20.

41. *Wenatchee Daily World* (June 14, 1951), p. 1; (December 4, 1951), p. 6.

42. Norwood, *Columbia River Power for the People*, p. 247; *Wenatchee Daily World* (March 19, 1952), p. 6.

43. A. L. Walker, et. al., *The Economic Significance of the Columbia Basin Project Development*, Washington Agricultural Experiment Station, College of Agriculture, Bulletin 669, September 1966, p. 48; Richard L. Neuberger, "Power Struggle on Canadian Border," *Harper's Monthly* 215 (December 1957): 49; *Wenatchee Daily World* (February 20, 1958), p. 4, editorial.

44. Floyd Dominy to Henry M. Jackson, November 24, 1965, BOR Office Mss.

45. "Third Powerhouse for Grand Coulee," *Electrical West* 131 (August 1964): 28; *Wenatchee Daily World* (December 4, 1964), p. 1; (February 14, 1965), p. 2; "Congress Approves Grand Coulee Dam Third Powerplant," *Pacific Northwest Public Power Bulletin* 20 (May 1966): 3.

46. *Wenatchee Daily World* (June 2, 1966), p. 1.

47. Norwood, *Columbia River Power for the People*, p. 248; *Wenatchee Daily World* (April 20, 1966), p. 1.

48. For comment on these points see Doris Kearns, *Lyndon Johnson and the American Dream* (New York: Harper & Row, 1979); and William E. Leuchtenburg, *In the Shadow of FDR: From Harry Truman to Ronald Reagan* (Ithaca, N.Y.: Cornell University Press, 1983), esp. pp. 121-160.

49. *Wenatchee Daily World* (June 2, 1966), p. 1.

50. *Ibid.* (January 19, 1967), p. 2; "Grand Coulee will Recapture Crown, Area Plans for Massive Work Force," *Pacific Builder and Engineer* 73 (February 1967): 19.

51. *Wenatchee Daily World* (January 13, 1967), p. 1.

52. "Grand Coulee Project Stirs up a Tempest," *Business Week* (April 15, 1967): 59.

53. *Spokesman-Review* (June 17, 1967), CBD League Mss.; *Wenatchee Daily World* (May 17, 1967), p. 2; (May 19, 1967), p. 1; (July 14, 1967), p. 1.
54. *Wenatchee Daily World* (November 17, 1967), p. 10; (January 1, 1968), p. 9.
55. *Ibid.* (April 1, 1968), p. 1; (April 3, 1968), p. 2; (April 5, 1968), p. 1.
56. Bureau of Reclamation Press Release, February 18, 1968, BOR Office Mss.
57. Bureau of Reclamation Press Release, October 15, 1968, *ibid.*; *Wenatchee Daily World* (October 15, 1968), p. 11.
58. *Wenatchee Daily World* (February 20, 1969, p. 3.
59. Armstrong to Boise, February 2, 1973, BOR Office Mss.
60. Draft letter, Armstrong to Dill, May 3, 1973, *ibid.*
61. A representative of the Bureau of Reclamation said, "God, how these guys hate it when they run into the old concrete. It doubles the powder ratio and triples the time. Next to that stuff, granite is like butter. Would you believe it, after more than 30 years, that concrete is still hardening." "Columbia Epic—Grand Coulee's Third Powerhouse," *Pacific Builder and Engineer* 76 (December 18, 1970): 16.
62. *Wenatchee Daily World* (December 31, 1968), p. 2; (February 3, 1969), p. 7; (March 6, 1969), p. 20.
63. The *Wenatchee Daily World* said that the lake had been lowered 133 feet. See *Wenatchee Daily World* (May 13, 1969), p. 2. Also see "Grand Coulee Power Plant Moving Ahead on Schedule," *Pacific Builder and Engineer* 75 (April 4, 1969): 29.
64. *Wenatchee Daily World* (April 2, 1969), p. 1 and picture.
65. "$112 Million Contract for Third Powerplant: Reclamation's Largest, World's Largest," *Reclamation Era* 56 (May 1970): 11; *Wenatchee Daily World* (February 2, 1970), p. 1.
66. *Spokesman-Review* (February 11, 1970), CBD League Mss.; *Wenatchee Daily World* (February 11, 1970), p. 1; Project History—Grand Coulee Dam, 1975, vol. 1, p. 4.
67. "First Concrete Placed at Grand Coulee Third Powerplant," *Vinnell* 13 (February 1971): 2.
68. *Wenatchee Daily World* (October 21, 1970), p. 1; (October 22, 1970), p. 1.
69. "Women's Lib at Grand Coulee," *Pacific Builder and Engineer* 76 (December 4-8, 1970): 28.
70. *Wenatchee World* (May 16, 1973), p. 1.
71. *Ibid.* (December 7, 1973), p. 2; (December 17, 1973), p. 1; (December 21, 1973), p. 1; (January 20, 1974), p. 1; (February 14, 1974), p. 1.
72. *Ibid.* (March 21, 1974), p. 2.
73. The time set for the opening of the forebay had been 3:00 p. m. on April 24, but it had been put off until the next morning. See *Ibid.* (April 24, 1974), p. 1; (April 25, 1974), p. 11; Project History—Grand Coulee Dam, 1975, vol. I, p. 5.
74. Project History—Grand Coulee Project Office, 1975, Vol. 1; Project History—Grand Coulee Third Powerhouse, 1976, Vol. 2, p. 10.
75. *Wenatchee Daily World* (May 26, 1968), p. 9.
76. "Grand Coulee DEW Turbine Breaks Records," *Engineering Journal* 58 (November/December 1975): 90.
77. "Buy American Policy Touches Off a Fuss," *Business Week* (July 27, 1968): 52. For a discussion of the "Buy American Policy" see Robert B. Reich, *The Work of Nations: Preparing Ourselves for the 21st Century* (New York: Random House, Inc., 1991), pp. 136-153.

78. *Wenatchee World* (September 6, 1973), p. 1; Project History—Grand Coulee Project Office, Columbia Basin Project, 1975, Vol. I, p. 3.
79. J. R. Granger, "Blasting Mass Concrete at Grand Coulee Dam," *Civil Engineering* 41 (August 1971): 28.
80. Project History—Grand Coulee Dam Left and Right Powerhouse Modification Final Construction Report, 1972, p. 18.
81. *Wenatchee World* (August 26, 1975), p. 1.
82. *Ibid.* (October 13, 1975), pp. 1-2.
83. Project History—Grand Coulee Dam Third Powerhouse, 1976, Vol. II, p. 4; *Wenatchee World* (November 19, 1976), p. 2.
84. Neil Stressman to Chief, Division of Design, Boise, July 24, 1978, BOR Office Mss.; *Wenatchee World* (May 14, 1975), p. 3; (November 3, 1978), p. 1; (November 5, 1978), p. 1; Project History—Grand Coulee Project Office, 1979, Vol. V, p. 6.
85. *Wenatchee World* (March 1, 1982), p. 2; (October 13, 1982), p. 3; (October 15, 1982), p. 1.
86. Vandalism delayed the inauguration of the second and third 700,000 Kw units. Someone, never identified and for reasons undetermined, damaged insulation on wiring coils in the generators. The Federal Bureau of Investigation looked into the matter briefly. Despite increased security, the ongoing incidents stretched from late October through December 1978, and damage to the machines and the loss of potential generating power amounted to over $1.5 million. *Wenatchee World* (March 4, 1980), p. 1; (November 10, 1978), p. 1; (November 14, 1978), p. 1.
87. Weekly Report on Construction Progress, Third Powerplant, May 23, 1980, BOR Office Mss.
88. *Wenatchee World* (December 30, 1980), p. 1.
89. Project History—Final Construction Report, Third Powerhouse, Left and Right Powerplant and Switchyard, 1972, p. 17; Final Construction Report, Grand Coulee Dam Third Powerhouse Modification, 1972, p. 17; *Wenatchee Daily World* (January 27, 1971), p. 7.
90. *Wenatchee World* (October 13, 1975), p. 1.
91. Project History—Columbia Basin Project, 1969, p. 11; *Wenatchee Daily World* (May 27, 1969), p. 1; (May 28, 1969), p. 20.
92. *Wenatchee World* (April 28, 1976), p. 2.
93. From 1933 through 1976, over six million people visited the dam.
94. Bureau of Reclamation, *The Story of the Columbia Basin Project*, (Washington, D.C.: U. S. Government Printing Office, 1978), p. 20.
95. Lee, et. al., *Electric Power and the Future of the Pacific Northwest*, p. 101.
96. *Wenatchee World* (March 15, 1984), p. 33.
97. I attribute the comment to University of Oregon history professor Richard Maxwell Brown because it is from him that I first heard the phrase, although in a more limited context. It may well have been used by others.

## Notes for Afterword

1. Rufus Woods, *The Twenty-Three Years' Battle for Grand Coulee Dam* (Wenatchee: *Wenatchee Daily World*, 1944), p. 6.
2. Richard Lowitt, *The New Deal and the West* (Bloomington: Indiana University Press, 1984), p. 152.

3. Alistair Cooke, *Alistair Cooke's America* (New York: Alfred A. Knopf, 1973), p. 323.

4. Phil Hamburger (?) "Notes for a Gazetter—XLIII" *New Yorker* 39 (October 1963): 204. C. C. Dill suffered no such attack of modesty. In his autobiography he wrote that he is "known as the Father of Grand Coulee dam." Clarence C. Dill, *Where Water Falls* (Spokane, Wash.: C. W. Hill, Printers, 1970), p. iv.

5. Hugh Gregory Gallagher, *FDR's Splendid Deception* (New York: Dodd, Mead & Company, 1985), p. 211.

6. In saying that the private power companies had started construction, Billingsley means that they had begun preliminary investigations. See "Great Rivers of the West," *Electrical West* 129 (August 1962): 344-372.

7. Roy M. Robbins, *Our Landed Heritage: The Public Domain 1776-1936* (New York: Peter Smith, 1950), pp. 368-370, 393-396.

8. Gene Tollefson, *BPA and the Struggle for Power at Cost* (Portland: Bonneville Power Administration, 1987), p. 194.

9. Donald C. Swain, "The Bureau of Reclamation and the New Deal, 1933-1940," *Pacific Northwest Quarterly* 61 (July 1970): 143.

10. *Ibid.*

11. Gerald Nash points out that national or regional planning was never popular in the West. See Gerald D. Nash, *World War II and the West: Reshaping the Economy* (Lincoln: University of Nebraska Press, 1990), p. xi.

12. Robert G. Athearn, *The Mythic West in Twentieth Century America* (Lawrence: University of Kansas Press, 1986), pp. 50, 70, 109. Also see Nash, *World War II and the West*, p. 20.

13. William Cronon, George Miles, and Jay Gitlin, *Under An Open Sky: Rethinking America's Western Past* (New York: Norton & Co., 1992), p. 9.

14. William E. Warne, *The Bureau of Reclamation* (New York: Praeger Publishers, 1973), p. 16.

15. Stewart H. Holbrook, *The Columbia* (New York: Rinehart and Co., 1956), p. 323.

16. *Wenatchee World* (April 18, 1983), p. 8; (June 27, 1983), p. 2; (July 3, 1983), p. 2; (July 8, 1983), p. 1; (July 18, 1983), p. 1.

17. Warne, *The Bureau of Reclamation*, p. 16; *Oregonian* (October 13, 1990), p. A1.

18. Kai N. Lee, Donna Lee Klemka, and Marion E. Marts, *Electric Power and the Future of the Pacific Northwest* (Seattle: University of Washington Press, 1980), *passim.*

19. Data board at Grand Coulee Information Center, June 24, 1989.

20. Lowitt, *The New Deal and the West*, pp. 162-163. Note that the placement of the aluminum industry during World War II depended on the availability of power and not its cost. It was not cheap power that brought in factories but simply the fact that the power existed in large blocks. The largest aluminum plant in the United States was located in Queens, Long Island because the New York City public utility systems had a surplus of power at that time. See Edwin J. Cohn Jr., *Industry in the Pacific Northwest and the Location Theory* (New York: King's Crown Press, Columbia University, 1954), p. 123.

21. William Ira Davisson, "The Impact of Electric Power on the Economic Development of the Pacific Northwest" (unpublished Ph.D. dissertation, Cornell University, 1961), p. 97. Early project planners anticipated that between 350,000 and 400,000 people would live in the project area when all of

the 1,029,000 acres received irrigation. As the project stands half finished, the population today, then, should be around 175,000 to 200,000. In 1970, 65,753 people lived in the area, and that reflected an unanticipated group of workers at Hanford and their families. See Ronald Albert Weinkauf, "The Columbia Basin Project, Washington: Concept and Reality, Lessons for Public Policy" (unpublished Ph.D. dissertation, Oregon State University, 1973), p. 161.

22. As of 1973, the 2,290 farms operated on 5,700 units. The units were established when the project was laid out in the late 1940s. Many units have been combined to make single farms, either through direct consolidation or through renting. The presence of the military during and since World War II at Moses Lake and at Hanford has made it impossible to assess exactly the demographic effect of the Columbia Basin Project. See Bureau of Reclamation, Columbia Basin Project Data Sheet, 1973; Weinkauf, "The Columbia Basin Project, Washington," p. 92.

23. Subsidy is a theme that appears to run through Western history. Robert Athearn, in his book *The Mythic West in Twentieth-Century America*, p. 206, quotes historian Joe Franz as stating that the West is the "child of subsidy." Patricia Limerick also comments on the role of subsidy in building the West. See Patricia Nelson Limerick, *The Legacy of Conquest: The Unbroken Past of the American West* (New York: W. W. Norton & Co., 1987), pp. 78-96.

24. In their study on the Bureau of Reclamation in 1973, a Ralph Nader group reported that without electric power and industrial water benefits, very few reclamation projects would be economically feasible. See Richard L. Berkman and W. Kip Viscusi, *Damming the West: Ralph Nader's Study Group Report on the Bureau of Reclamation* (New York: Grossman Publishers, 1973), p. 90.

25. *Oregonian* (June 24, 1976), p. B7.

26. In their study on the Bureau of Reclamation in 1973, a Ralph Nader group reported that reclamation projects have generally been built for special interest groups, and not for the general welfare of the nation. They also concluded that Native Americans are used to help get appropriations passed for reclamation but are almost always left out or receive only token aid. See Berkman and Viscusi, *Damming the West*, p. 151.

27. United States House of Representatives, *Columbia River and Minor Tributaries* House Document No. 103, 73rd Congress, 1st Session (Washington, D.C.: U.S. Government Printing Office, 1933), pp. 537-538. In the same government document, Secretary of Agriculture Arthur M. Hyde is represented by a long letter in which he points out that only businessmen will benefit from the project. See pp. 538-544. The section quoted here also appears in Cornelis Walterus Johannes Maria Crossmit, "An Analysis of Social Overhead Capital Expenditures on the Columbia Basin Irrigation Project, 1950-1970" (unpublished Ph.D. dissertation, Washington State University, 1973), p. 25.

28. In 1980 farmers on the Columbia Basin Project grew seventy different crops on 514,390 acres for a value of $275.1 million. That represented 13.5 percent of the total crop valuation for the state of Washington. See Herbert R. Hinman, M. Anthony Wright, and Gayle S. Willett, *1982 Crop Enterprise Budgets for the Columbia Basin, Washington*, Extension Bulletin 1019, College of Agriculture, Cooperative Extension Division (Pullman: Washington

State University, January 1982), p. 1. Also see, *Oregonian* (June 24, 1976), p. B7.

29. Warne, *The Bureau of Reclamation*, p. 139.
30. Weinkauf, "The Columbia Basin Project, Washington: Concept and Reality, Lessons for Public Policy," p. 172.
31. Craig Lynn Infanger, "Income Distributional Consequences of Publicly Provided Irrigation: The Columbia Basin Project" (unpublished Ph.D. dissertation, Washington State University, 1974), p. 131. Infanger stated flatly that the income redistribution pattern was not in favor of the poor, pp. 141, vi. For a similar conclusion, see Yahaya Doka, "Policy Objectives, Land Tenure, and Settlement Performance: Implications for Equity and Economic Efficiency in the Columbia Basin Irrigation Project" (unpublished Ph.D. dissertation, Washington State University, 1979), p. 132.
32. Otis L. Graham Jr., *Toward A Planned Society: From Roosevelt to Nixon* (New York: Oxford University Press, 1976), p. 67.
33. Total investment in the irrigation aspect of the project in 1986 was tabulated by the Bureau of Reclamation to be $496 million. The 20,000 people include actual farm operators, people who both owned and rented land at the same time, and those who owned land but did not farm it themselves. See Infanger, "Income Distributional Consequences of Publicly Provided Irrigation," p. 75.
34. For a discussion of the costs involved, see CH2M Hill, *Draft Environmental Impact Statement: Continued Development of the Columbia Basin Project, Washington* (Boise, Ida.: Bureau of Reclamation, September 1989), p. III-78-81. For a full evaluation of the cost, the cost potential of the power lost because of water removed for irrigation, plus the direct subsidy to irrigation costs, must be computed. Should the project be completed, the CH2M Hill study estimates that for Public Utility District ratepayers in the project area the average rate increase could reach $15.65 per household.
35. Anthony Netboy, *The Columbia River Salmon and Steelhead Trout: Their Fight for Survival* (Seattle: University of Washington Press, 1980), p. 87.
36. *Draft Environmental Impact Statement: Continued Development of the Columbia Basin Project, Washington*, pp. viii, II-1-30. Craig Infanger points out that efforts like the Columbia Basin Project, built over many years, and financed piecemeal, result in taking money from one generation and giving it to succeeding generations. See Infanger, "Income Distributional Consequences of Publicly Provided Irrigation, p. 120.
37. Eric Ernest Elder, "Economic Impacts of Irrigation Development in Washington" (unpublished Ph.D. dissertation, Washington State University, 1985), p. 1.
38. *Oregonian* (November 2, 1993), p. A10; "Interior Reorganizes Bureau," X•Press Information Services, Ltd., April 13, 1994.

# Bibliographical Note

Considering the volume of material written over the years about Grand Coulee Dam, surprisingly little is readily available on it or on the irrigation and power it provides. Out of print are George Sundborg's 1954 work, *Hail Columbia* (Macmillan), which details the efforts to build the dam and the Columbia Basin Project, and Murray Morgan's *The Dam* from the same year (Viking Press), which gives a thrilling minute-by-minute account of the March 1952 disaster that nearly ruined the powerhouses. Readily available is L. Vaughn Downs's recent book, *The Mightiest of Them All* (1986, Ye Galleon Press). It is notable for its numerous, well-selected photographs.

There is as yet no complete, broad study of public and private power development in the Pacific Northwest. Daniel Ogden's 1949 dissertation, "The Development of Federal Power Policy in the Pacific Northwest" (Chicago University) and Franklyn Mahar's 1968 dissertation "Douglas McKay and the Issues of Power Development in Oregon" (University of Oregon), would be starting points for such a work. Both Gus Norwood, with his 1981 *Columbia River Power For the People* (U.S. Government Printing Office) and Gene Tollefson's 1987 *BPA and the Struggle for Power At Cost* (Bonneville Power Administration), have chronicled the birth and growth of the Bonneville Power Administration. Ken Billington's *People, Politics and Public Power* (Washington Public Utility District's Association, 1988) offers valuable information. Among the studies of private power companies are Craig Wolner's *Electrifying Eden* (Oregon Historical Society Press, 1990) on Portland General Electric. A short history of the Washington Water Power Company appeared in the August 1962 edition of *Electrical West* along with a second article that surveys, from the perspective of private power, water power development in the eleven Western states. On the history of various dams and their construction, among the recent items available are Joseph Stevens's *Hoover Dam* (University of Oklahoma Press, 1988), and Russell Martin's *A Story That Stands Like A Dam* (Henry Holt and Company, 1989). Reporter Hu Blonk's personal memoir, *Behind the By-Line Hu* (Clark Printing, 1992), offers insightful reminiscences about the Grand Coulee Dam construction period.

There is also no comprehensive treatment of irrigation in the region. Rose Boening's pioneering article, "History of Irrigation in the State of Washington" (*Washington Historical Quarterly*, October 1918), and Emmett VandeVere's 1948 dissertation, "History of Irrigation in Washington" (University of Washington), are places to begin such an overdue book. Anyone looking into this or power development should consult Charles McKinley's epic 1952 study, *Uncle Sam in the Pacific Northwest* (University of California Press). Criticism of irrigation in general and the Bureau of Reclamation in particular is found in Donald Worster's *Rivers of Empire* (Pantheon Books, 1985) and, in a more journalistic vein, in Marc Reisner's *Cadillac Desert* (Viking, 1986), although both of these deal mostly with California.

As for the fascinating geology of northeastern Washington, the most recent and best book available is *Cataclysms on the Columbia* by John Allen, Marjorie Burns, and Sam Sargent (Timber Press, 1986). Almost two works in one, it details the heroic life of J Harlen Bretz, as well as outlining, in words the untrained reader can understand, the forces that carved the Grand Coulee and the unique features of eastern Washington.

For this book I drew from many of the rich archives available. The James O'Sullivan papers at Gonzaga University are badly organized and rapidly deteriorating, but still provide a wealth of material on both power and irrigation. The papers of the Columbia Basin Commissions in the Washington State Archives in Olympia are in excellent condition and readily accessible along with the papers of the various state governors. Senators Warren Magnuson's and Henry Jackson's papers at the University of Washington, along with those of Willis Batcheller, are of particular interest. Washington State University holds the papers of Frank Banks and other collections related to the Columbia Basin Project, including some material from Osmar Waller and Roy Gill of the Spokane Chamber of Commerce.

Documents from the Bureau of Reclamation are in a number of locations. Processed and categorized are those held by the National Archives Branch in Denver, Colorado. In the same building are boxes in storage held by the Federal Records Center. These are unculled, disorganized, and require permission for access, as do those still held by the Bureau itself in the basement storage facility of its Denver headquarters building. The caveat here, familiar to many historians who work with government documents, is that researchers must request permission to view each specific file, yet one often cannot know what is in a file (and hence if he or she really wants to see it) until it is actually viewed. This is because of the rather Byzantine and always enigmatic system of codes used to identify and store material. The same is true for papers held at the Bureau's regional office in Boise, Idaho.

Annually the Bureau of Reclamation compiles a so-called "Project History" for each of its facilities. Like the Anglo-Saxon Chronicle of old, these detail, without comment or interpretation, the happenings of the year. They are illustrated with maps, photographs, and technical descriptions. Collectively they represent the in-house memory of the organization through which it keeps track of the physical growth of each operation and the details of changes implemented over time. They are collages of minutia. But each usually includes a helpful chronology of the year's events and a retrospective for that project. A few copies are produced and one is stored at the headquarters in Denver, another at the project's regional office, and a third at the site itself.

For technical material related to dam and project-building, the trade journal *Pacific Builder and Engineer* is excellent. Over the years it carried extensive articles and photographs detailing the construction of just about everything of note in the Northwest. Anyone dealing with physical engineering work erected in Oregon, Washington, Idaho, or British Columbia during the twentieth century should consult its pages. Another of the same genre is *Western Construction News*, covering much the same ground. This wealth of articles is enhanced by additional material in the national publications *Engineering News-Record, Electrical West*, and the Bureau of Reclamation's publication *Reclamation Era* (until 1924, *Reclamation Record*).

A list of all periodical articles related to Grand Coulee Dam and the Columbia Basin Project (well over 1,000) located by this author, along with books, dissertations, government documents, and other references, can be found in the manuscript "Grand Coulee—the Bibliography." It is a companion to "Grand Coulee—The Struggle," "Grand Coulee—The Dam," and "Visions, Plans, and Realities: A History of the Columbia Basin Project" (unpublished Ph.D. dissertation, University of Oregon, 1990). These are the four considerably more detailed works from which this book was drawn. Copies are available at: the Washington State University Library in Pullman, Washington; the Washington State Library in Olympia, Washington; The Bureau of Reclamation in Denver, Colorado; The Grand Coulee Project Office in Grand Coulee, Washington; and the Bonneville Power Administration Library in Portland, Oregon.

# Acknowledgments

*Feci Quod Potui, Facient meliora potentes* *

One summer in the mid-1950s my parents took my sister and me on a trip that included a stop at Grand Coulee Dam. I remember looking out of our motel window, seeing the dam lighted, and being impressed with its size. I also remember the long expanse of hot, dry land that we crossed on our way through eastern Washington. At Dry Falls, I stood amazed that so much water had ever existed on what was then such a desolate place. Years later, I spent five summers working for Seattle City Light as a tour guide at its Skagit River hydroelectric project in northwestern Washington. I remember how often we had to admit to tourists that while our three dams were large, the dam at the Grand Coulee was greater than all of them put together. Grand Coulee had a mystique that haunted us and dwarfed Seattle's sizable undertaking, and it intrigued me.

I began this work while attending the University of Oregon, studying under the direction of Beekman Professor of Northwest and Pacific Northwest History, Richard Maxwell Brown (now retired). His time, his encouragement, and his availability have supported the effort throughout. Dick Brown's wide knowledge of the field and formidable bibliographic expertise, willingly and frequently shared, were invaluable. He is not only a gentleman and a scholar but also a teacher, and I am in his debt.

Along the way, a rather sizable group of people helped me. In the Bureau of Reclamation I found ready assistance from James V. Cole, Columbia Basin Project Manager in Ephrata, Washington; Cline Sweet, also in Ephrata; Craig L. Sprankle, Public Affairs Officer at

---

* I have done what I could—let those who can, do better.

Grand Coulee Dam; Judy Calhoun and the late Bert Holmes at the Grand Coulee Project Office; Edward J. Lenhart, Regional Property and Services Director, Boise, Idaho; Gertrude E. Schalow, Records Division, Denver, Colorado; Carolyn McNee, Reclamation Librarian, Denver, Colorado; and L. Vaughn Downs, Ephrata, Washington.

At the National Archives branch in Denver, Assistant Director of Archives Daniel Nealand and Eileen Bolger guided me. Their knowledge of the files entrusted to their care expedited my time there, and I greatly appreciate their willingness to help. David W. Hastings, Chief of Archival Services with the State of Washington in Olympia, and the staff with whom he works, spent considerable time leading me through the mass of material they have accumulated. The same was also true at Washington State University, the University of Washington, and Gonzaga University. I am indebted to the people at all of these institutions for their time and patience.

In Wenatchee, I met Wilfred Woods, son of Rufus Woods and current publisher of *Wenatchee World*, and his son, Rufus Woods, now managing editor. Their willingness to accommodate my needs went beyond anything I might have hoped for, and I am grateful to them and to Bruce Mitchell, Hu Blonk, Janelle Hamm, and Greg Hartgrave, all of whom went out of their way to answer questions, make explanations, or provide materials.

Those who read all or part of the manuscript and commented or who offered encouragement include Daniel A. Pope, Jack P. Maddex, Jr. (who alerted me to the failings of the passive voice and browbeat me into avoiding it), Patricia F. McDowell, Carman Morgan, Gordon Dodds, Bureau of Reclamation senior historian Brit Allan Storey (who also helped locate and provide photographs), Peter G. Boag, Sally Morita, and George Sundborg. Robert E. Ficken of Issaquah, currently working on a biography of Rufus Woods, graciously went over the entire manuscript and offered helpful suggestions and comments. My student assistant, Andrea VanBemmel, spent hours sorting and arranging my notes.

At Washington State University Press I have received help, encouragement, and expertise from my editor Keith Petersen, along with others on the staff: marketing and promotions coordinator Beth DeWeese, copyeditor Nancy Grunewald, designer David Hoyt, typesetter Wes Patterson, director Thomas Sanders, assistant director Mary Read, order fulfillment coordinator Arline Lyons, and editor

Glen Lindeman. They are an excellent group and I have found working with them a pleasure and a welcome learning experience.

None of those I have mentioned are in any way responsible for whatever errors I may have made.

Without question, I am in the debt of a small army of interlibrary loan librarians across the country. Without this service and their efforts, I could not have looked at the quantity and variety of materials that ultimately found their way into my hands. Especially I must thank Reginald B. Sullivan, librarian at Aloha High School in Beaverton, Oregon, where I teach American History. Reggie supported my every request for materials and went out of his way to locate periodicals, dissertations, and public documents. My employer, the Beaverton Public Schools, annually allows ten of its teachers a sabbatical with pay. I was fortunate enough to obtain one of these and, after twenty years in the classroom, it proved a welcome and refreshing break. I congratulate school districts that have the wisdom to make such enrichment possibilities available. Too often those who teach at elementary and high schools feel separated and even alienated from the academic community, and these opportunities provide a way for us to interact with others in our fields of interest.

Finally, I understand now why so many acknowledgments to books make mention of the author's spouse and relatives. My parents, Lloyd and Eleanor Pitzer, and the rest of my family, including Margaret Galley, have been supportive throughout. But without question, I am most indebted to my wife, Grace. She has put up with my being preoccupied or out-of-sorts now and then. She has understood when I have told her to leave me alone. She accompanied me to dark and dusty archives and sat patiently while I ignored her for hours, if not days on end. She has encouraged me when I needed it and she made me learn how to use a computer whether I wanted to or not. She has loved me in spite of all of this, and I am hard pressed to ask for more.

To all of these, and some whom I have undoubtedly forgotten, I offer my sincere thanks.

# About the Author

Paul C. Pitzer has taught American history at Aloha High School in Beaverton, Oregon, since 1969. He spent two years with the Peace Corps in Iranian Azarbaijan, and holds a Ph.D. in history from the University of Oregon. Among his publications are a number of articles on Northwest history and the book, *Building the Skagit*, published in 1978.

# Index